Pattern Recognition
and Machine Learning

Pattern Recognition and Machine Learning

Proceedings of the Japan–U.S. Seminar on the Learning Process in Control Systems, held in Nagoya, Japan August 18-20, 1970

Edited by K. S. Fu
School of Electrical Engineering
Purdue University
Lafayette, Indiana

PLENUM PRESS · NEW YORK–LONDON · 1971

Library of Congress Catalog Card Number 77-163287

SBN 306-30546-1

A Division of Plenum Publishing Corporation
227 West 17th Street, New York, N.Y. 10011

United Kingdom edition published by Plenum Press, London
A Division of Plenum Publishing Company, Ltd.
Davis House (4th Floor), 8 Scrubs Lane, Harlesden, NW10, 6SE, England

Printed in the United States of America

PREFACE

This book contains the Proceedings of the US-Japan Seminar on Learning Process in Control Systems. The seminar, held in Nagoya, Japan, from August 18 to 20, 1970, was sponsored by the US-Japan Cooperative Science Program, jointly supported by the National Science Foundation and the Japan Society for the Promotion of Science. The full texts of all the presented papers except two are included.† The papers cover a great variety of topics related to learning processes and systems, ranging from pattern recognition to systems identification, from learning control to biological modelling. In order to reflect the actual content of the book, the present title was selected. All the twenty-eight papers are roughly divided into two parts--Pattern Recognition and System Identification and Learning Process and Learning Control. It is sometimes quite obvious that some papers can be classified into either part. The choice in these cases was strictly the editor's in order to keep a certain balance between the two parts.

During the past decade there has been a considerable growth of interest in problems of pattern recognition and machine learning. In designing an optimal pattern recognition or control system, if all the a priori information about the process under study is known and can be described deterministically, the optimal system is usually designed by deterministic optimization techniques. If all or a part of the a priori information can only be described statistically, for example, in terms of probability distribution or density functions, then stochastic or statistical design techniques will be used. However, if the a priori information required is unknown or incompletely known, in general, an optimal design cannot be achieved. Two different approaches have been taken to solve this class of problems. One approach is to design a system based only upon the amount of information available. In that case, the unknown information is either ignored or is assumed to be some known values from the designer's best guess. The second approach is to design a system which is capable of estimating the unknown information during its operation and an optimal decision or control action will be determined on the basis of the estimated information.

†Two papers, which were not received in time to meet the publication deadline, have not been included.

In the first case, a rather conservative design criterion (for example, minimax criterion) is often used; the systems designed are in general inefficient and suboptimal. In the second case, if the estimated information gradually approaches the true information as time proceeds, then the system thus designed will eventually approach the optimal system. Here the optimal system means that the performance of the system designed will be equally as good as in the case where all required a priori information is known. Because of the gradual improvement of performance due to the improvement of the estimated unknown information, this class of systems may be called learning systems. The system learns the unknown information during operation, and the learned information is, in turn, used as an experience for future decisions or controls.

The research on learning process in control systems started first in the U.S.A. about 1958. The aim was to analyze the psychological learning function from an engineering point of view and to introduce it into the control system. A few years later (in about 1960), the research started also in Japan. At present, we can list a number of research projects in both countries. However, the research is still in the early stage of development, and there are many essential problems left to be solved. Although there are relatively only a small number of researchers in this area, each has been concentrating on their specific area of interest, and they are very eager to exchange information on the topic. With the opportunity of the US-Japan Seminar, the researchers in both Japan and the U.S.A. could meet together to discuss their work and development in the area. The results of the seminar has formed a close bond between the researchers in Japan and the U.S.A. and, in addition, has contributed to future developments in systems engineering and control engineering in both countries.

Credit for any success in this Seminar must be shared with many prople who contributed significantly of their time and talents. In organizing the Seminar the co-chairmen, Kahei Nakamura and K. S. Fu received considerable help from J. E. O'Connell of the National Science Foundation; the planning committee including K. Nakamura, T. Katagawa, J. Nagumo, B. Kondo, K. Tanaka, S. Tsuji; and the staff of the National Science Foundation/Tokyo and the Japan Society for the Promotion of Science. It is the authors of the indivisual papers whose contributions made possible the Seminar and the subsequent post-seminar proceedings. As the editor of the proceedings, I wish to express my heartfelt appreciation for the help received from J. T. Tou, K. Nakamura, G. N. Saridis, Kathy Mapes and J. M. Mendel during the preparation of this publication.

K. S. Fu
Lafayette, Indiana
April 1971

CONTENTS

PART I: PATTERN RECOGNITION AND SYSTEM IDENTIFICATION

PART I
PATTERN RECOGNITION AND SYSTEM IDENTIFICATION

SOME STUDIES ON PATTERN RECOGNITION WITH NONSUPERVISED LEARNING PROCEDURES

Kokichi Tanaka

University of Osaka

Osaka, Japan

1. INTRODUCTION

Generally speaking, the pattern recognition techniques must perform two basic functions, that is, the process of characterizing a class of the common pattern of inputs that belong to the same class and the process of classifying any input as a member of one of several classes. Now, the process of classification is based on a procedure of decision making in which a set of sample patterns may be attributed to the corresponding class. Let a set of sample patterns belonging to the same class be characterized by the set of parameters accompanied by the probability distribution which governs how each pattern involved in the same class is generated. This is the method of Parametric Statistics. In the parametric methods the training set whose statistical structures are in some cases known correctly and in other cases unknown is used for the purpose of obtaining estimates of the parameter values on the basis of the Bayes' estimation theory, and the discriminant function which may implement the process of classification is then determined by these estimates. This is the so-called Bayes' Machine.

There are two types of Bayes' Machine which are of interest as follows. (i) The first type of system is designated "Nonsupervised Learning Machine," in which the parameters to be estimated are determined by the use of the set of sample patterns in the learning feature without knowing their classes. The typical example of the present system is "Decision-Directed Adaptive Pattern Recognizer" [1],and "Adaptive Filter" [2] may be viewed as the special case for this. (ii) On the contrary, those systems where the correct classification of the sample patterns is supplied are designated "Supervised Learning Machine." A pioneering paper by N. Abramson [3] is an excellent introduction to the subject. The well-known "Matched Filter" may be viewed as the limiting case with degenerate random variables.

Methods of nonparametric statistics are also introduced when little or nothing is known a priori about the functional forms of the probability distributions. There are two types of learning procedures, that is, "Supervised Learning" and Nonsupervised Learning", in the same way as in the parametric case. As is well known, the typical examples of the former are "Perceptron" [4],"ADALINE" [5], and other layered machines [6]; and examples of the latter are the "Adaptive Waveform Recognizer [7], and the Adaptive Correlating Filter" [8].

In the present paper some extensions of existing works on the "Nonsupervised Adaptive Recognizer" [1,2,7,8,9,10] are developed and several machines are proposed to achieve improved performances in comparison with the results obtained hitherto. In addition, the proposed approaches are also applied to recognize a set of time-varying patterns [11,12] in a noisy environment such as in a noisy channel with fading. A computer experiment is conducted to verify these results.

2. SOME IMPROVEMENTS ON THE EXISTING NONSUPERVISED MACHINES

Let us now assume that an ℓ-dimensional pattern vector $\underline{X}$, which is fixed but unknown, is a sample taken from a normal population $\mathcal{N}(\underline{M},\underline{\Lambda})$, and a noise vector $\underline{N}$ is the one from $\mathcal{N}(0,\Sigma)$. Then the input vector at the nth interval will be denoted by $\underline{Y}_n=\underline{\theta}_n X+\underline{N}_n$, where θ_n is a binary variable independent of n, and it takes either 1 with a probability p or 0 with a probability 1-p. These parameters $\underline{M}$, $\underline{\Lambda}$, $\underline{\Sigma}$, and p are given to the machine a priori. The decision-directed machine (DDM) has been proposed as one type of the discriminators with learning for such a case. The decision rule of the DDM which is a self-taught learning machine using its own output is given as follows:

$$Q_n \begin{cases} > \delta_n & \hat{\theta}_n=1 \text{ (input contains a pattern)} \\ \leqq \delta_n & \hat{\theta}_n=0 \text{ (input does not contain a pattern)} \end{cases} \tag{1}$$

where

$$Q_n = \underline{Y}_n^t \underline{\Sigma}^{-1} \underline{Y}_n-(\underline{Y}_n-\underline{H}_n)^t (\underline{\Xi}_n+\underline{\Sigma})^{-1} (\underline{Y}_n-\underline{H}_n) \tag{2}$$

$$\delta_n = \log \frac{(1-P)^2}{P^2} + \log \frac{|\underline{\Xi}_n+\underline{\Sigma}|}{|\underline{\Sigma}|} \tag{3}$$

with the a posteriori probability given as

$$P(\theta_n=1/\underline{Y}^n,\ \hat{\theta}^{n-1}) = [1 + \frac{1-p}{p}\ \frac{|\underline{\Xi}_n+\underline{\Sigma}|^{1/2}}{|\underline{\Sigma}|^{1/2}}\exp(-\frac{1}{2}Q_n)]^{-1} \quad (4)$$

where p=a priori probability that the input contains a pattern; $\underline{Y}^n=(\underline{Y}_1,\underline{Y}_2,\ldots,\underline{Y}_n)$ is the input sequence; and $\hat{\theta}^n=(\hat{\theta}_1,\hat{\theta}_2,\ldots,\hat{\theta}_n)$ is the teaching sequence. In addition, $\underline{H}_n$ and $\underline{\Xi}_n$ are the iteratively obtained estimator of the unknown pattern $\underline{X}$ and its covariance, respectively, by the following rule:

$$\underline{H}_{n+1} = \underline{H}_n + \hat{\theta}_n\underline{\Sigma}^{-1}\underline{\Xi}_{n+1}(\underline{Y}_n-\underline{H}_n)\ ;\ \text{a posteriori mean of pattern } \underline{X} \quad (5)$$

$$\underline{\Xi}_{n+1} = (\underline{\Xi}_n^{-1} + \hat{\theta}_n\underline{\Sigma}^{-1})^{-1}\ ;\ \text{a posteriori covariance of } \underline{X}\ . \quad (6)$$

Let us now obtain the distribution of Q_n to determine the probability of error. Define

$$\begin{aligned} Q_n^{(1)} &= Q_n + \underline{H}_n^t\underline{\Xi}_n^{-1}\underline{H}_n \\ &= (\underline{Y}_n+\underline{\Sigma}\underline{\Xi}_n^{-1}\underline{H}_n)^t(\underline{\Sigma}\underline{\Xi}_n^{-1}\underline{\Sigma}+\underline{\Sigma})^{-1}\ (\underline{Y}_n+\underline{\Sigma}\underline{\Xi}_n^{-1}\underline{H}_n) \\ &= \underline{Z}_n^t\underline{F}\underline{Z}_n \end{aligned} \quad (7)$$

$$\delta_n^{(1)} = \delta_n + \underline{H}_n^t\underline{\Xi}_n^{-1}\underline{H}_n \quad (8)$$

where

$$\underline{Z}_n = \underline{Y}_n + \underline{\Sigma}\underline{\Xi}_n^{-1}\underline{H}_n;\qquad \underline{F} = (\underline{\Sigma}\underline{\Xi}_n^{-1}\underline{\Sigma}+\underline{\Sigma})^{-1} = \underline{\Sigma}^{-1}-(\underline{\Sigma}+\underline{\Xi}_n)^{-1}\ . \quad (9)$$

Now assuming that

$$\underline{\Sigma} = \text{diag.}(\sigma_1^2,\sigma_2^2,\ldots,\sigma_\ell^2);\qquad \underline{\Lambda} = \text{diag.}(\lambda_1,\lambda_2,\ldots,\lambda_\ell) \quad (10)$$

then $\underline{\Xi}_n$ is always reduced to a diagonal form as

$$\underline{\Xi}_n = \text{diag.}(\xi_{n1},\xi_{n2},\ldots,\xi_{n\ell})\ .$$

Consequently, $\underline{F}$ is also reduced to a diagonal form. In this particular case, eqs. (7) and (8) may be written in terms of the components of each vector and the elements in each diagonal matrix as

$$\begin{aligned} Q_n^{(1)} &= Q_n + \sum_{i=1}^{\ell}\frac{h_{ni}^2}{\xi_{ni}} = \sum_{i=1}^{\ell}C_{ni}(\frac{y_{ni}}{\sigma_i} + \frac{\sigma_i h_{ni}}{\xi_{ni}})^2 \\ &= \sum_{i=1}^{\ell}C_{ni}(n_h' + \frac{\theta_n\xi_{ni}X_i+\sigma_i^2 h_{ni}}{\sigma_i\xi_{ni}})^2 \\ &= \sum_{i=1}^{\ell}C_{ni}u_{ni} = \underline{U}_n\underline{C}_n\underline{U}_n^t \end{aligned} \quad (11)$$

and

$$\delta_n^{(1)} = \delta_n + \sum_{i=1}^{\ell} \frac{h_{ni}^2}{\xi_{ni}} ,$$

respectively, where

$$\underline{H}_n = \begin{bmatrix} h_{n1} \\ h_{n2} \\ \vdots \\ h_{n\ell} \end{bmatrix} ; \quad \underline{Y}_n = \begin{bmatrix} y_{n1} \\ y_{n2} \\ \vdots \\ y_{n\ell} \end{bmatrix} ; \quad \underline{N}_n = \begin{bmatrix} n_{n1} \\ n_{n2} \\ \vdots \\ n_{n\ell} \end{bmatrix} ; \quad \underline{X} = \begin{bmatrix} x_1 \\ x_2 \\ \vdots \\ x_\ell \end{bmatrix}$$

$$\underline{U}_n = \begin{bmatrix} u_{n1} \\ u_{n2} \\ \vdots \\ u_{n\ell} \end{bmatrix} = \text{a normal random vector with a unit covariance matrix*}$$

$$\underline{C}_n = \text{diag.}(c_{n1}, c_{n2}, \ldots, c_{n\ell})^* ; \quad n'_{ni} = \frac{n_{ni}}{\sigma_i} = \text{a standard normal random variable}$$

$$c_{ni} = \frac{\xi_{ni}}{\sigma_i^2 + \xi_{ni}} ; \quad u_{ni} = n'_{ni} + \frac{\theta_n \xi_{ni} x_i + \sigma_i^2 h_{ni}}{\sigma_i \xi_{ni}} \tag{12}$$

From eq. (10) the distribution of $Q_n^{(1)}$ may easily be found to be a distribution of the variable which consists of the weighted sum of ℓ independent variables according to a noncentral chi-squared distribution law. However, the cumulative distribution of $Q_n^{(1)}$ may not be expressed in a simple form. For example, employing a series expansion in terms of a chi-squared distribution, it may be written as

$$P_{\ell,\varepsilon}(Q_n^{(1)}) = \sum_{j=0}^{\infty} K_j(\rho) G_{\ell+2j}\left(\frac{Q_n^{(1)}}{\rho}\right) \tag{13}$$

where

$$K_j(\rho) = \prod_{i=1}^{\ell} \left(\frac{\rho}{c_{ni}}\right)^{1/2} \exp\left(-\frac{\varepsilon}{2}\right) \frac{1}{j!} \int_{-\infty}^{\infty} \left[\frac{1}{2} F(\underline{V})\right]^j (2\pi)^{-\ell/2} e^{-(1/2)\underline{V}^t \underline{V}} d\underline{V} \tag{14}$$

$$F(\underline{V}) = \sum_{i=1}^{\ell} \left\{\left(1 - \frac{\rho}{c_{ni}}\right)^{1/2} v_i + \left(\frac{\rho}{c_{ni}}\right)^{1/2} E(U_{ni})\right\}^2 \tag{15}$$

$$\underline{V} = (v_1, v_2, \ldots, v_\ell)^t$$

$$E(U_{ni}) = \frac{\hat{\theta}_n \xi_{ni} x_i + \sigma_i^2 h_{ni}}{\sigma_i \xi_{ni}} \tag{16}$$

$G_m(\cdot)$ = chi-squared cumulative distribution function of m degree of freedom.

ρ = arbitrary positive constant satisfying the condition $\rho \leqq \min_{i=1,2,\ldots,\ell} c_{ni}$

$\varepsilon = \sum_{i=1}^{\ell} \{E(U_{ni})\}^2$ = noncentrality parameter.

Thus the probability of error at nth time becomes

$$W_{en} = (1-p)P_e(\hat{\theta}_n=1|\theta_n=0)+pP_e(\hat{\theta}_n=0|\theta_n=1) \tag{17}$$

where

$$\left.\begin{aligned} P_e(\hat{\theta}_n=1|\theta_n=0) &= 1-P_{\ell,\varepsilon_o}(\delta_n^{(1)}) \\ P_e(\hat{\theta}_n=0|\theta_n=1) &= P_{\ell,\varepsilon_1}(\delta_n^{(1)}) \end{aligned}\right\} \tag{18}$$

ε_o and ε_1 give the value of ε corresponding to $\theta_n=0$ and $\theta_n=1$ in eq. (16), respectively. For the special case of $\underline{\Lambda} = \lambda\underline{I}$, $\underline{\Sigma} = \sigma\underline{I}$, where $\underline{I}$ is the identity matrix, $c_{ni} = c_n$ holds. Thus, putting $q_n = Q_n^{(1)}/c_n$, eq. (13) reduces to a well-known noncentral chi-squared distribution $P^{(1)}_{\ell,\varepsilon_r}(q_n)$. Then, $P_{\ell,\varepsilon_r}(\delta_n^{(1)})$ in eq. (18) may also be substituted by $P^{(1)}_{\ell,\varepsilon_r}(\delta_n^{(1,1)})$, where $\delta_n^{(1,1)}=\delta_n^{(1)}/c_n (r=0,1)$.

For large values of n, eq. (2) becomes approximately

$$Q_n \simeq 2\underline{Y}_n^t\underline{\Sigma}^{-1}\underline{H}-\underline{H}^t\underline{\Sigma}^{-1}\underline{H} \tag{2'}$$

which is a linear function of $\underline{Y}_n$, and eq. (3) becomes

$$\delta_n = \log\,(1-p)^2/p^2 \tag{3'}$$

since $\underline{\Xi}_n \to 0$ (zero matrix) and $\underline{H}_n \to \underline{H}$ as $n \to \infty$. Accordingly, the decision rule given as eq. (1) may be expressed as

$$\hat{\theta}_n=1, \text{ if } Q_n^{(2)} > \delta_n^{(2)} \quad \text{or} \quad \hat{\theta}_n=0, \text{ if } Q_n^{(2)} \leqq \delta_n^{(2)} \tag{1'}$$

where

$$Q_n^{(2)} = \underline{Y}_n^t\,\underline{\Sigma}^{-1}\underline{H} \tag{7'}$$

$$\delta_n^{(2)} = 1/2\,\{\log \frac{(1-p)^2}{p^2} + \underline{H}^t\underline{\Sigma}^{-1}\underline{H}\} \tag{8'}$$

Now, because the covariance $\underline{\Sigma}$ is symmetric, let us consider an orthogonal matrix $\underline{A}$ which diagonalizes $\underline{\Sigma}$ such that

$$\underline{A}^t\underline{\Sigma}\underline{A} = \underline{D} = \text{diag.}(\sigma_{d1}^2,\sigma_{d2}^2,\ldots,\sigma_{d\ell}^2) \tag{19}$$

Then define

$$\underline{Y}'_n = \underline{A}^t \underline{Y}_n, \quad \underline{W} = \underline{A}^t \underline{\Sigma}^{-1} \underline{H}, \quad \text{and} \quad \underline{X} = \underline{A}^t \underline{A} \tag{20}$$

we have

$$Q_n^{(2)} = \underline{Y}'^{\,t}_n \underline{W} \tag{21}$$

Thus, the conditional probability distribution of the ith component y'_{ni} of the vector $\underline{Y}'_n$ may be expressed as

$$P(y'_{ni}/\theta_n=1,\theta_n=r) = \frac{1-\Phi(d^i_r)}{1-\Phi(d_r)} \frac{1}{\sqrt{2\Pi}\,\sigma_{di}} \exp -\left[\frac{(y'_{ni}+rx'_i)^2}{2\sigma^2_{di}}\right] \tag{22}$$

where $\Phi(\cdot)$ = cumulative distribution function of standardized normal random variable;

$$d_r = \frac{\delta_n^{(2)} - r\underline{X}'^{\,t}\underline{W}}{\sqrt{\underline{W}^t \underline{D}\underline{W}}} \quad ; \quad d^i_n = \frac{\delta_n^{(2)} - r\underline{X}'^{\,t}\underline{W} - w_i y'_{ni}}{\sqrt{\underline{W}^{it}\underline{D}\underline{W}^i}}$$

$$\underline{W}^i = (w_1, w_2, \ldots, w_{i-1}, 0, w_{i+1}, \ldots, w_\ell)^t .$$

Consequently, the probability of error at the nth time for a linear discriminant case will be given by the same expression as eq. (17) obtained by the substitution of eq. (18').

$$P_e(\hat{\theta}_n=1/\theta_n=0) = 1-\Phi(d_o); \quad P_e(\hat{\theta}_n=0/\theta_n=1) = \Phi(d_1) \tag{18'}$$

instead of eq. (18). Making use of eq. (22), $H = \lim_{n\to\infty} \underline{H}_n$ may be given as

$$E(\underline{Y}_n/\theta_n=r,\hat{\theta}_n=1) = \underline{A}^t E(\underline{Y}'_n/\theta_n=r,\hat{\theta}_n=1) = \underline{A}^t(\underline{X}'+\beta_r(\sqrt{\underline{W}^t\underline{D}\underline{W}})^{-1}\underline{D}\underline{W})$$

$$= r\underline{X} + \alpha_r(\sqrt{\underline{H}^t\underline{\Sigma}^{-1}\underline{H}})^{-1}\,\underline{H} \tag{23}$$

$$\underline{H} = E(\underline{Y}_n/\hat{\theta}_n=1)$$

$$= P(\theta_n=1/\hat{\theta}_n=1)\underline{X} + P(\theta_n=1/\hat{\theta}_n=1)\frac{\alpha_1\underline{H}}{\sqrt{\underline{H}^t\underline{\Sigma}\underline{H}}} + P(\theta_n=0/\hat{\theta}_n=1)\frac{\alpha_o\underline{H}}{\sqrt{\underline{H}^t\underline{\Sigma}^{-1}\underline{H}}} \tag{24}$$

$$P(\theta_n=1/\hat{\theta}_n=1) = \frac{p\{1-\Phi(d_1)\}}{p\{1-\Phi(d_1)\}+(1-p)\{1-\Phi(d_o)\}}$$

$$P(\theta_n=0/\hat{\theta}_n=1) = 1-p(\theta_n=1/\hat{\theta}_n=1) \tag{25}$$

where

$$\alpha_r = \frac{\Phi'(d_r)}{1-\Phi(d_r)} \; ; \quad E(.) = \text{ensemble avg.}; \quad \underline{H}^t = \text{transposed matrix of } \underline{H}$$

$$\Phi'(x) = \frac{d}{dx}\Phi(x) .$$

Then, $\underline{H}$ may also be expressed as

$$\underline{H} + k\underline{X} \tag{26}$$

where

$$k = P(\theta_n=1/\hat{\theta}_n=1)\ [1 + (\frac{\alpha_1-\alpha_o}{\sqrt{\underline{X}^t\underline{\Sigma}^{-1}\underline{X}}})] + \frac{\alpha_o}{\sqrt{\underline{X}^t\underline{\Sigma}^{-1}\underline{X}}}$$

is a function of signal-to-noise ratio $\underline{X}^t\underline{\Sigma}^{-1}\underline{X}$. The results thus obtained may be more general than those given in the case of $\underline{\Sigma}=\psi\underline{I}$ by H. J. Scudder [1]. Consequently, the DDM does not converge to the template machine. Hence, we shall now propose several machines which may converge better than DDM to the template machine under keeping the SN ratio small.

a. Consistent-Estimator Type DDM (CDDM)

The CDDM is the machine which seeks a consistent estimator in the form of $\underline{H} = aE(\underline{Y}_n/\hat{\theta}_n=1)$, where $E(\cdot/\cdot)$ denotes a conditional expectation and a is a scalar constant given as a function of SN ratio as follows.

$$\underline{\hat{H}} = a\{P(\theta_n=1/\hat{\theta}_n=1)\underline{X}+P(\theta_n=1/\hat{\theta}_n=1)\frac{\alpha_1}{\sqrt{\underline{X}^t\underline{\Sigma}^{-1}\underline{X}}}\underline{H}+P(\theta_n=0/\hat{\theta}_n=1)\frac{\alpha_o}{\sqrt{\underline{X}^t\underline{\Sigma}^{-1}\underline{X}}}\underline{\hat{H}}\} \tag{27}$$

Eliminating $\underline{\hat{H}}(=\underline{X})$ on both sides of eq. (27), we can obtain a given as follows:

$$a = 1/\left[p(\theta_n=1/\hat{\theta}_n=1)(1+\frac{\alpha_1}{\sqrt{\underline{X}^t\underline{\Sigma}^{-1}\underline{X}}}) + P(\theta_n=0/\hat{\theta}_n=1)\frac{\alpha_o}{\sqrt{\underline{X}^t\underline{\Sigma}^{-1}\underline{X}}}\right] \tag{28}$$

In the special case where the probability of occurrence of patterns in the input is equal to 1/2 (i.e., p=1/2), a is expressed as

$$a = 1/\left[\Phi(\sqrt{\underline{X}^t\underline{\Sigma}^{-1}\underline{X}}/2) + (2/\sqrt{\underline{X}^t\underline{\Sigma}^{-1}\underline{X}}\ \Phi')\ (\sqrt{\underline{X}^t\underline{\Sigma}^{-1}\underline{X}}/2)\right] \tag{29}$$

Table 1 shows the value of a as a function of SN ratio.

The learning rule of this machine is given as follows.

$$\begin{aligned} &\underline{H}_1 = \underline{M}, \quad \underline{H}_{n+1} = \underline{H}_n + \hat{\theta}_n\underline{\Xi}_{n+1}\ \underline{\Sigma}^{-1}(a\underline{Y}_n-\underline{H}_n) \\ &\underline{\Xi}_1 = \underline{\Lambda}, \quad \underline{\Xi}_{n+1} = (\underline{\Xi}_n^{-1} + \hat{\theta}_n\ \underline{\Sigma}^{-1})^{-1} \end{aligned} \tag{30}$$

This rule differs from that of DDM only in the point that $\underline{Y}_n$ is modified as $a\underline{Y}_n$.

Table 1

SN ratio $\underline{X}^t\Sigma^{-1}\underline{X}$	Equivalent SN ratio obtained when the signal used was uniformly distributed over L=20 dimensions	k	a	Error prob. attained at terminal phase of DDM	Error prob. for template machines
0.5	-16 dB	2.050	0.5889	0.3675	0.3619
1	-13	1.589	0.7165	0.3155	0.3085
2	-10	1.291	0.8336	0.2444	0.2398
5	-6	1.081	0.9441	0.1328	0.1318
10	-3	1.019	0.9849	0.0569	0.0569
20	0	1.0021	0.9980	0.0127	0.0127

b. Modified DDM (MDDM)

The DDM computes a posteriori probability $P(\theta_n/\underline{Y}^n,\hat{\theta}^{n-1})$ only for a teaching sequence $\hat{\theta}^{n-1}$ which the machine has actually decided, and uses it instead of $P(\theta_n/\underline{Y}^n)$, where $\underline{Y}^n$ is a sequence of $\underline{Y}_1,\underline{Y}_2,\ldots,\underline{Y}_n$. On the other hand, the modified DDM(MDDM) proposed here computes a posteriori probabilities for all possible sequences in the past m intervals, within the extent that the machine structure is not too large, to approximate $P(\theta_n/\underline{Y}^n)$ well, and makes the decision by employing thus obtained probabilities. That is, $P(\theta_n=1/\underline{Y}^n)$ is approximated by $P(\theta_n=1/\underline{Y}^n,\hat{\theta}^{*n-m-1})$ as

$$P(\theta_n=1/\underline{Y}^n,\hat{\theta}^{*n-m-1}) = \sum_{\theta_m^{n-1}} P(\theta_m^{n-1})P(\theta_n=1/\underline{Y}^n,\hat{\theta}^{*n-m-1},\theta_m^{n-1}) \qquad (30)$$

where $\hat{\theta}^{*n-m-1}$ is a sequence of $\hat{\theta}_1^*,\hat{\theta}_2^*,\ldots,\hat{\theta}_{n-m-1}^*$, and $\hat{\theta}_k^*$ is a discrimination result of $\underline{Y}_k$ decided at the end of the (k+m)th interval. Consequently, the probability of mis-discrimination of $\hat{\theta}_k^*$ is less than that of $\hat{\theta}_k$ decided without the time delay. From these reasons described above, the MDDM seems to operate better than DDM.

c. Non-Decision-Directed Machine (NDDM)

As is shown previously, the decision-directed machine will not make a decision quite well and consequently will not converge to the true pattern vector at a signal-to-noise ratio much smaller than one. Let us now suggest a modified system designated as a non-decision-directed machine appropriate for the case of small signal-to-noise ratio [9]. The estimated pattern vector at the (n+1)th learning interval is now given by

$$\hat{\underline{\eta}}_{n+1} = \frac{1}{np} \sum_{i=1}^{n} \underline{Y}_i \qquad (31)$$

where p is the a priori probability that the input contains a pattern instead of $(\Sigma_{i=1}^{n} \hat{\theta}_i \underline{Y}_i)/(\Sigma_{i=1}^{n} \hat{\theta}_i)$ for the decision-directed machine. Since

$$E = [\hat{\underline{\eta}}_{n+1}] = \frac{1}{np} (np\underline{X} + \underline{0}) = \underline{X} \tag{32}$$

where $E[\cdot]$ = ensemble average and $\underline{0}$ = zero vector, $\hat{\underline{\eta}}_{n+1}$ is an unbiased and also consistent estimator by the law of large numbers. However, the optimal (maximum a posteriori probability at each time interval) system which makes use of only an unbiased and consistent estimator $\hat{\underline{\eta}}_{n+1}$ obtained from the set of all past input data $\underline{Y}^n$ will become complicated, because $\hat{\underline{\eta}}_{n+1}$ is not a Gaussian variable. Then, let us consider the case where $\hat{\underline{\eta}}_{n+1}$ is nearly Gaussianly distributed as in the case of the decision-directed machine. Under such a condition, covariance of $\hat{\underline{\eta}}_{n+1}$ may be expressed as

$$\text{Covar.}[\hat{\underline{\eta}}_{n+1}] = E[(\hat{\underline{\eta}}_{n+1}-\underline{X})(\hat{\underline{\eta}}_{n+1}-\underline{X})^t] = \underline{K}/n \tag{33}$$

where

$$\underline{K} = 1/p^2 \; [(p-p^2)(\underline{\Lambda} + \underline{M}\underline{M}^t) + \underline{\Sigma}] \tag{34}$$

Thus, after the example of the decision-directed machine, the learning rule equations for the non-decision-directed machine may be given by

$$\left.\begin{aligned} \underline{H}_{n+1} &= \underline{H}_n + \underline{K}^{-1}\underline{\Xi}_{n+1}\left(\frac{\underline{Y}_n}{p} - \underline{H}_n\right), \quad \underline{H}_1 = \underline{M} \\ \underline{\Xi}_{n+1} &= (\underline{\Xi}_n^{-1} + \underline{K}^{-1})^{-1}, \quad \underline{\Xi}_1 = \underline{\Lambda} \end{aligned}\right\} \tag{35}$$

This means that in the non-decision-directed machine $\underline{H}_{n+1}$ and $\underline{\Xi}_{n+1}$ are always updated in an iterative manner by using the input vector $\underline{Y}_n$ as an aid in making its next estimate, no matter whether the pattern was actually contained or not contained in $\underline{Y}_n$. On the other hand, the decision-directed machine applies the input vector $\underline{Y}_n$ to update only when the pattern was estimated to be present.

d. Experimental Results

A computer simulation of the DDM, CDDM, MDDM and NDDM, combining those comparative studies with experimental tests, were performed. The results of the computer simulation are shown in Figures 1-(a) and 1-(b). Both the pattern and noise were assumed to have white Gaussian distributions, that is, $\underline{\Lambda} = \lambda\underline{I}$, $\underline{\Sigma} = \sigma^2\underline{I}$, $\underline{M} = \underline{0}$, and the same fixed 20-dimensional pattern was used for all the tests, the probability p that the patterns occured was 1/2. The input SN ratios (=10 log λ/σ^2) used were -13 dB, -6 dB, and 0 dB. Then each curve is plotted as the average of the 10 runs at each desired time instant. ε_n is the normalized-squared-error of $\underline{H}_n$ given as $\varepsilon_n = [(\underline{H}_n-\underline{X})^t(\underline{H}_n-\underline{X})]/(\underline{X}^t\underline{X})$, and $\blacktriangleleft$ indicates ε_n at the

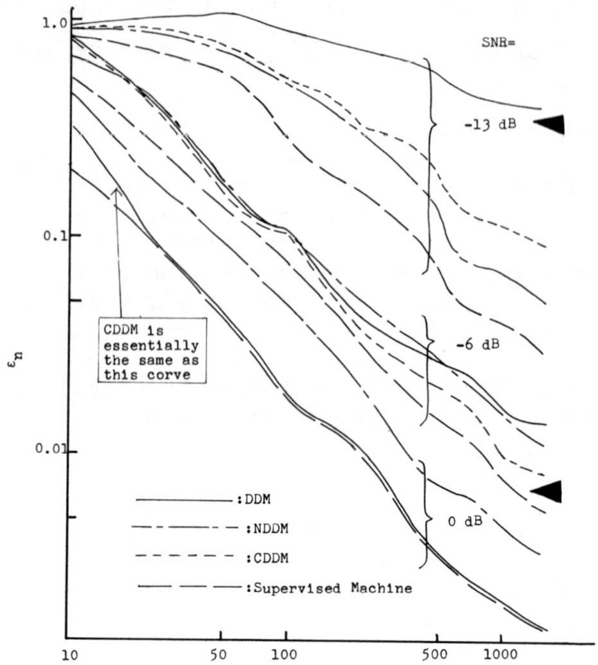

Figure 1-(a). Normalized-Squared-Error ε_n as a Function of Time in Learning Processes for Four Machines Used.

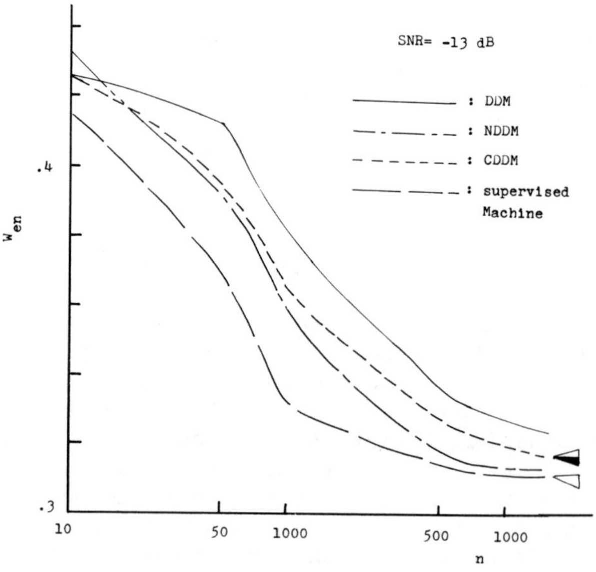

Figure 1-(b). Probability of Error W_{em} as a Function of Time in Learning Processes for Four Machines Used.

terminal phase of the DDM, that is, $\lim_{n \to \infty} \varepsilon_n = (k-1)^2$. W_{en} is a probability of error. $\triangleleft$ and $\blacktriangleleft$ indicate the probability of error of the machine matched to the template and that of the machine achieved at the terminal phase of DDM, respectively. As is seen in Figures 1-(a) and 1-(b), the probability of error of the NDDM is fairly improved than that of the DDM at low SN ratios.

3. SOME CONSIDERATIONS ON RECOGNITION OF TIME-VARYING PATTERNS

In this section we shall analyze a pattern recognizer for a set of time-varying patterns.

a. 1st-and 2nd-Order Autoregressive Processes

Let us now consider a pattern class given as a 1st-order autoregressive time series.

$$(\underline{X}_n - \underline{M}) + \underline{a}(\underline{X}_{n-1} - \underline{M}) = \underline{Z}_n \tag{36}$$

where $\underline{Z}_n$ is a Gaussian random variable and

$$\left.\begin{array}{l} E[\underline{Z}_n] = \underline{0}, \quad E[\underline{X}_{n-j}\underline{Z}_n^t] = \underline{0} \quad \text{for } j=1,2,\ldots \\ E[\underline{Z}_n \underline{Z}_k^t] = \underline{0} \quad \text{for } n \neq k, \quad = \underline{\psi} \text{ for } n = k \end{array}\right\} \tag{37}$$

Then the distribution of $\underline{X}_n$ will be approximated by the Gaussian law of $\mathcal{N}(\underline{M},\underline{\Phi})$. The a posteriori mean $\underline{H}_n$ and covariance $\underline{\Xi}_n$ of $\underline{X}_n$ are given as follows [14]:

$$\underline{H}_1 = \underline{M}, \quad \underline{\Xi}_1 = \underline{\Phi} \tag{38-1}$$

$$\underline{H}_{n+1} = \underline{M} - \underline{a}\{(\underline{H}_n - \underline{M}) + \theta_n(\underline{\Xi}_n^{-1} + \underline{\Sigma}^{-1})^{-1}\underline{\Sigma}^{-1}(\underline{Y}_n - \underline{H}_n)\} \tag{38-2}$$

$$\underline{\Xi}_{n+1} = \underline{\psi} + \underline{a}(\underline{\Xi}_n^{-1} + \theta_n \underline{\Sigma}^{-1})^{-1}\underline{a}^t \tag{38-3}$$

Unlike the fixed pattern case, $\underline{H}_n$ does not converge to $\underline{X}_n$, and $\underline{\Xi}_n$ does not vanish even if n becomes large enough. Consequently, the machine does not converge to the template machine, but remains at the quadric discriminator even for large n.

On the other hand, from the viewpoint of estimation theory, the pattern recognition may be viewed as an estimation problem of the state vector in the (n+1)th interval when the measurements in the 1,2,...,nth intervals are given. A value of $\theta_n = 0$ may be regarded as the lack of the measurement for that value of n. Here we will show how our approach ties to a Kalman filter. Let $\underline{x}_n = \underline{X}_n - \underline{M}$, $y_n = \underline{Y}_n - \underline{M}$, and $\underline{h}_n = \underline{H}_n - \underline{M}$. Then the system considered here is rewritten as

$$\underline{X}_n = -\underline{a}\underline{X}_{n-1} + \underline{Z}_n \tag{39}$$

and the measurement is given by

$$\underline{Y}_n - \theta_n \underline{M} = \theta_n \underline{X}_n + \underline{N}_n \tag{40}$$

The optimal estimation of this system may be given by [15]

$$\hat{\underline{X}}_{n/n} = -\underline{a}\hat{\underline{X}}_{n-1/n-1} + \theta_n \underline{P}_{n/n}\underline{\Sigma}^{-1}(\underline{Y}_n + \underline{a}\hat{\underline{X}}_{n-1/n-1}) \tag{41}$$

$$\underline{P}_{n/n-1} = \underline{a}\underline{P}_{n-1/n-1}\underline{a}^t + \underline{\psi} \tag{42}$$

$$\underline{P}_{n/n} = [\underline{P}^{-1}_{n/n-1} + \theta_n \underline{\Sigma}^{-1}]^{-1} \tag{43}$$

where $\hat{\underline{X}}_{i/j}$ is the best estimate of $\underline{X}_i$ when the measurements in the 1,2,...,jth intervals are given, and $\underline{P}_{i/j}$ is its covariance. In our pattern recognition problem, $\underline{h}_n$ corresponds to $\hat{\underline{X}}_{n/n-1}$, and $\underline{\Xi}_n$ corresponds to $\underline{P}_{n/n-1}$. Further, the single-stage optimal predicted estimate is given by

$$\underline{h}_n = \hat{\underline{X}}_{n/n-1} = -\underline{a}\hat{\underline{X}}_{n-1/n-1} \tag{44}$$

Substituting (43) nto (42), and replacing $\underline{P}_{n/n-1}$ with $\underline{\Xi}_n$, we obtain

$$\underline{\Xi}_n = \underline{\psi} + a(\underline{\Xi}^{-1}_{n-1} + \theta_{n-1}\underline{\Sigma}^{-1})^{-1}\underline{a}^t$$

This is the same as (38-3). Multiplying (41) by $-\underline{a}$ from the left side, and using (44), we obtain

$$\underline{h}_{n+1} = -\underline{a}\{\underline{h}_n + \theta_n \underline{P}_{n/n}\underline{\Sigma}^{-1}(\underline{Y}_n - \underline{h}_n)\}$$

Then

$$\underline{H}_{n+1} - \underline{M} = -\underline{a}\{(\underline{H}_n - \underline{M}) + \theta_n(\underline{\Xi}_n^{-1} + \underline{\Sigma}^{-1})^{-1}\underline{\Sigma}^{-1}(\underline{Y}_n - \underline{H}_n)\}$$

This is the same as (38-2). The behavior of $\underline{H}_n$ and $\underline{\Xi}_n$ as $n \to \infty$ are well known in the optimal estimation theory, though few explicit statements have been done.

When the patterns vary slowly from time to time according to a 2nd-order autoregressive process as

$$\underline{Z}_n = (\underline{X}_n - \underline{M}) + \underline{a}_1(\underline{X}_{n-1} - \underline{M}) + \underline{a}_2(\underline{X}_{n-2} - \underline{M})$$

we can also obtain the same type of recursive relations as eq. (38) [15]. That is,

$$\underline{H}_1 = \underline{M}, \quad \underline{\Xi}_1 = \underline{\Phi}$$

$$\underline{H}_n = \underline{M} + \underline{\Xi}_n[\underline{g}_{n-1} + \underline{\beta}^t_{n-1}(\underline{\alpha}_{n-1} + \theta_{n-1}\underline{\Sigma}^{-1})^{-1}\{\underline{f}_{n-1} + \theta_{n-1}\underline{\Sigma}^{-1}(\underline{Y}_{n-1} - \underline{M})\}]$$

$$\underline{\Xi}_n = \{\underline{\gamma}_{n-1} - \underline{\beta}^T_{n-1}(\underline{\alpha}_{n-1} + \theta_{n-1}\underline{\Sigma}^{-1})^{-1}\underline{\beta}_{n-1}\}^{-1}, \quad n=2,3,..., \tag{45}$$

where

$$\underline{\alpha}_{k+1} = \underline{\gamma}_k + \underline{a}_1^t\underline{\psi}^{-1}\,\underline{a}_1 - \underline{\beta}_k'^{\,t}\underline{\alpha}_k'\underline{\beta}_k' \;;\quad \underline{\beta}_{k+1} = -(\underline{a}_1^t + \underline{\beta}_k'^{\,t}\underline{\alpha}_k'\underline{a}_2^T)\underline{\psi}^{-1}$$

$$\underline{\gamma}_{k+1} = \underline{\psi}^{-1} - \underline{\psi}^{-1}\underline{a}_2\underline{\alpha}_k'\underline{a}_2^t\underline{\psi}^{-1} \;;\quad \underline{f}_{k+1} = \underline{g}_k + \underline{\beta}_k'^{\,t}\underline{\alpha}_k'\{\underline{f}_k + \theta_k\underline{\Sigma}^{-1}(\underline{Y}_k - \underline{M})\}$$

$$\underline{g}_{k+1} = -\underline{\psi}^{-1}\underline{a}_2\underline{\alpha}_k'\{\underline{f}_k + \theta_k\underline{\Sigma}^{-1}(\underline{Y}_k - \underline{M})\};\quad \underline{\alpha}_k' = (\underline{\alpha}_k + \underline{a}_2^t\underline{\psi}^{-1}\underline{a}_2 + \theta_k\underline{\Sigma}^{-1})^{-1}$$

$$\underline{\beta}_k' = \underline{\beta}_k - \underline{a}_2^t\underline{\psi}^{-1}\underline{a}_1 \qquad (46)$$

Each of the parameters $\underline{\alpha}_k$, $\underline{\beta}_k$, and $\underline{\gamma}_k$ is an inverse of a covariance matrix. $\underline{f}_k$ is a parameter conveying information about $\underline{X}_k$, and $\underline{g}_k$ is one about $\underline{X}_{k+1}$.

b. Computer-Experimental Results and Considerations

When $\underline{\Phi} = \Phi\underline{I}$ and $\underline{\Phi}_n = \Phi_n\underline{I}$, the normalized autocorrelation function of the 1st-order autoregressive process is an exponential function $\rho_n = \Phi_n/\Phi = \rho_1^n$. For the 2nd-order autoregressive process, the normalized autocorrelation function is $\rho_n = b\alpha^n + (1-b)\beta^n$, where b is a constant, and α and β are two roots of the equation $t^2 + a_1t + a_2 = 0$. We show some examples of the normalized autocorrelation function of the 1st-order and the 2nd-order autoregressive processes in Figure 2.

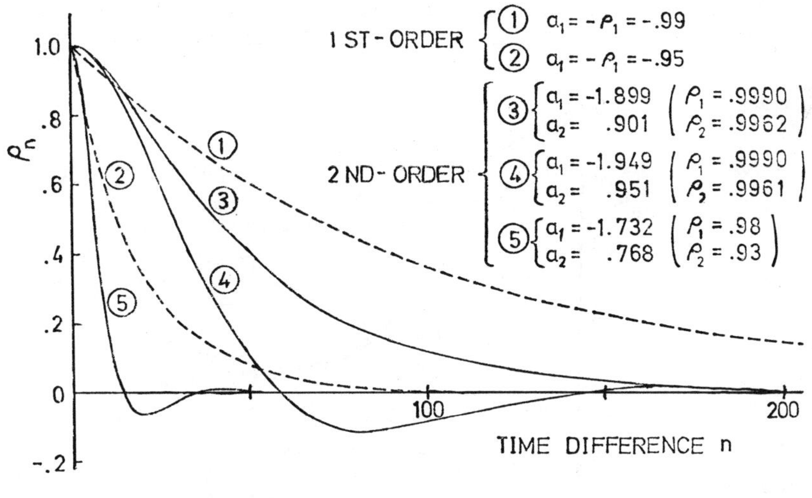

Figure 2. Normalized Autocorrelation Functions of Autoregressive Processes.

Under the conditions of L=20, p=1/2, $\underline{M}=\underline{0}$, $\underline{\Sigma}=\sigma^2\underline{I}$, $\underline{\Phi}=\Phi\underline{I}$, $\underline{\Phi}_n=\Phi_n\underline{I}$, $\underline{\psi}=\psi\underline{I}$, m=5, and $\underline{a}_i=a_i\underline{I}$, we have simultaneously made the computer simulations of the three types of the machines above discussed; i.e., the DD machine, the MDD machine, and the supervised machine, for some of the processes having the autocorrelations shown in Fig. 2. The averaged results of 100 runs are shown in Figure 3-(a) and 3-(b). PNR is a nominal energy ratio of pattern to noise defined by PNR = 10 $\log(\Phi/\sigma^2)$ [dB]. Therefore, the actual pattern to noise ratio fluctuates momentarily around the (nominal) PNR according to the fluctuation of the pattern. Symbols used in Figure 3-(a) and 3-(b) are as follows:

$$\varepsilon_n = [(\underline{H}_n-\underline{X}_n)^t(\underline{H}_n-\underline{X}_n)] \,/\, \underline{X}_n^t\underline{X}_n \; :$$

normalized squared error

$$W_{en} = P(\theta_n=1)P(\hat{\theta}_n=0/\theta_n=1)+P(\theta_n=0)P(\hat{\theta}_n=1/\theta_n=0) \; :$$

probability of dichotomization error

◁ : $\varepsilon=\varepsilon_n$ in the stationary region of 1st-order autoregressive process

Since all the three machines have the function to discriminate the measurements by the normalized distance between the estimate $\underline{H}_n$ of the pattern and the measurement $\underline{Y}_n$, we used the squared error of $\underline{H}_n$ in addition to the probability of error as the performance index of the machines.

The MDD machine has $\underline{H}_n$ and $\underline{\Xi}_n$ of 2^5. However, we have calculated ε_n and W_{en} using only one pair of $\underline{H}_n$ and $\underline{\Xi}_n$ corresponding to $\underline{\theta}_m^{n-1}$ that has the maximum a posteriori probability. Since these W_{en} are theoretical ones of the quadric dichotomizer that employs only

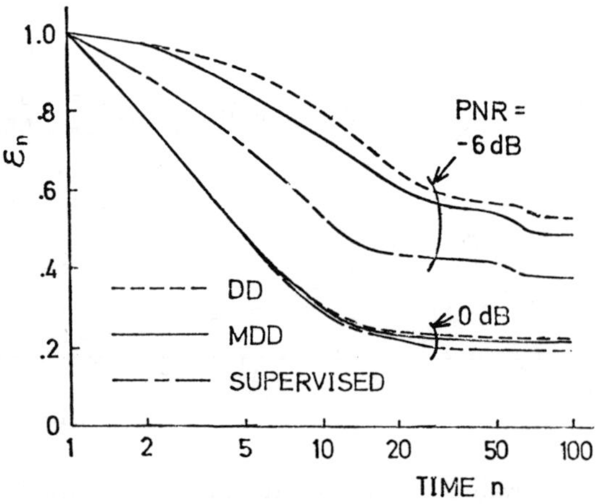

Figure 3-(a). Normalized Squared Error of Estimates for 2nd-Order Autoregressive Process ④ in Figure 2.

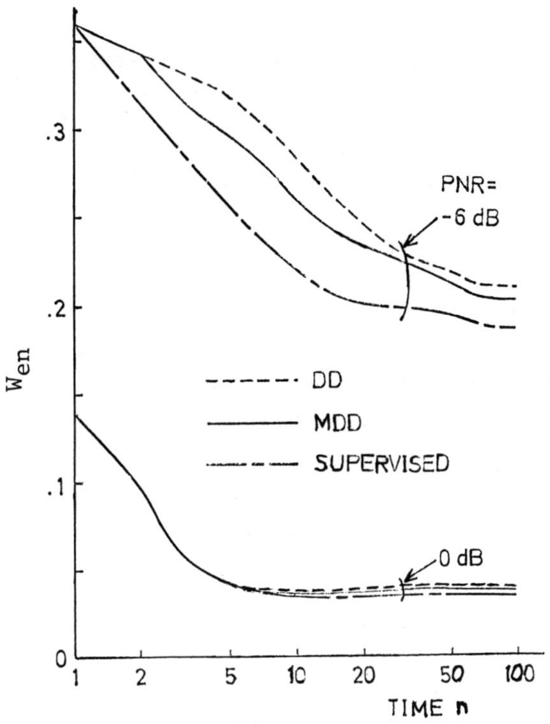

Figure 3-(b). Probability of Error of Three Machines for 2nd-Order Autoregressive Process (4) in Figure 2.

one pair of $\underline{H}_n$ and $\underline{\Xi}_n$, the actual probabilities of the dichotomization error are slightly better than those values.

From these results, we can see that the effect of the learning appears within the time having comparatively large autocorrelation coefficient ρ_i, and as ρ_i becomes smaller, it comes to the stationary state, where the variation of the pattern is balancing with the quantity of the learning. For example, in Figure 2, the autocorrelation of (4) becomes very small after the time difference 50. Corresponding to this fact, in Figure 3-(a) and 3-(b), the performances of the machines are poorly improved after the time 50. We can also see that the performance of the MDD machine is better than the DD machine especially when the noise is large. This is because of two reasons. First, the structure of the MDD machine is closer to that of the optimum nonsupervised machine than that of the DD machine. Second, the DD machine is compelled to use biased measurements to update $\underline{H}_n$. The bias accompanies the DD approach because misclassified samples of $\underline{Y}_n$ are included in the estimator $\underline{H}_n$.

4. CONCLUSION

We have analyzed several parametric-type pattern recognizers based on nonsupervised learning and proposed the three types of machines; i.e., CDDM, MDDM, and NDDM. The superiority of the proposed machines was demonstrated by computer simulation for time-varying patterns as well as fixed patterns.

REFERENCES

1. H. J. Scudder, "Probability of Error of Some Adaptive Pattern-Recognition Machines," IEEE Trans. on Information Theory, Vol. IT-11, No. 4, July 1965, p. 336.

2. E. M. Glaser, "Signal Detection by Adaptive Filters," Trans. IRE, Vol. IT-7, No. 2, April 1961, p. 87.

3. N. Abramson, et.al., "Learning to Recognizer Patterns in a Random Environment," Trans. IRE, Vol. IT-8, Sept. 1962, pp. 58-63.

4. F. Rosenblatt, "The Perceptron, A Perceiving and Recognizing Automaton," Cornell Aeronautical Lab. Rept., No. 85-460-1, Jan. 1957.

5. B. Widrow, "Generalization and Information Storage in Networks of ADALINE "Neurons", in Self-Organizing Systems edited by Yovits, Jacobi, and Goldstein (eds.), pp. 435-461, Spartan Books, Washington, D.C., 1962.

6. E. G. Henrichon and K. S. Fu, "A Nonparametric Partitioning Procedure for Pattern Classification," IEEE Trans. on Computers, Vol. C-18, No. 7, July 1969, pp. 615-624.

7. C. V. Jackowatz, R. L. Shuey and G. M. White, "Adaptive Waveform Recognition," 4th Intl. Symp. on Inf. Theory, London, in Information Theory edited by C. Cherry, pp. 317-326, Butterworths Book, 1961.

8. K. Tanaka, et.al., "An Identification Method of System Characteristics Using a New Type of Adaptive Correlating Filter," Proc. IFAC, Tokyo Symp., August 1965, p. 245.

9. D. B. Cooper and P. W. Cooper, "Nonsupervised Adaptive Signal Detection and Pattern Recognition," Information and Control, Vol. 7, No. 3, Sept. 1964, p. 416.

10. K. S. Fu and C. H. Chen, "Sequential Decision, Pattern Recognition and Machine Learning," Rept. TR-EE 65-6, Purdue Univ., April 1965.

11. C. G. Hilborn and D. G. Lainiotis, "Optimal Unsupervised Learning Multicategory Dependent Hypotheses Pattern Recognition," IEEE Trans. on Information Theory, Vol. IT-14, May 1968, pp. 468-470.

12. Y. T. Chien and K. S. Fu, "Stochastic Learning of Time-Varying Parameters in Random Environment," IEEE Trans. on Systems Science and Cybernetics, Vol. SSC-5, No. 3, July 1969, pp. 237-246.

13. K. Tanaka and S. Tamura, "Some Considerations on a Type of Pattern Recognition Using Nonsupervised Learning Procedure," IFAC Intl. Symp. on Tech. and Biol. Problems of Control, Yerevan, Armenia, Sept. 1968. Also Trans., IECE of Japan, Vol. 52, No. 2, pp. 165-173, Feb. 1969.

14. S. Tamura and K. Tanaka, "On the Recognition of Time-Varying Patterns Using Learning Procedures", IEEE Trans. on Information Theory, (to appear July 1971).

15. R. C. K. Lee, Optimal Estimation, Identification and Control, MIT Press, New York, 1969, pp. 43-49.

LINEAR AND NONLINEAR STOCHASTIC APPROXIMATION ALGORITHMS FOR LEARNING SYSTEMS*

Y. T. Chien

The University of Connecticut

Storrs, Connecticut, U.S.A.

I. INTRODUCTION

Frequently in the design of on-line learning systems for pattern recognition or system identification, there is a need to construct successive estimates for the unknown parameters of some underlying probability distribution. One of the most widely used methods for this purpose is stochastic approximation. It is well known that stochastic approximation is concerned with the successive estimation algorithms which converge to the true value of some sought (unknown) parameter when, due to the random nature of the system environment, the measurements are inevitably noisy. The algorithms of most interest to on-line pattern recognition or system identification are those which have following properties: (1) they are self-correcting, that is, the error of estimates tends to vanish in the limit, and (2) their convergence to the true value of an unknown parameter is of some specific nature, for example, in mean square or with probability one. This paper will discuss two types of algorithms which have been developed to generate recursive estimates that are linear or nonlinear functions of the past measurements. These algorithms will be discussed in terms of their computational structure as well as their statistical properties such as mean square error. In particular, comparison will be made between linear and nonlinear algorithms on the basis of specific assumptions about the unknown inputs and parameters characterizing the learning environment.

---

*This work was supported in part by National Science Foundation Grant GK-4696.

Although the treatment here will be in the context of parameter estimation and pattern recognition, the concepts and algorithms developed are applicable to other learning systems involving identification or recognition.

II. LEARNING TO RECOGNIZE PATTERNS--LINEAR ALGORITHMS

One of the problems in learning to recognize patterns in a random environment is to estimate the unknown parameters of a conditional probability distribution characterizing a class of patterns. The estimated parameters for each class of patterns then determine the structure as well as the performance of a recognition device.

Let us suppose that a pattern sample to be recognized has been adequately described by a set of measurements $(x_1, x_2, \ldots, x_p)$ as a column vector X. One can interpret the vector X as a point in a p-dimensional space. The function of a recognition device is to classify the vector X as belonging to one of m possible classes, say $C_1 C_2, \ldots, C_m$, or equivalently, to partition the p-space of all possible pattern measurements into subspaces. In most applications it is desired that the partitioning of the p-space be made in an optimal fashion, i.e., with minimum probability of misclassification. To be specific, let the m classes of patterns be described by m conditional probability densities, denoted by $P(X/C_1), P(X/C_2), \ldots, P(X/C_m)$. It has been shown that under certain conditions, an optimal recognition device is to decide X as belonging to class C_i if

$$P(X/C_i) \geqq P(X/C_j) \quad \text{for all } j \tag{1}$$

Since the choice is determined by the maximum of the m values $P(X/C_j)$, one can just as well compute an arbitrary monotone function of $P(X/C_j)$, the likelihood function, to select C_j. This is the well-known maximum likelihood procedure. What is implied in this procedure is that one has been given a complete statistical description of the m classes of patterns in terms of the m probability densities. This is of course an unrealistic assumption in practical applications. A more realistic case in learning to recognize patterns is to assume that a complete statistical description is not given. In fact, through a learning process, the conditional probability densities should be allowed to change on the basis of pattern vectors that have been previously classified. These vectors are commonly called learning samples, denoted by $X_1, X_2, \ldots, X_n$, with subscript indicating the sequence of classification. Since these vectors are classified samples, we can assume a sequence of learning samples $X_1, X_2, \ldots, X_n$ to be the n learning samples for each and every conditional density under consideration.

In order to see a more detailed structure of the learning process, let us assume a particular form for the conditional density $P(X/C_j)$. We assume that the vectors X_i, $i = 1,2,\ldots,n$, are samples from a multivariate normal density with mean vector M and covariance matrix K. That is,

$$P(X) = [(2\pi)^P|K|]^{-1/2}[\exp - 1/2(X-M)^T K^{-1}(X-M)] \quad (2)$$

where M is a column vector consisting elements $m_1, m_2, \ldots, m_p$ and $E(x_i) = m_i$. Note that the class notation C_j has been dropped from the conditional density as it will be understood that this process is repeated for each and every class. Now consider the learning process in which the mean vector M is unknown. Abramson and Braverman [1] have shown that the successive estimates for M, M_n, can be obtained by computing the conditional expectation of M given the learning samples. The resulting estimate M_n is a weighted average of an initial guess for M, M_0, and the sample mean vector. One can also arrive at an equivalent result by considering the learning samples to be composed of a constant mean M and an independent gaussian noise N_i with zero mean and covariance matrix K. That is,

$$X_i = M + N_i \quad (3)$$

It has been shown by Chien and Fu [2] that the successive estimate M_n for M can be obtained by using a stochastic approximation algorithm of the following type:

$$M_n = M_{n-1} + \gamma_n (X_n - M_{n-1}) \quad (4)$$

where γ_n is a sequence of numbers satisfying certain conditions. The recursive estimate M_n defined in (4) will converge to the true mean M in mean square and with probability one. In addition, if γ_n is chosen according to the initial guess M_0 as in [1], the mean square error for M_n computed for each n is minimized. In terms of our recognition problem, we see that the recursive algorithm defined in (4) enables the recognition device to lock into the unknown patterns by learning the mean vector for each and every class. The optimal recognition performance (with minimum probability of misclassification) is achieved in the long-run as the number of learning samples increases indefinitely.

It should be pointed out that the relationship in (4) defines the simplest form of a stochastic approximation algorithm for estimating the unknown parameter M. This algorithm is <u>linear</u> in that only the first order difference $(X_n - M_{n-1})$ is used in modifying the previous estimate M_{n-1} in order to compute the current estimate M_n. In essence, M_n is a linear combination of the past learning samples. Linear algorithms of this type have proved to be very useful for learning process which involves reliable measurements representing the input pattern vectors.

III. A NONLINEAR ALGORITHM WITH THRESHOLD EFFECT

Let us now consider a natural extension to the linear algorithm defined in (4) and make the transition from linear to nonlinear relationship in the following way:

$$M_n = M_{n-1} + F_n(D_n) \tag{5}$$

where

$$D_n = X_n - M_{n-1}$$

The term $F_n(D_n)$ in (5) is in general a nonlinear function in D_n. Our objective here is to determine the characteristics of this type of nonlinear algorithms in relation to a learning process that involves the use of unreliable pattern samples. Unreliable samples may result from the inherent variability of patterns or from the imperfect measurements due to excessive noise encountered in practice. In either case, a simple linear algorithm may not work satisfactorily.

In order to see how a nonlinear algorithm may improve the situation, we must describe the unreliable samples in a mathematically meaningful way. The following assumptions are made regarding the composition of learning samples.

<u>Assumption (i).</u> Let each learning sample X_i again be composed of two independent components as in (3). That is, $X_i = M + N_i$, where N_i is the noise component and M is the unknown mean vector.

<u>Assumption (ii).</u> Each N_i is attributed to two types of gaussian noise. Type I noise is associated with the ordinary measurement variation that carries the reliable information for M. Type II noise characterizes the unreliable information in regard to the unknown M. Let type I noise be described by a multivariate gaussian

density with zero mean and covariance matrix K_1 and type II noise be described by a similar gaussian density but with a covariance matrix K_2. Note that this assumption implies that the two types of noise differ only in their covariance matrices.

We further assume that K_1 and K_2 differ by a proportional constant such that $K_1 = K_2/\beta$, $\beta >> 1$.

Assumption (iii). For each N_i, type I noise can occur with a probability $(1 - \alpha)$ and type II noise can occur with a probability α where $\alpha << 1$. Thus if $P_1(N)$ denotes the probability density of type II, then the noise component of X_i can be described by a mixture distribution, with probability density $P(N)$:

$$P(N) = (1 - \alpha)\, P_1(N) + \alpha P_2(N) \tag{6}$$

Returning to the nonlinear algorithm defined in (5), we are interested in determining the functional form $F_n(\cdot)$ so that M_n can be obtained in some optimal fashion, say, minimum square error. To achieve this, let us define an error vector at the nth approximation, $e_n = M_n - M$. The recursive relation in (5) can then be rewritten in terms of the error vectors.

$$e_n = e_{n-1} + F_n(D_n) \tag{7}$$

To determine the optimal form of $F_n(D_n)$, it is only necessary to compute the conditional expectation of e_{n-1}, given D_n. After some calculation based on our three assumptions, we have the following approximate form for $F_n(D_n)$:

$$F_n(D_n) \approx \frac{K_{n-1}[(1-\alpha)\, P_1(D_n)K_1^{-1} + \alpha P_2(D_n)K_2^{-1}]}{(1-\alpha)\, P_1(D_n) + \alpha P_2(D_n)}\, D_n \tag{8}$$

Let

$$d(D_n) = D_n^T\, K_1^{-1}\, D_n \tag{9}$$

Note that $d(D_n)$ can be considered as a measure of the distance between the current learning sample X_n and the previous estimate for the mean, M_{n-1}. We can discuss the behavior of $F_n(D_n)$ in (8) by finding its asymptotes in terms of the measure $d(D_n)$.

Case I. $d(D_n)$ is small. Since $K_1 = K_2/\beta$ where $\beta \gg 1$, a good approximation for $F_n(D_n)$ can be shown to be

$$F_n(D_n) \approx K_{n-1}K_1^{-1}D_n \tag{10}$$

Case II. $d(D_n)$ is large such that $P_1(D_n) \ll P_2(D_n)$ and $P_1(D_n) \to 0$ but $P_2(D_n) \neq 0$. This is the most important case of interest to us insofar as dealing with unreliable samples is concerned. Then, after some calculation, we have

$$F_n(D_n) \approx K_{n-1}\, K_2^{-1}\, D_n \tag{11}$$

Combining case (I) and case (II), we can characterize the asymptotic behavior of $F_n(D_n)$ in the following way: If the current learning sample X_n is sufficiently close to the previous estimate M_n, $F_n(D_n)$ takes the form of D_n, weighted by a factor $K_{n-1}\, K_1^{-1}$. If X_n is far away from M_{n-1} (which most likely signifies an unreliable sample), $F_n(D_n)$ is essentially D_n but weighted by a factor $K_{n-1}K_2^{-1}$. Since $K_{n-1}\, K_2^{-1}$ is negligible ($\beta \gg 1$), a good approximation to $F_n(D_n)$ can be mathematically described by selecting a threshold, say θ, so that

$$F_n(D_n) \approx \begin{cases} K_{n-1}K_1^{-1}D_n & \text{if } d(D_n) \leqq \theta \\ 0 & \text{if } d(D_n) > \theta \end{cases} \tag{12}$$

This is the threshold effect that characterizes the approximation to the optimal nonlinear transformation $F_n(D_n)$. It implies that one can simply discard the learning samples that are found unreliable and at the same time carry out a linear transformation on the remaining samples that seem reliable.

IV. A NUMERICAL EXAMPLE AND A RECOGNITION EXPERIMENT

Several other aspects must be investigated before the nonlinear algorithm derived in Section III can be applied to numerical calculations or physical experiments. For example, one must have the means of computing the error covariance K_n at each stage of the learning process. In addition, a proper selection of the threshold θ is also important for a desirable learning performance. The interested reader is referred to [3] for further discussions on these subjects. For illustration purpose, we give a numerical example and describe a character recognition experiment in which both linear and nonlinear algorithms are used in learning the unknown parameters.

(a) A Numerical Example

This example deals with the successive calculation of the mean value m of a normal distribution, with variance σ_1^2, which is mixed with a second normal distribution with the same mean value but a variance ten times as large ($\beta = 10$). The mixing probability α was chosen to be 0:1 so that the assumptions made in Section III should hold. Several threshold values were chosen to allow the calculation of and plot of mean square errors for a number of performance curves as a function of learning samples. The mean square errors for both linear and nonlinear algorithms were plotted in Figure 1 in which the "effective" number of samples refers to the actual number of samples used in the successive updating of the mean values. Note that the mean square error of nonlinear algorithm was smaller than that of linear algorithm and the errors could be controlled by proper thresholding in the learning process.

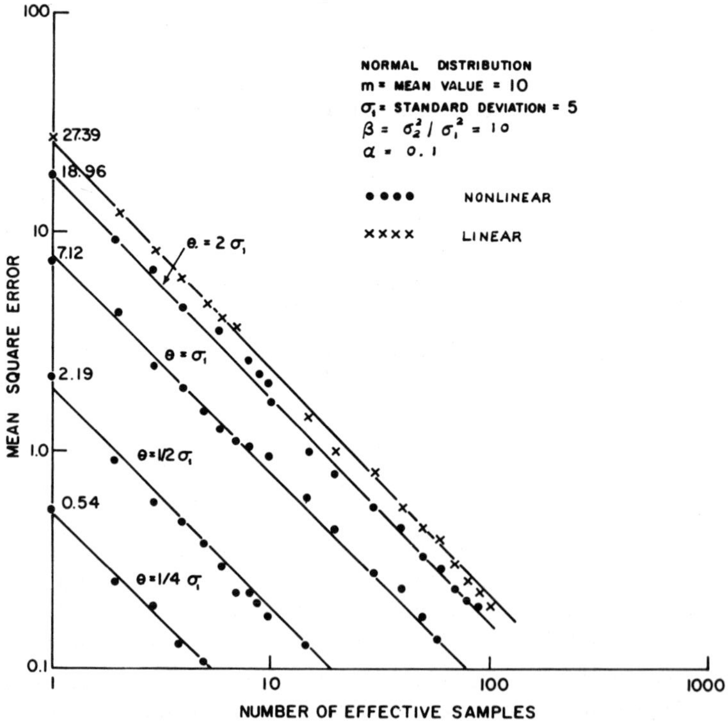

Figure 1. Mean Square Errors as a Function of Learning Samples.

(b) A Recognition Experiment

This experiment deals with a computer-simulated recognition of character samples. A selection of four English letters was chosen, each letter had two distinct categories of samples. Samples of category I belong to those which carry reliable information about the letter in question and samples of category II were generated to represent those that carry extremely noisy information. A computer program was written to simulate two primary functions in the recognition experiment--the computation of recursive estimates for the mean vectors of the four letters and the recognition function based on the estimation. Both linear and nonlinear algorithms were simulated in the program for the purpose of comparison in their learning performance. Figure 2 shows the recognition rates (percentage of correction recognition) as a function of effective learning samples. Note that for the nonlinear algorithm, samples of category II were found to be mostly discarded by the thresholding effect, and consequently the recognition rate showed a marked improvement.

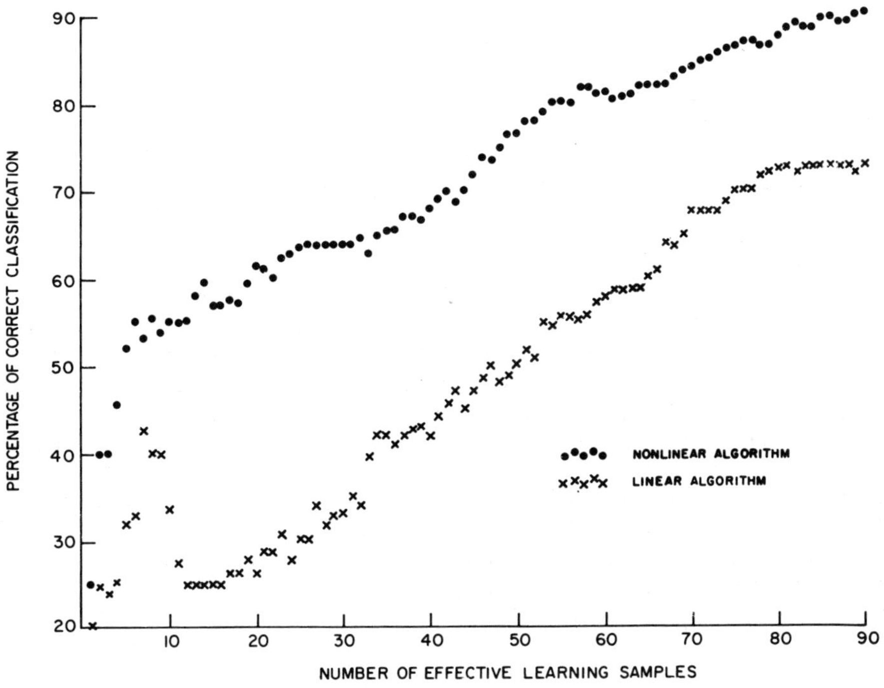

Figure 2. Recognition Rates as a Function of Learning Samples.

V. DISCUSSIONS AND FURTHER REMARKS

During the seminar at Nagoya, Dr. Widrow of Stanford University pointed out that he and his colleagues had experienced similar problems which motivated the study of nonlinear learning algorithm in pattern recognition. Dr. Widrow commented that when sample size was small, a noisy input data sample would drive the weight vector (e.g., mean values) far off. One of Dr. Widrow's students, Dr. Lee R. Talbert, has derived a <u>linear</u> algorithm which has similar objectives to the nonlinear approach discussed in the paper. Talbert [16] allows the desired response associated with the given input pattern sample to be adapted as well as the weight vector is adapted. This is an extension of the least mean square (LMS) algorithm which causes the weight vector to converge to a vector linear combination of the LMS solution and the matched filter solution. The sample statistics determine the mixture rates. It is found in [16] that the solution vector tends to LMS or matched filter, depending on which is more advantageous.

Although the problem treated in this paper is in the context of parameter estimation and pattern recognition, a number of related works have attacked the problem in other applications. Gilbo and Chelpanov [6] considered the problem of syntherizing a nonlinear characteristic for an inertialess component in the presence of pulse noise. The threshold effect of an optimal (LMS) nonlinear algorithm was first derived by Rozov and Chelpanov [7] in connection with highly distorted signals characterized by univariate Gaussian noise. The nonlinear algorithm derived in this paper is essentially an extension of Rozov and Chelpanov's algorithm. Also related are problems of detecting outlying samples in a statistical population. Karlin and Truax [17] examined the nature of this problem from the Bayesian viewpoint and showed that the Bayes solution was also characterized by a threshold effect.

A different but related nonlinear algorithm should be mentioned here for a possible further study and comparison. This is the truncation algorithm characterized by a "clipping" effect. In the context of pattern recognition and learning, extremely noisy samples detected by this algorithm are not to be discarded altogether from the learning population but are used to modify the unknown parameters with a vector of constant norm, regardless of the actual measurements. Albert and Gardner [14] have discussed the convergence property of this type of algorithms and several theorems have been conjectured. Further studies, however, would have to be made in regard to its learning performance so that the effect of a clipping can be compared with that of a threshold discussed in this paper.

REFERENCES

1. N. Abramson and D. Braverman, "Learning to Recognize Patterns in a Random Environment," IRE Trans. Information Theory, Vol. IT-8, September 1962.

2. Y. T. Chien and K. S. Fu, "On Bayesian Learning and Stochastic Approximation," IEEE Trans. on Systems Science and Cybernetics, Vol. SSC-3, No. 1, June 1967.

3. Y. T. Chien, "The Threshold Effect of a Nonlinear Learning Algorithm for Pattern Recognition," J. Information Science, 2, July 1970.

4. R. Grimsdale, F. Sumner, C. Tunis and T. Kilburn, "A System for the Automatic Recognition of Patterns," in Pattern Recognition (edited by L. Uhr), John Wiley and Sons, 1966, pp. 317-338.

5. R. Bakis, N. Herbst and G. Nagy, "An Experimental Study of Machine Recognition of Hand-printed Numerals," IEEE Trans. on Systems Science and Cybernetics, Vol. SSC-4, No. 2, July 1968, pp. 119-132.

6. E. P. Gilbo and B. Chelpanov, "Optimal Nonlinear Transformation of Signals from Several Instruments, Taking the Unreliability of Their Operation into Account," Automation and Remote Control, Vol. 27, No. 2, 1966.

7. Yu. L. Rozov and I. B. Chelpanov, "The Use of the Stochastic Approximation Method When the Possibility Arises of Errors with the 'miss' Type," Automation and Remote Control, Vol. 29, No. 12, 1968.

8. T. Marill and D. Green, "Statistical Recognition Functions and the Design of Pattern Recognizers," IRE Trans. Electronic Computers, Vol. EC-9, No. 4, December 1960, pp. 472-477.

9. N. J. Nilsson, Learning Machines, McGraw-Hill, 1965, Chapter 3.

10. C. K. Chow, "On Optimum Recognition Error and Reject Tradeoff," IEEE Trans. on Information Theory, Vol. IT-16, No. 1, January 1970, pp. 41-46.

11. D. G. Keehn, "A Note on Learning for Gaussian Properties," IEEE Trans. Information Theory, Vol. IT-11, January 1965, pp. 126-132.

12. V. S. Pugachev, Theory of Random Functions, Pergamon Press, 1965.

13. T. W. Anderson, Introduction to Multivariate Statistical Analysis, John Wiley and Sons, 1958, p. 347.

14. A. E. Albert and L. A. Gardner, Stochastic Approximation and Nonlinear Regression, The MIT Press, 1967.

15. R. B. Hennis, "Recognition of Unnurtured Characters in Multifont Application," in Pattern Recognition (edited by L. Kanal), Thompson Book Company, 1968.

16. L. R. Talbert, "The Sum-line Extrapolative Algorithm and its Application to Statistical Classification Problems," IEEE Trans. on Systems Science and Cybernetics, Vol. SSC-6, No. 3, July 1970, pp. 229-239.

17. S. Karlin and Truax, "Slippage Problems," Ann. Math. Stat., Vol. 31, 1960, pp. 296-324.

ACKNOWLEDGMENT

The computational part of this work was carried out in the Computer Center of the University of Connecticut, which is supported in part by NSF Grant GJ-9.

MULTI-CATEGORY PATTERN CLASSIFICATION USING A NONSUPERVISED LEARNING ALGORITHM

Iwao Morishita and Ryuji Takanuki

University of Tokyo

Tokyo, Japan

I. INTRODUCTION

The purpose of this paper is to present a nonsupervised learning algorithm for multi-category pattern classification and to make it clear what types of classifications are obtained by using this algorithm.

During the last decade a number of nonsupervised learning algorithms for pattern classification or signal detection have been reported in the literature [1-8]. In the previous work, however, little attention has been paid to the steady-state behavior of such algorithms. In the case of supervised learning, if the convergence of an algorithm is proved, the algorithm is guaranteed to yield a unique solution, and hence there is no need to investigate its steady-state behavior.

In the problem of nonsupervised learning, the situation changes radically, because available data are restricted to a sequence of unclassified patterns. Recently, Ya. Z. Tsypkin [9] has formulated this problem as an optimization problem and developed a general method for solving the optimization problem. According to his method, we can obtain a recursive algorithm which converges to a minimum of a criterion function given. On the other hand, even if a recursive algorithm has been obtained from a heuristic consideration, we may consider a criterion function which is minimized by the algorithm. In either case, we should not assume, in general, that a recursive algorithm obtained will yield a unique steady-state solution, because criterion functions are often multimodal. Thus, in the case of nonsupervised learning, the following questions arise naturally. How many steady-state solutions will the algorithm yield?

With what type of classification is each solution associated? If we wish to answer to these questions, we must investigate the steady-state behavior of the algorithm.

In the algorithm proposed, weight vectors are represented as time functions and the rule for their adjustments is described by a differential equation. First, on the basis of the assumption that patterns in the input sequence are those obtained from a distribution by random sampling, analytical techniques are developed for finding the steady-state solutions of the differential equation and for determining the stability of each solution found. Next, by applying the techniques, the steady-state behavior of the algorithm to patterns sampled from a four-cluster distribution is analyzed in detail. It is shown that there exists several stable steady-state solutions and that each of them is associated with a reasonable classification such that one cluster or one pair of adjacent clusters in the distribution corresponds to one category.

II. THE ALGORITHM

Let a pattern be characterized by N features $x_1,x_2,\ldots,x_N$. These features can be represented by an N-dimensional vector $\underline{x}$, and the vector $\underline{x}$ can be interpreted as a point in the N-dimensional feature space X. The problem of classifying patterns into M categories $C_1,C_2,\ldots,C_M$ is equivalent to the problem of partitioning the space X into M disjoint subspaces $X_1,X_2,\ldots,X_M$.

As is well known, M-category classification can be performed by employing M discriminant functions. In this paper we use the following form of discriminant functions:

$$y_i(\underline{x}) = \underline{w}_i'\underline{x} \qquad i=1,2,\ldots,M, \tag{1}$$

where $\underline{w}_i$ is an N-dimensional vector called the weight vector and the prime denotes a transpose operation. The subspace X_i is defined as a region in which the following inequality holds:

$$\underline{w}_i'\underline{x} \geqq \underline{w}_j'\underline{x} \qquad \text{for all } j \neq i. \tag{2}$$

If two subspaces X_i and X_j share a common boundary B_{ij}, it is a segment of the hyperplane described by the equation

$$(\underline{w}_i-\underline{w}_j)'\underline{x} = 0 . \tag{3}$$

Clearly, this hyperplane passes through the origin and is perpendicular to the vector $\underline{v}_{ij} = \underline{w}_i-\underline{w}_j$. In general, X_i and B_{ij} are determined from the M weight vectors $\underline{w}_1,\underline{w}_2,\ldots,\underline{w}_M$. This may be shown explicitly by writing as

$$\begin{aligned} X_i &= X_i(\underline{w}_1,\underline{w}_2,\ldots,\underline{w}_M) && \text{for all } i\ , \\ B_{ij} &= B_{ij}(\underline{w}_1,\underline{w}_2,\ldots,\underline{w}_M) && \text{for all } i \text{ and for all } j \neq i. \end{aligned} \tag{4}$$

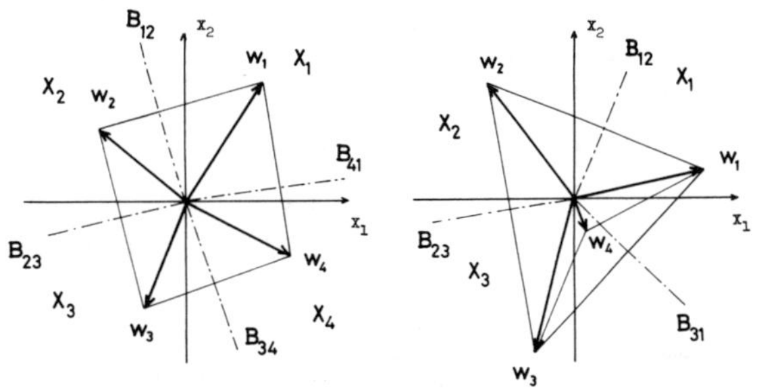

Figure 1. Partition of the Two-Dimensional Feature Space X by the Four Discriminant Functions $\underline{w}_i'\underline{x}$, i=1,2,3,4.

Figure 1 shows two examples of the partition of X for the case of N=2 and M=4. The distribution of patterns in X is described by a probability density function $p(\underline{x})$. In the following, we assume that the distribution consists of clear clusters of pattern points and that the clusters can be separated from each other by the use of the discriminant functions given by eq. (1).

Figure 2 shows a schematic diagram representation of the algorithm. Let patterns $\underline{x}(1),\underline{x}(2),\ldots,\underline{x}(k),\ldots$ be presented sequentially at intervals of Δ. Then we introduce a time function $\underline{x}(t)$ defined as

$$\underline{x}(t) = \underline{x}(k) \quad \text{for } (k-1)\Delta \leq t < k\Delta, \quad k=1,2,\ldots.$$

Each pattern in the input sequence $\underline{x}(t)$ is assigned to one of the M categories by comparing the M values

$$y_i(t) = y_i(\underline{x}(t)) = \underline{w}_i(t)'\underline{x}\ , \quad i=1,2,\ldots,M. \tag{5}$$

If $y_k(t)$ is the largest, the pattern is assigned to the category C_k. The outputs $f_1(t),f_2(t),\ldots,f_M(t)$ are two-valued functions such that

$$f_i(t) = \begin{array}{ll} 1 & \text{if } y_i(t) \geq y_j(t) \quad \text{for all } j \neq i, \\ 0 & \text{otherwise} \end{array} \tag{6}$$

Thus, the assignment to the category C_k is represented by the following values of the outputs:

$$f_k(t) = 1, \qquad f_j(t) = 0 \qquad \text{for all } j \neq k\ . \tag{7}$$

The rule for the adjustment of the weight vector $\underline{w}_i(t)$ is given by the differential equation

$$T\,\frac{d\underline{w}_i(t)}{dt} + \underline{w}_i(t) = f_i(t)\underline{x}(t) \qquad i=1,2,\ldots,M. \tag{8}$$

The coefficient T is assumed to be positive and much greater than Δ.

In the case of M=2, the algorithm can be implemented by the following form of a threshold logic unit:

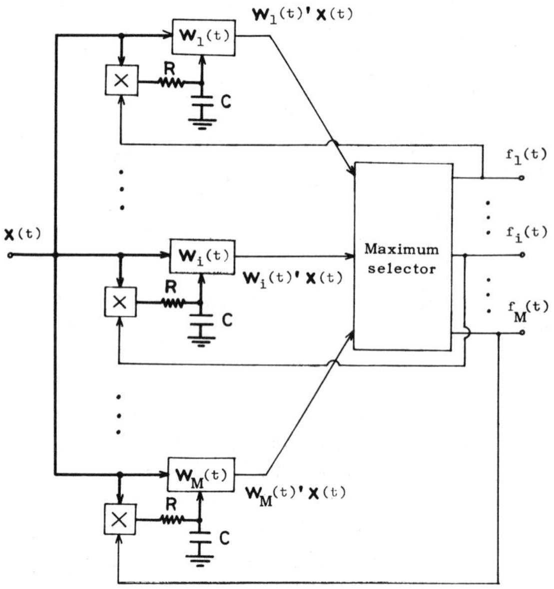

Figure 2. Schematic Diagram Representation of the Algorithm.

$$f(t) = \mathrm{sgn}[y(t)] = \mathrm{sgn}[\underline{w}(t)'\underline{x}] \ , \quad T\,\frac{d\underline{w}(t)}{dt} + \underline{w}(t) = f(t)\underline{x}(t) \ .$$

The behavior of this unit was investigated intensively in a previous paper [10].

III. STEADY-STATE SOLUTIONS OF THE ALGORITHM

Assume that there exists a time interval h such that

$$T \gg h \gg \Delta \ . \tag{9}$$

Then, from (8) it is clear that $\underline{w}_i(t)$ does not change appreciably for the interval h. In other words, we can assume that during the interval h the weight vector $\underline{w}_i(t)$ remains approximately constant. Thus, an approximation for $\underline{w}_i(t)$ is given by

$$\tilde{\underline{w}}_i(t) = \frac{1}{h}\int_{t-h}^{t} \underline{w}_i(t)dt \ , \qquad i=1,2,\ldots,M. \tag{10}$$

Averaging (8) from t-h to t yields

$$T\,\frac{d\tilde{\underline{w}}_i(t)}{dt} + \tilde{\underline{w}}_i(t) = \frac{1}{h}\int_{t-h}^{t} f_i(t)\underline{x}(t)dt, \qquad i=1,2,\ldots,M. \tag{11}$$

Assume that patterns in the input sequence $\underline{x}(t)$ are those obtained from a distribution by random sampling. Since during the interval h a large number of patterns are presented, we can put

$$\frac{1}{h}\int_{t-h}^{t} f_i(t)\underline{x}(t)dt = \int_{X_i(\tilde{\underline{w}}_1(t),\tilde{\underline{w}}_2(t),..,\tilde{\underline{w}}_M(t))} \underline{x}p(\underline{x})d\underline{x}. \quad (12)$$

Substituting this into (11) yields:

$$T\frac{d\tilde{\underline{w}}_i(t)}{dt} + \tilde{\underline{w}}_i(t) = \int_{X_i(\tilde{\underline{w}}_1(t),\tilde{\underline{w}}_2(t),..,\tilde{\underline{w}}_M(t))} \underline{x}p(\underline{x})d\underline{x}, \quad i=1,2,..,M. \quad (13)$$

Strictly speaking, the algorithm described by (8) never converges to a solution, because T is a constant. In the staedy-state, however, the weight vectors will remain in a neighborhood of a stable steady-state solution of the approximate differential equation (13). In the following, we shall investigate the steady-state solutions of (13).

In contrast to (8), eq. (13) is an autonomous type of differential equation. The condition for determining its steady-state solutions is given by

$$\tilde{\underline{w}}_i = \int_{X_i(\tilde{\underline{w}}_1,\tilde{\underline{w}}_2,..,\tilde{\underline{w}}_M)} \underline{x}p(\underline{x})d\underline{x}\ , \quad i=1,2,...,M. \quad (14)$$

Let a solution of this equation be denoted by $\underline{w}_1^*,\underline{w}_2^*,...,\underline{w}_M^*$. In order to determine whether the solution is stable or not, we transfer it to the origin by the change of coordinates

$$\delta\underline{w}_i(t) = \tilde{\underline{w}}_i(t)-\underline{w}_i^*\ , \quad i=1,2,...,M. \quad (15)$$

Then, by neglecting the higher order terms of $\delta\underline{w}_i(t)$, we obtain the linearized differential equation

$$T\frac{d\delta\underline{w}_i(t)}{dt} + \delta\underline{w}_i(t) = \sum_{j\neq i} P_{ij}(\delta\underline{w}_i(t)-\delta\underline{w}_j(t)), \quad i=1,2,..,M. \quad (16)$$

where

$$P_{ij} = \int_{B_{ij}(\underline{w}_1^*,\underline{w}_2^*,..,\underline{w}_M^*)} \frac{\underline{x}\underline{x}'p(\underline{x})}{|\underline{w}_i^*-\underline{w}_j^*|}d\sigma, \quad \text{for all } i \text{ and for all } j\neq i. \quad (17)$$

Let us define a N×M-dimensional column vector $\delta\underline{w}(t)$ as

$$\delta\underline{w}(t) = (\delta\underline{w}_1(t)',\delta\underline{w}_2(t)',...,\delta\underline{w}_M(t)')' \quad (18)$$

and a (N×M)×(N×M) matrix P as

$$P = \begin{bmatrix} \sum_{j\neq 1} P_{1j} & -P_{12} & \cdots & -P_{1i} & \cdots & -P_{1M} \\ -P_{21} & \sum_{j\neq 2} P_{2j} & \cdots & -P_{2i} & \cdots & -P_{2M} \\ \vdots & \vdots & & \vdots & & \vdots \\ -P_{i1} & -P_{i2} & \cdots & \sum_{j\neq i} P_{ij} & \cdots & -P_{iM} \\ \vdots & \vdots & & \vdots & & \vdots \\ -P_{M1} & -P_{M2} & \cdots & -P_{Mi} & \cdots & \sum_{j\neq M} P_{Mj} \end{bmatrix} \quad (19)$$

Then, (16) can be rewritten as

$$T\frac{d\delta\underline{w}(t)}{dt} + \delta\underline{w}(t) = P\delta\underline{w}(t) \quad \text{or} \quad T\frac{d\delta\underline{w}(t)}{dt} = (P-I)\delta\underline{w}(t) . \tag{20}$$

Since $P_{ij}= P_{ji}$, the matrix Q=P-I is symmetric and hence its eigenvalues are all real. The solution $\underline{w}_1^*,\underline{w}_2^*,\ldots,\underline{w}_M^*$ is stable if and only if the eigenvalues are all negative.

A sufficient condition for asymptotic stability is that $p(\underline{x})$ is zero for all $\underline{x}$ on every boundary $B_{ij}^*= B_{ij}(\underline{w}_1^*,\underline{w}_2^*,\ldots,\underline{w}_M^*)$. In fact, from the definition (17), $P_{ij}= 0$ for all i and for all j≠i, and hence all the eigenvalues are equal to -1.

IV. A CRITERION FUNCTION MINIMIZED BY THE ALGORITHM

Let us consider a criterion function of the form

$$J(\underline{w}_1,\underline{w}_2,\ldots,\underline{w}_M) = \int_X \sum_{i=1}^{M} (\underline{x}f_i(\underline{x};\underline{w}_1,\underline{w}_2,\ldots,\underline{w}_M)-\underline{w}_i)^2 p(\underline{x})d\underline{x} \tag{21}$$

or

$$J(\underline{w}_1,\underline{w}_2,\ldots,\underline{w}_M) = E\{\sum_{i=1}^{M} (\underline{x}f_i(\underline{x};\underline{w}_1,\underline{w}_2,\ldots,\underline{w}_M)-\underline{w}_i)^2\} . \tag{22}$$

The gradient of J with respect to $\underline{w}_j$ is given by

$$\nabla_{\underline{w}_j} J = E\{\sum_{i=1}^{M} (\underline{x}^2-2\underline{w}_i'\underline{x})\nabla_{\underline{w}_j} f_i\} - E\{2(f_j\underline{x}-\underline{w}_j)\} . \tag{23}$$

It turns out that the first term in the right-hand side of the above equation vanishes. Thus, we obtain

$$\nabla_{\underline{w}_j} J = - E\{2(f_j\underline{x}-\underline{w}_j)\} = - 2(\int_{X_j} \underline{x}p(\underline{x})d\underline{x}-\underline{w}_j) . \tag{24}$$

Putting

$$2T\frac{d\underline{w}_j(t)}{dt} = - \nabla_{\underline{w}_j(t)}J \tag{25}$$

yields

$$T\frac{d\underline{w}_j(t)}{dt} + \underline{w}_j(t) = \int_{X_j(\underline{w}_1(t),\underline{w}_2(t),\ldots,\underline{w}_M(t))}\underline{x}p(\underline{x})d\underline{x} . \tag{26}$$

It has been shown that the algorithm minimizes the criterion function J. Substituting a steady-state solution into (22) yields a minimum value of J:

$$J^* = J(\underline{w}_1^*,\underline{w}_2^*,\ldots,\underline{w}_M^*) = E\{\underline{x}^2\} - \sum_{i=1}^{M} \underline{w}_i^{*2} . \tag{27}$$

V. AN EXAMPLE

In this section the analytical techniques developed in section III are applied to the case of N=2 and M=4. Employing a polar coordinate system

$$r = \sqrt{x_1^2+x_2^2} \qquad \theta = \tan^{-1}(x_2/x_1) \ , \tag{28}$$

we can write the probability density $p(\underline{x})$ as

$$p(\underline{x})d\underline{x} = p(r,\theta)r\ dr\ d\theta \ . \tag{29}$$

Let

$$p(\theta) = a \sin^2 2\theta \ , \tag{30}$$

and let $p(r)$ be an arbitrary function of r attaining its maximum value at $r=r_o$. In the following, we assume that the input patterns are sampled from the distribution

$$p(r,\theta) = p(r)p(\theta) \ . \tag{31}$$

Clearly, this distribution consists of four clusters; their centers are

$$(r_o,\tfrac{1}{4}\pi),\ (r_o,\tfrac{3}{4}\pi),\ (r_o,\tfrac{5}{4}\pi) \text{ and } (r_o,\ \tfrac{7}{4}\pi) \ .$$

The two-diemsnional space X is partitioned into four subspaces X_1^*, X_2^*, X_3^*, X_4^* by a steady-state solution $\underline{w}_1^*,\underline{w}_2^*,\underline{w}_3^*,\underline{w}_4^*$. First, let us assume that any of the four subspaces is not null. Then, there exists four boundaries B_{12}^*, B_{23}^*, B_{34}^*, B_{41}^*. We may describe them by the equation

$$\underline{x} = \begin{bmatrix} r\cos\theta_i^* \\ r\sin\theta_i^* \end{bmatrix} \qquad i=1,2,3,4. \tag{32}$$

The weight vectors $\underline{w}_i^*$ and the angles θ_i^* must satisfy the equations

$$\underline{w}_i^* = \int_{X_i^*} \underline{x}p(\underline{x})d\underline{x} = a\overline{r^2}\int_{\theta_i^*}^{\theta_{i+1}^*} \begin{bmatrix}\cos\theta\\ \sin\theta\end{bmatrix} \sin^2 2\theta d\theta,\ i=1,2,3,4, \tag{33}$$

$$(\underline{w}_i^*-\underline{w}_{i+1}^*)' \begin{bmatrix}\cos\theta_i^*\\ \sin\theta_i^*\end{bmatrix} = 0 \qquad i=1,2,3,4, \tag{34}$$

where

$$\overline{r^2} = \int_0^\infty r^2 p(r)dr \ , \quad \theta_5^* = \theta_1^* \ , \quad \underline{w}_5^* = \underline{w}_1^* \ . \tag{35}$$

Then, we obtain the two solutions

$$\underline{w}_1^*=a\overline{r^2}\begin{bmatrix}8/15\\8/15\end{bmatrix},\ \underline{w}_2=a\overline{r^2}\begin{bmatrix}-8/15\\8/15\end{bmatrix},\ \underline{w}_3^*=a\overline{r^2}\begin{bmatrix}-8/15\\-8/15\end{bmatrix},\ \underline{w}_4^*=a\overline{r^2}\begin{bmatrix}8/15\\-8/15\end{bmatrix}, \tag{36a}$$

$$\theta_1^* = \tfrac{1}{2}\pi, \quad \theta_2^* = \pi, \quad \theta_3^* = \tfrac{3}{2}\pi, \quad \theta_4^* = 0 \ , \tag{36b}$$

or

$$\underline{w}_1^*=a\overline{r^2}\begin{bmatrix}7\sqrt{2}/15\\0\end{bmatrix},\ \underline{w}_2^*=a\overline{r^2}\begin{bmatrix}0\\7\sqrt{2}/15\end{bmatrix},\ \underline{w}_3^*=a\overline{r^2}\begin{bmatrix}-7\sqrt{2}/15\\0\end{bmatrix},\ \underline{w}_4^*=a\overline{r^2}\begin{bmatrix}0\\-7\sqrt{2}/15\end{bmatrix}, \tag{37a}$$

$$\theta_1^* = \frac{1}{4}\pi, \quad \theta_2^* = \frac{3}{4}\pi, \quad \theta_3^* = \frac{5}{4}\pi, \quad \theta_4^* = \frac{7}{4}\pi, \tag{37b}$$

which are shown in Figure 3(a) and (b), respectively.

Next, let us assume that one subspace is null. Without loss of generality we can put

$$\underline{w}_4^* = \underline{0} \qquad \theta_4^* = \theta_3^* \ . \tag{38}$$

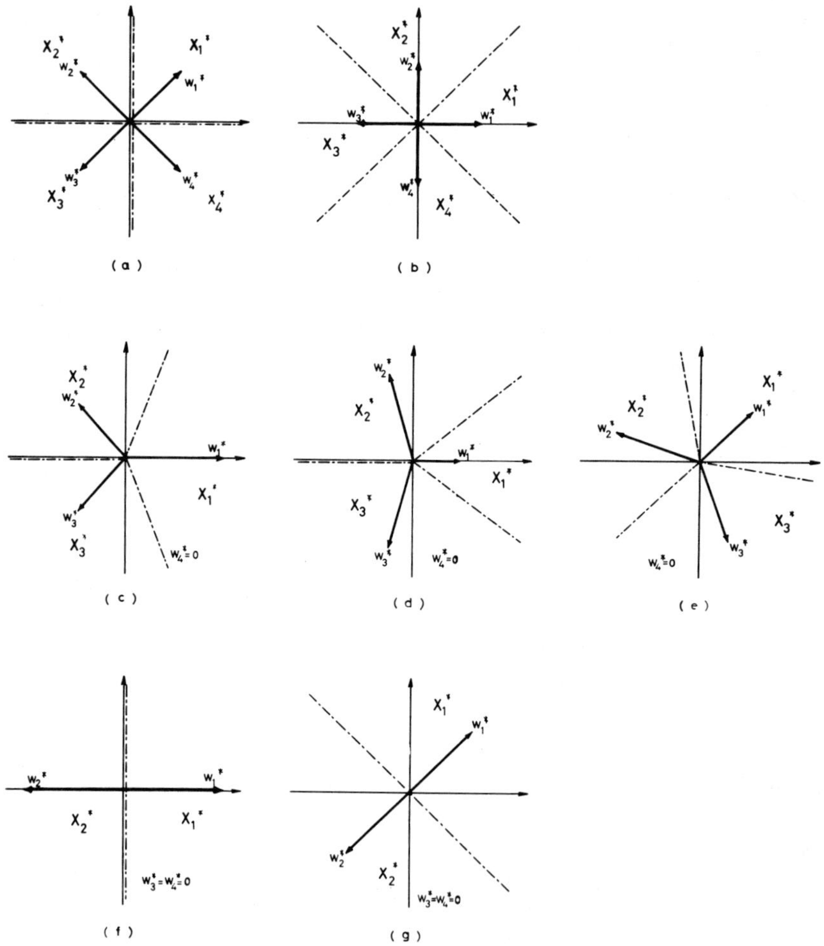

Figure 3. Steady-State Solutions of the Differential Equation (13) for the Case that M=4, N=2, and $p(\underline{x})d\underline{x}=a\ \sin^2 2\theta\cdot p(r)r\ dr\ d\theta$.

Assuming that $\theta_2^* = \pi$, we obtain

$$\underline{w}_1^* = \overline{ar^2}\begin{bmatrix}1.037\\0\end{bmatrix}, \quad \underline{w}_2^* = \overline{ar^2}\begin{bmatrix}-0.513\\0.589\end{bmatrix}, \quad \underline{w}_3^* = \overline{ar^2}\begin{bmatrix}-0.513\\-0.589\end{bmatrix}, \tag{39a}$$

$$\theta_1^* = 1.209, \quad \theta_2^* = \pi, \quad \theta_3^* = -1.209 \ , \tag{39b}$$

or

$$\underline{w}_1^* = \overline{ar^2}\begin{bmatrix}0.500\\0\end{bmatrix}, \quad \underline{w}_2^* = \overline{ar^2}\begin{bmatrix}-0.250\\0.898\end{bmatrix}, \quad \underline{w}_3^* = \overline{ar^2}\begin{bmatrix}-0.250\\-0.898\end{bmatrix}, \tag{40a}$$

$$\theta_1^* = 0.676, \quad \theta_2^* = \pi, \quad \theta_3^* = -0.676 \ , \tag{40b}$$

and assuming that $\theta_2^* = \frac{5}{4}\pi$, we obtain

$$\underline{w}_1^* = \overline{ar^2}\begin{bmatrix}0.537\\0.537\end{bmatrix}, \quad \underline{w}_2^* = \overline{ar^2}\begin{bmatrix}-0.863\\0.326\end{bmatrix}, \quad \underline{w}_3^* = \overline{ar^2}\begin{bmatrix}0.326\\-0.863\end{bmatrix}, \tag{41a}$$

$$\theta_1^* = 1.721, \quad \theta_2^* = \frac{5}{4}\pi, \quad \theta_3^* = 0.150 \ . \tag{41b}$$

The three solutions are shown in Figure 3(c), (d) and (e), respectively. Clearly, rotating each solution by the angle 1/2 π, π or 3/2 π yields another solution.

Now let

$$\underline{w}_3^* = \underline{w}_4^* = \underline{0} \ . \tag{42}$$

Then we obtain

$$\underline{w}_1^* = \overline{ar^2}\begin{bmatrix}16/15\\0\end{bmatrix}, \quad \underline{w}_2^* = \overline{ar^2}\begin{bmatrix}-16/15\\0\end{bmatrix}, \tag{43a}$$

$$\theta_1^* = \frac{1}{2}\pi, \quad \theta_2^* = -\frac{1}{2}\pi \ , \tag{43b}$$

or

$$\underline{w}_1^* = \overline{ar^2}\begin{bmatrix}7\sqrt{2}/15\\7\sqrt{2}/15\end{bmatrix}, \quad \underline{w}_2^* = \overline{ar^2}\begin{bmatrix}-7\sqrt{2}/15\\-7\sqrt{2}/15\end{bmatrix}, \tag{44a}$$

$$\theta_1^* = \frac{3}{4}\pi, \quad \theta_2^* = -\frac{1}{4}\pi. \tag{44b}$$

The two solutions are shown in Figure 3(f) and (g), respectively. Clearly, rotating each solution by the angle 1/2 π yields another solution.

In order to determine the stabilities of the solutions obtained above, the matrix Q=P-I and its eigenvalues have been calculated for each solution. The results show that the solutions (36), (39) and (43) are stable and that (37), (40) (41) and (44) are unstable. Therefore, the algorithm can converge to any of the three stable solutions; the initial conditions determine which of them will be yielded in the steady-state.

As Figure 3 shows, the stable solutions (36), (39) and (43) are associated with four-category, three-category and two-category classifications, respectively. In these three types of classification, either one cluster or one pair of adjacent clusters in the distribution corresponds to one category. On the other hand, no a priori information has been given about the number of categories. Thus, it seems natural to conclude that these three types of classifications are all reasonable.

Substituting the stable solutions into (27) yields

$$\begin{aligned} J^* &= E\{\underline{x}^2\} -2.276(\overline{ar^2})^2 \quad \text{for (36)} , \\ J^* &= E\{\underline{x}^2\} -2.296(\overline{ar^2})^2 \quad \text{for (39)} , \\ J^* &= E\{\underline{x}^2\} -2.276(\overline{ar^2})^2 \quad \text{for (43)} . \end{aligned} \tag{45}$$

On the other hand, substituting an unstable solution, for example, (37) into (27) yields

$$J^* = E\{\underline{x}^2\} -1.742(\overline{ar^2})^2 . \tag{46}$$

VI. DIGITAL COMPUTER SIMULATION

In order to verify the theoretical results obtained in the above section, simulation experiments were performed. The algorithm was implemented on a FACOM 270/20 computer and the experimental

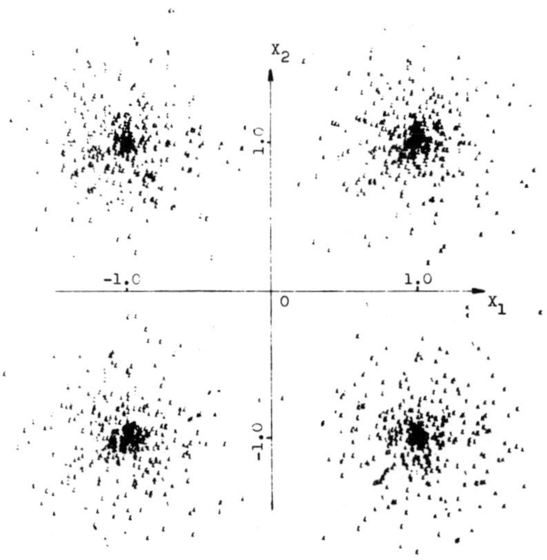

Figure 4. Four-Cluster Pattern Distribution Used in the Simulation Experiments.

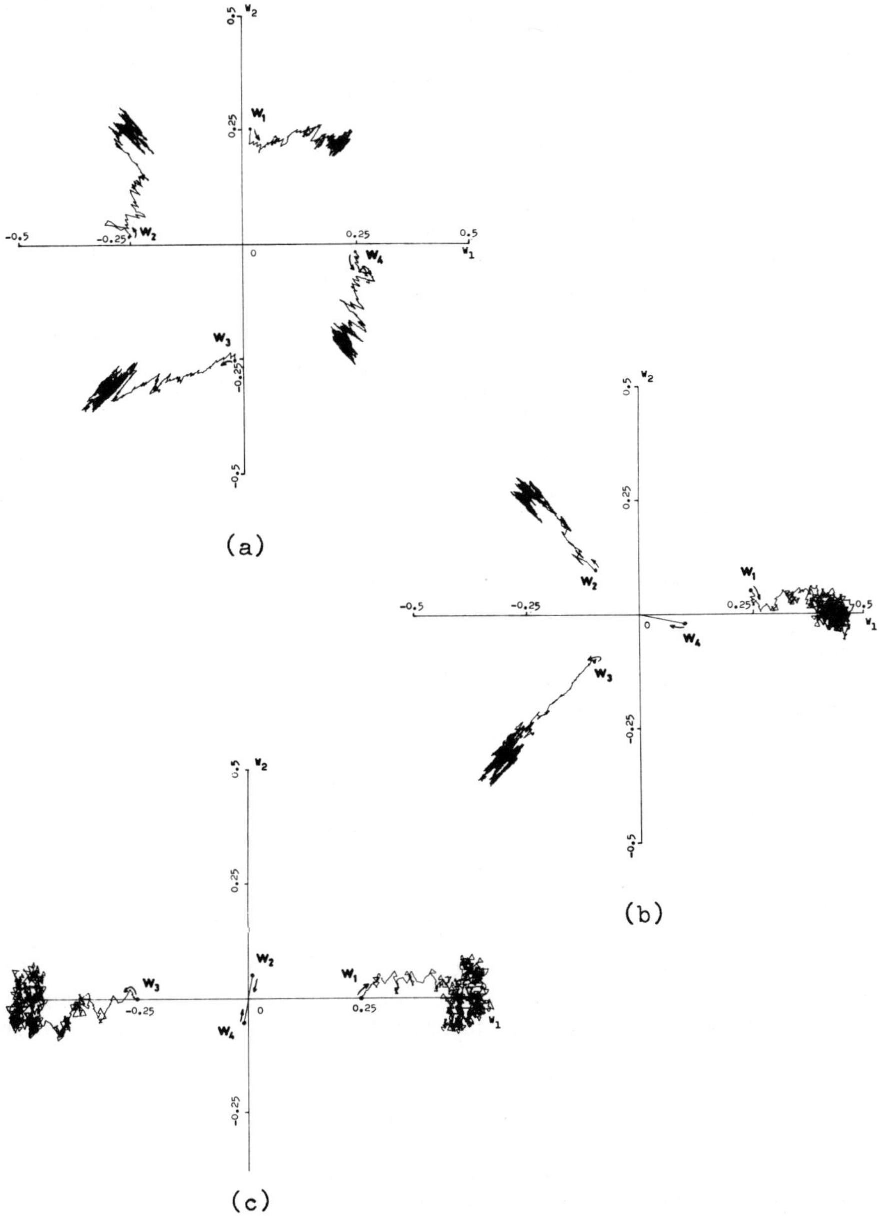

Figure 5. Three Experimental Results Obtained By Presenting Patterns Sampled from the Distribution Shown in Figure 4.

results were recorded by a CalComp digital plotter. The pattern distribution used in the simulation is shown in Figure 4. It consists of four clusters. Every member of each cluster was generated by adding zero-mean pseudo-random Gaussian noise to a prototype pattern vector. The number of the weight vectors used was four, the experimental value for T was 100 Δ, and the time duration of an input sequence was 8T = 800 Δ.

When the weight vectors started from

$$\underline{w}_1(0)= \begin{bmatrix}0.02\\0.25\end{bmatrix},\ \underline{w}_2(0)= \begin{bmatrix}-0.25\\0.02\end{bmatrix},\ \underline{w}_3(0)= \begin{bmatrix}-0.02\\-0.25\end{bmatrix},\ \underline{w}_4(0)= \begin{bmatrix}0.25\\-0.02\end{bmatrix},$$

they moved along the trajectories shown in Figure 5(a). When the initial position was selected as

$$\underline{w}_1(0)= \begin{bmatrix}0.25\\0.05\end{bmatrix},\ \underline{w}_2(0)= \begin{bmatrix}-0.10\\0.10\end{bmatrix},\ \underline{w}_3(0)= \begin{bmatrix}-0.10\\-0.10\end{bmatrix},\ \underline{w}_4(0)= \begin{bmatrix}0.10\\-0.02\end{bmatrix},$$

the phase-plane trajectories shown in Figure 5(b) were obtained. Figure 5(c) shows the trajectories obtained with the initial position

$$\underline{w}_1(0)= \begin{bmatrix}0.25\\0\end{bmatrix},\ \underline{w}_2(0)= \begin{bmatrix}0.01\\0.05\end{bmatrix},\ \underline{w}_3(0)= \begin{bmatrix}-0.25\\0\end{bmatrix},\ \underline{w}_4(0)= \begin{bmatrix}-0.01\\-0.05\end{bmatrix}.$$

Thus, the stability of the three solutions shown in Figure 3(a), (c) and (f) has been verified.

When the initial position was selected very close to the solution shown in Figure 3(b), (d), (e) or (g), the weight vectors did not remain in a neighborhood of the initial position. Thereby, the instability of these solutions has been verified.

VIII. CONCLUSIONS

A nonsupervised learning algorithm for multi-category pattern classification has been presented, and its steady-state behavior has been investigated by means of analysis and simulation. It has been shown that although the algorithm does not necessarily yield a unique solution, every solution yielded is associated with a reasonable classification on the distribution of the input patterns.

REFERENCES

1. C. V. Jakowatz, et.al., "Adaptive Waveform Recognition," Proc. 4th London Symp. on Information Theory, edited by C. Cherry, Washington, D. C., Butterworths, 1961, pp. 317-326.

2. E. M. Glaser, "Signal Detection by Adaptive Filters," IRE Trans. on Information Theory, Vol. IT-7, April 1961, pp. 87-98.

3. H. D. Block, et.al., "Analysis of a Four-Layer Series-Coupled Perceptron II," Rev. Mod. Phys., Vol. 34, Jan. 1962, pp. 135-142.

4. D. B. Cooper and P. W. Cooper, "Adaptive Pattern Recognition and Signal Detection Without Supervision," 1964 Int. Conv. Rec., Pt. I, pp. 246-257.

5. G. Nagy and G. L. Shelton, Jr., "Self-Corrective Character Recognition System," IEEE Trans. on Information Theory, Vol. IT-12, April 1966, pp. 215-222.

6. S. C. Fralic, "Learning to Recognize Patterns Without a Teacher," IEEE Trans. on Information Theory, Vol. IT-13, Jan. 1967, pp. 57-64.

7. Y. T. Chien and K. S. Fu, "On Bayesian Learning and Stochastic Approximation," IEEE Trans. on Systems Science and Cybernetics, Vol. SSC-3, June 1967, pp. 28-38.

8. E. M. Braverman, "The Method of Potential Functions in the Problem of Training Machines to Recognize Patterns Without a Trainer," Automation and Remote Control, Vol. 27, Oct. 1967. pp. 1728-1736.

9. Ya. Z. Tsypkin, "Self-Learning--What is it?," IEEE Trans. on Automatic Control, Vol. AC-13, Dec. 1968, pp. 608-612.

10. I. Morishita, "Analysis of an Adaptive Threshold Logic Unit," IEEE Trans. on Computers, Vol. C-19, Dec. 1970, pp. 1181-1192.

ACKNOWLEDGMENT

The authors wish to thank Prof. T. Isobe for his encouragement.

A MIXED-TYPE NON-PARAMETRIC LEARNING MACHINE WITHOUT A TEACHER

Masamichi Shimura

Osaka University

Osaka, Japan

I. INTRODUCTION

In this paper, we consider the design of a non-parametric learning machine without a teacher. Most pattern recognition problems may be categorized as parametric or non-parametric on the basis of knowledge that we have concerning the conditional densities of the input patterns. Problems in which the densities are completely unknown are called non-parametric. In addition, the learning machine can be further classified into two types. One is a supervised machine, that is, a machine with an external teacher. In this case, the teacher gives the information regarding the category to which the input pattern belongs and the information regarding the correctness of the machine's action. The second type is an unsupervised machine.

Some examples of well-known models of the supervised parametric machines are the Adaline by Professor B. Widrow and the Perceptron by Professor F. Rosenblatt. These machines are linear pattern classifiers with learning mechanisms. In general, the main advantage of the linear machine is that the structure of the system is comparatively simple and that no information regarding the input patterns is needed. The major shortcoming of the linear machine is that the input patterns must be linearly separable. If the input patterns are linearly separable, the machine can be trained by an external teacher so as to respond correctly even if it is given no information about the input patterns.

Next, we will consider an unsupervised machine. The unsupervised machine has no teacher and so it must learn by itself. That is, it must improve its performance according to its own decision instead of the teacher. Therefore, it is very difficult for the machine to learn by itself.

However, variable references on this problem of the parametric learning method are works of Dr. J. Scudder, Drs. Cooper and Cooper, Drs. Patrick and Hancock, and Dr. Fu. In the case of the non-parametric unsupervised machine, in turn, the learning process is quite difficult to accomplish. This remains true even if the input patterns are assumed to be linearly separable. This difficulty increases especially when the set of input patterns has intersections and the signal-to-noise ratio is low. Jakowatz's model is one approach to realize such a machine. However, the analysis of his method is still incomplete.

Let us consider the practical problem of signal detection. Signal detection can be considered as a type of pattern recognition. An easy method to detect a signal embedded in noise is to use a simple filter such as an energy detector. However, if no information about the signal and noise is available, we can not get good performance by such a simple filter. This is because the threshold can not be set to a suitable value.

In this paper, we propose a new machine in which a linear machine is combined with such a filter to obtain a mixed-type non-parametric machine without an external teacher. Our primary purpose in this discussion is to analyze the learning process in the machine and to show that this model is very reasonable.

II. THE MODEL

The proposed model consists of a well-known linear learning classifier with a simple signal detector as shown in Figure 1. The detector operates as an acting teacher of the machine and the machine learns according to the decision made by this detector in

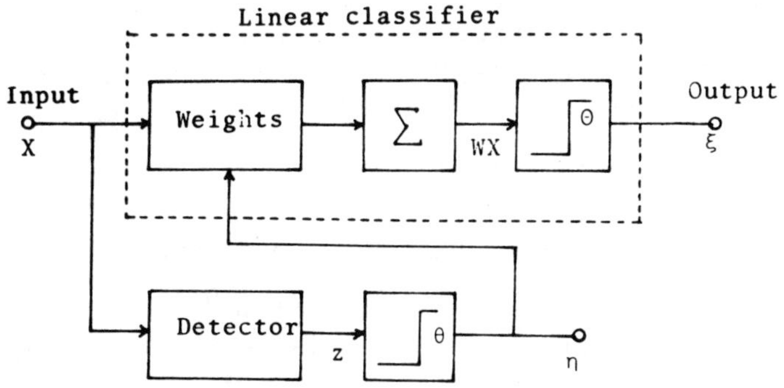

Figure 1. Block Diagram of the Mixed-type Learning Machine Without a Teacher.

the learning period. Although this detector is not a good teacher as was mentioned before, it may give more correct information about the category of the input patterns than that obtained by the classifier itself at an early stage. One reason why learning is difficult for the non-parametric unsupervised machine is that the decision made by itself can not be expected to be always correct. In our machine, this difficulty is circumvented by using an acting teacher.

As an example of such a non-parametric problem, consider the detection of a signal with unknown amplitude embedded in noise. The problem can be considered as a type of hypothesis-testing.

In the case of coherent detection, where the times when signals occur are known, the detector decides at time t_k between the two hypotheses

$$H_o^k \; : \; X_k = N_k$$

and

$$H_1^k \; : \; X_k = S + N_k \tag{1}$$

where X_k is the received waveform in the k-th sampline interval, N_k is the k-th sample of a noise waveform, and S is an unknown signal. Assume that we have a received waveform X(t) by the vector of m samples

$$X(t) = (x_1, x_2, \ldots, x_m) = (x(t_o), x(t_o + \tau), \ldots, x(t_o+(m-1)\tau))$$

At first, let us consider the simplest case, in which a mean value detector is used as the imperfect teacher. In this case, the output of the detector becomes

$$z = \frac{1}{m} \sum_i^m x_i \tag{2}$$

If the actual signal is of alternating waveform (i.e., containing negative as well as positive excursions), then the mean value detector can not always detect it. In such a case, an energy detector must be used and its output is

$$z = \frac{1}{m} \sum_i^m x_i^2 \tag{3}$$

The decision of these detectors is made as follows:

$$\eta = 1 \qquad \text{if } |z| \geqq \theta$$

and

$$\eta = -1 \qquad \text{if } |z| < \theta \tag{4}$$

where θ is the threshold value and η is the output of the threshold element.

The decision rule of the linear classifier is

$$d(X_k) = \begin{cases} H_o^k & \text{if } \Sigma\, w_i x_i \geqq \Theta \quad (\Theta>0) \\ H_1^k & \text{otherwise} \end{cases} \tag{5}$$

and the output ξ is -1 or +1 when the machine selects H_o^k or H_1^k, respectively.

III. LEARNING PROCEDURE WITHOUT A TEACHER

The machine shown in Figure 1 studies according to the following algorithm. Let us introduce new pattern vector and weight vector,

$$W = (w_1, w_2, \ldots, w_m, -\Theta)$$

$$\chi = (x_1, x_2, \ldots, x_m, 1)^t$$

Using these vectors, the decision rule of the machine analogous to that in Eq. (5) is stated in the form

$$d(\chi_k) = \begin{cases} H_o^k & \text{if} \quad W_k\chi_k < 0 \\ H_1^k & \text{if} \quad W_k\chi_k \geqq 0 \end{cases} \tag{6}$$

Therefore, we have four cases:

$$\begin{array}{llll} \text{a)}\ \xi = 1, & \eta = 1 & \text{b)}\ \xi = 1, & \eta = -1 \\ \text{c)}\ \xi = -1, & \eta = 1 & \text{d)}\ \xi = -1, & \eta = -1 \end{array} \tag{7}$$

The cases a) and d) are those in which the linear classifier responds identically to the detector, so that the weighting factors are not changed. Otherwise, the (k+1)st member of the weight vector sequence is adjusted in order to give the same response as that of the detector. That is, the (k+1)st member is given by

$$W_{k+1} = W_k \qquad \text{if} \quad \xi_k\eta_k = 1$$

and

$$W_{k+1} = W_k + c\eta_k\chi_k^t \qquad \text{if} \quad \xi_k\eta_k = -1 \tag{8}$$

where c is a positive number, and ξ_k, η_k are the responses of the classifier and the detector, respectively, at the k-th step.

Next, we discuss the convergence of the above mentioned learning procedure. Assume that there exists statistically a solution vector, say W, such that

$$\begin{aligned} W\chi_k^1 &\geqq 0 \\ W\chi_k^o &< 0 \end{aligned} \tag{9}$$

where χ_k^1 is the k-th "signal" pattern, that is, the input pattern in which an actual signal exists with noise and χ_k^o is the k-th "no signal" pattern.

Now we omit from the training sequence any pattern χ_k for which $W_{k+1} = W_k$. Relabel the remaining patterns in the sequence by the symbols $Y_1, Y_2, \ldots, Y_k, \ldots$ and call this a reduced training sequence. At the same time, we introduce a reduced weight vector sequence $V_1, V_2, \ldots, v_k, \ldots$. Thus, the (k+1)st weight vector is changed by

$$V_{k+1} = V_k + c\eta_k Y_k^t \tag{10}$$

for each reduced pattern.

Consider the decrease in the squared distance to W, d_k, effected by the k-th step

$$d_k = (W - V_k)^2 - (W - V_{k+1})^2 \tag{11}$$

Without loss of generality, we set c=1 in the following analysis. Substituting Eq. (10) into Eq. (11), we have

$$d_k = -2\eta_k V_k Y_k - \eta_k^2 Y_k^t Y_k + 2\eta_k W Y_k \tag{12}$$

It is clear that d_k is not always positive according to the imperfectness of the detector. In order to circumvent this problem, let us consider the υ sum of d_k, D_h

$$D_h = \frac{1}{\upsilon} \sum_{j=h}^{h+\upsilon-1} d_j \tag{13}$$

If D_h is positive for an arbitrary number h and for an appropriate number υ, then the weight vector V_k approaches the vector W. The solution will converge even in an oscillatory case, as is shown in Figure 2. From Eqs. (12) and (13), we obtain

$$\begin{aligned} D_h &> \frac{1}{\upsilon} \sum\{2\eta_j W Y_j - Y_j^t Y_j\} \\ &= \frac{2}{\upsilon} \{\sum_i^{\alpha_1}(-WY_i^o) - \sum_j^{\alpha_2}(-WY_j^o) + \sum_k^{\beta_1}(WY_k^1) - \sum_\ell^{\beta_2}(WY_\ell^1)\} - \frac{1}{\upsilon}\sum_n^{\upsilon} Y_n^t Y_n \qquad (14) \\ &\equiv Q \end{aligned}$$

where α_1, α_2 are the numbers of "no signal" patterns Y_j^o when $\eta_j = -1$, $+1$, respectively, and β_1, β_2 are the numbers of "signal" patterns Y_ℓ^1 when $\eta_\ell = +1$, -1, respectively.

As the input patterns are statistical, Q is a probabilistic value. For example, if the noise is assumed to be Gaussian with

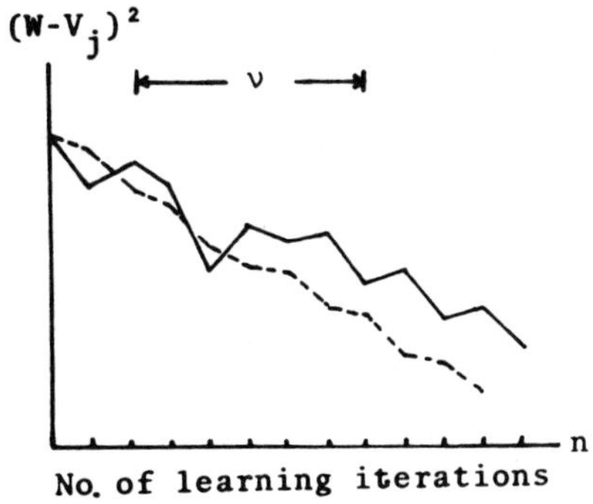

Figure 2. Learning Process

variance σ^2, then the probability density of Q when $\alpha \equiv \alpha_1+\alpha_2 = \beta_1+\beta_2 \equiv \beta$ becomes

$$p(Q) = \frac{\upsilon}{\sqrt{2\pi\Phi}} \exp \frac{-\upsilon^2(Q-U)^2}{2\Phi^2}$$
$$\Phi = 2\sqrt{\upsilon\Omega^2+\beta m s^2+m\upsilon\sigma^2/2} \tag{15}$$
$$U = (2p-1)(\sum w_i s_i-\Theta)+(2q-1)\Theta-ms^2/2-m\sigma^2$$

where $\Omega^2 = \sum^m w_i^2$, $s^2 = \frac{1}{m}\sum s_i^2$, $q=\alpha_1/\alpha$ and $p=\beta_1/\beta$.

Let ω be the solution region of weight vectors satisfying Eq. (9) and ω' be the solution region, lying in the interior of ω, having the property that

$$WY^1 \geqq \frac{m}{2(2p-1)}(s^2+2\sigma^2)+\frac{\Theta}{2}$$
$$WY^o < -\frac{\Theta}{2} \tag{16}$$

Combining these relations, we obtain the following inequality if $2p-1>0$

$$U > (2p-1)WY^1-(2p-1)\Theta/2+(2q-1)\Theta-m(s^2+2\sigma^2)/2$$
$$> (2q-1)\Theta \tag{17}$$

Therefore, if $p > 1/2$ and $q > 1/2$, then $U>0$. (18)

The probability that Q is positive, say p(Q 0), becomes

$$p(Q>0) = \int_0^\infty p(Q)dQ = \frac{1}{\sqrt{2\pi}}\int_{-\mu}^\infty \exp-\frac{t^2}{2}\,dt \tag{19}$$

where

$$\mu = \frac{\upsilon U}{2\sqrt{\upsilon\Omega^2+m\beta s^2+m\upsilon\sigma^2/2}}$$

Thus

$$\lim_{\upsilon \to \infty} p(Q>0) = 1 \tag{20}$$

In the actual case, however, the amplitude of the noise must be finite. Therefore

$$\lim_{\upsilon \to N} p(Q>0) = 1 \tag{21}$$

where N is a sufficient large integer.

From Eq. (21), we can obtain the following results. If υ is sufficiently large and the condition (18) is satisfied, D_h becomes positive with unit probability, and thus the weight vector approaches the solution vector lying in the region ω'. This result shows that the learning procedure converges statistically even if the input patterns are not linearly separable.

IV. PERFORMANCE OF DETECTORS

In the present system, the learning procedure converges if p and q are greater than 0.5. Note that the values of p and q are affected by the threshold value θ of the detector. In the case neither the waveform nor the power of the signal is known, the choice of θ is quite important, although the optimum value of θ is difficult to be obtained. A method to find the appropriate value θ which satisfies the condition Eq. (18) is the following:

i) Case of a mean value detector. In this case, p and q are given by the expressions

$$q = \text{prob.}\ (\eta=-1\,|\,\chi_j^o) = \frac{2}{\sqrt{2\pi}} \int_0^{\sqrt{m}\theta/\sigma} \exp - \frac{t^2}{2}\, dt \tag{22}$$

$$p = \text{prob.}\ (\eta=1\,|\,\chi_j^1) = 1 - \frac{1}{\sqrt{2\pi}} \int_{-\sqrt{m}(\theta+s_0)/\sigma}^{\sqrt{m}(\theta-s_0)/\sigma} \exp - \frac{t^2}{2}\, dt \tag{23}$$

where $s_0 = \frac{1}{m}\sum_i^m s_i$. Thus as θ increases, q also increases, but p becomes smaller.

If θ is chosen as $\theta=\sigma/\sqrt{m}$, then q is 0.63 and p is the following

$$p = 1 - \frac{1}{\sqrt{2\pi}} \int_{-1-\sqrt{m}s_0/\sigma}^{1-\sqrt{m}s_0/\sigma} \exp - \frac{t^2}{2}\, dt \tag{24}$$

The relation between p and m given by Eq. (24) is shown in Figure 3. From this figure, p is larger than 0.5 if $m > 0.84\sigma^2/s_0^2$.

ii) Case of an energy detector. If m is large enough, p and q can be approximated by use of the central limit theorem as follows:

$$q = \frac{1}{\sqrt{2\pi}} \int_{-\sqrt{m}/\sqrt{2}}^{\sqrt{m}(\theta-\sigma^2)/\sqrt{2}\sigma^2} \exp - \frac{t^2}{2}\, dt$$

$$p = \frac{1}{\sqrt{2\pi}} \int_{\sqrt{m}(\theta-s^2-\sigma^2)/\sigma\sqrt{4s^2+2\sigma^2}}^{\infty} \exp - \frac{t^2}{2}\, dt$$

Therefore if the value is chosen to be

$$\theta = (1 + 0.68/\sqrt{m})\sigma^2 \tag{25}$$

q is approximately 0.68 and

$$p = \frac{1}{\sqrt{2\pi}} \int_{(0.68-\sqrt{m}\lambda_e)/\sqrt{4\lambda_e+2}}^{\infty} \exp - \frac{t^2}{2}\, dt$$

as indicated in Figure 4, where $\lambda_e = s^2/\sigma^2$.

V. TWO SIGNALS

Consider the case in which there exists two signals S_1 and S_2. When the input patterns are classified into two classes: "signal S_1"

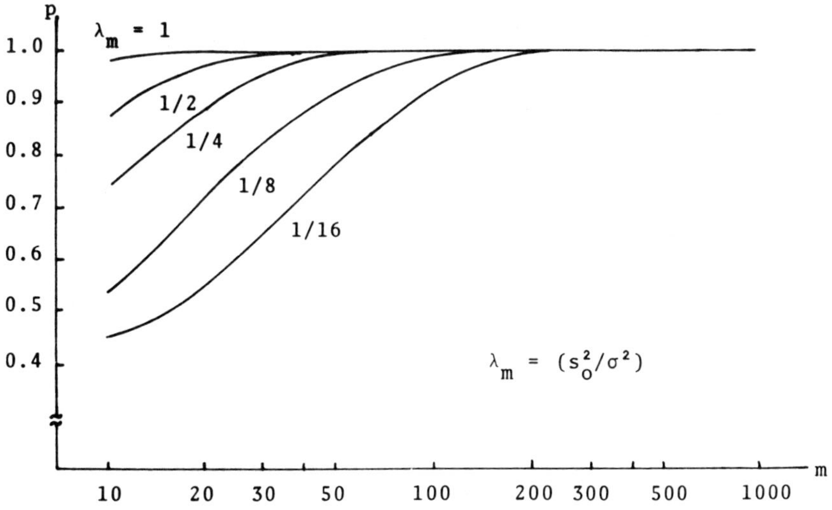

Figure 3. Relations Between the Probability of Correct Answers p and the Number of Dimensions m in a Mean Value Detector.

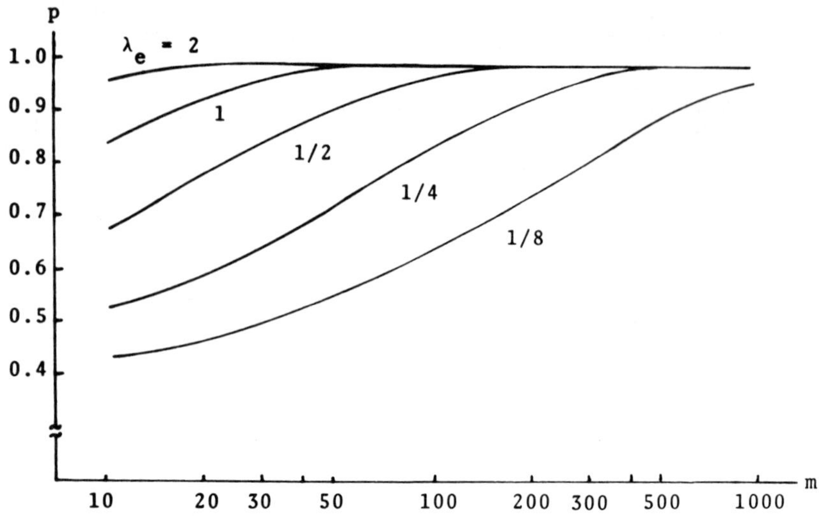

Figure 4. Relations Between the Probability of Correct Answers p and the Number of Dimensions m in an Energy Detector.

and "signal S_2" patterns, the decision rule of the linear classifier shown in Figure 1 is

$$d(X_k) = \begin{cases} H_1^k : X_k = S_1 + N_k & \text{if } \Sigma w_i x_{ki} \geqq \Theta \\ H_2^k : X_k + S_2 + N_k & \text{if } \Sigma w_i x_{ki} < \Theta \end{cases} \qquad (26)$$

In this case, we must assume that the power of signal S_1 is not equal to that of signal S_2. Under this assumption, we have four cases analogous to Eq. (7). The learning is performed by the similar algorithm to that given by Eq. (8). The statistical convergence of the learning process can be proved in the same manner as before.

The υ sum of the distance d_k, D_h corresponding to Eq. (14) becomes as follows:

$$D_h > \frac{2}{\upsilon}\left\{\sum_i^{\alpha_1}(-WY_i^1) - \sum_j^{\alpha_2}(-WY_j^1) + \sum_k^{\beta_1}(WY_k^2) - \sum_\ell^{\beta_2}(WY_\ell^2)\right\} - \frac{1}{\upsilon}\sum_n^{\upsilon} Y_n^t Y_n \qquad (27)$$

$$\equiv Q$$

where Y_k^i is the k-th pattern vector of "signal S_i" pattern. Therefore the probability density of Q corresponding to Eq. (15) becomes

$$p(Q) = \frac{1}{\sqrt{2\pi}\Phi}\exp-\frac{(Q-U)^2}{2\Phi^2}$$

$$\Phi = (2\sigma\sqrt{\upsilon\Omega^2+m\alpha s_{10}^2+m\beta s_{20}^2+m\upsilon\sigma^2/2})/\upsilon \tag{28}$$

$$U = (2p-1)(\Sigma w_i s_{1i}-\Theta) - (2q-1)(\Sigma w_i s_{2i}-\Theta)$$

$$-m\alpha s_{10}^2/\upsilon - m\beta s_{20}^2/\upsilon - m\sigma^2$$

where $s_{10}^2 = \frac{1}{m}\sum_i^m s_{1i}^2$ and $s_{20}^2 = \frac{1}{m}\sum_i^m s_{2i}^2$.

For the following region in the weight space,

$$WY^1 > \frac{m(s_{10}^2+s_{20}^2+\sigma^2)}{4(p+q-1)} + \Theta$$

$$WY^2 < -\{\frac{m(s_{10}^2+s_{20}^2+\sigma^2)}{4(p+q-1)} + \Theta\} \tag{29}$$

$$|WN| < \Theta$$

U becomes positive if $p > 0.5$ and $q > 0.5$. The probability that Q is positive can be written in the form

$$p(Q>0) = \frac{1}{\sqrt{2\pi}}\int_{-\mu}^{\infty}\exp-\frac{t^2}{2}\,dt$$

where

$$\mu = \frac{\upsilon U}{2\sqrt{\upsilon\Omega^2+m\alpha s_{10}^2+m\beta s_{20}^2+m\sigma^2/2}}$$

Thus we have

$$\lim_{\upsilon\to N} p(Q>0) = 1$$

VI. EXPERIMENTAL RESULTS

Some results of a computer experiment using an energy detector as an acting teacher are presented below. The learning process of the machine with 10 units as the probability of correct answers of the machine to the number of learning iterations is plotted in Figure 5. In these tests, the input signal-to-noise ratio was -6, -3, 0 and 3dB, and c was chosen as 1/10 of the mean amplitude of the input signal. From this figure no apparent difference can be seen among the three waveforms: triangle, sinusoidal and rectangular waveforms as shown in Figure 5.

In the case of a triangle waveform, the sharpest points of the waveform are the prominent feature and thus the learning rate is

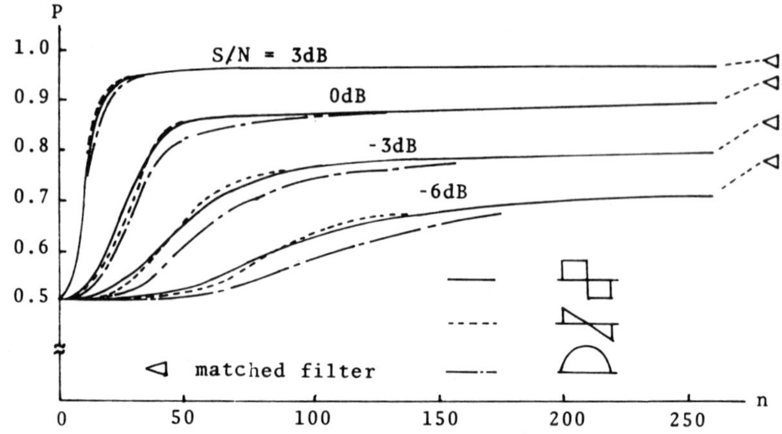

Figure 5. Learning Process. Probability of Correct Answers P of the System Vs. No. of Learning Iterations n.

comparatively slow at the beginning. In the case of a sinusoidal waveform, the learning rate is uniformly slow because the end points of the wave are zero. The probability of correct answers of a matched filter for an identical situation is included in Figure 5. Figure 6 shows the learning process when a machine is supervised by an external teacher. This figure shows that when the signal-to-noise ratio is low, the learning rate of the unsupervised machine is slower than that of the supervised machine in the early stages. However, when the signal-to-noise ratio is high, the unsupervised machine has almost the same learning performance as the supervised one. This is because the unsupervised machine receives incorrect information in learning when the signal-to-noise ratio is low.

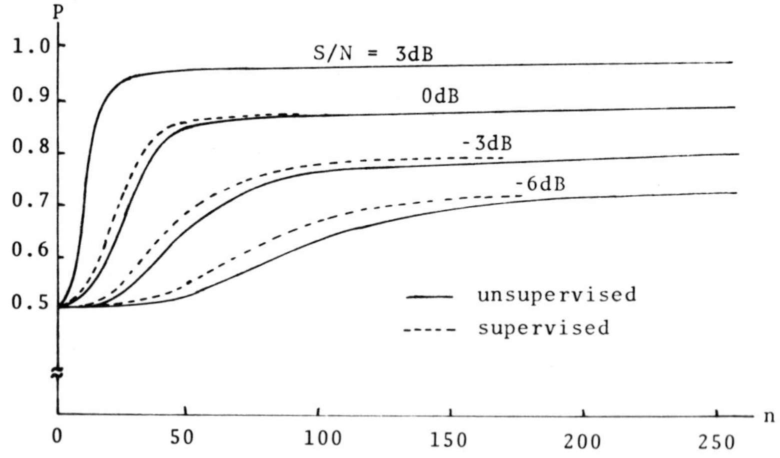

Figure 6. Learning Process in a Supervised and Unsupervised Machines.

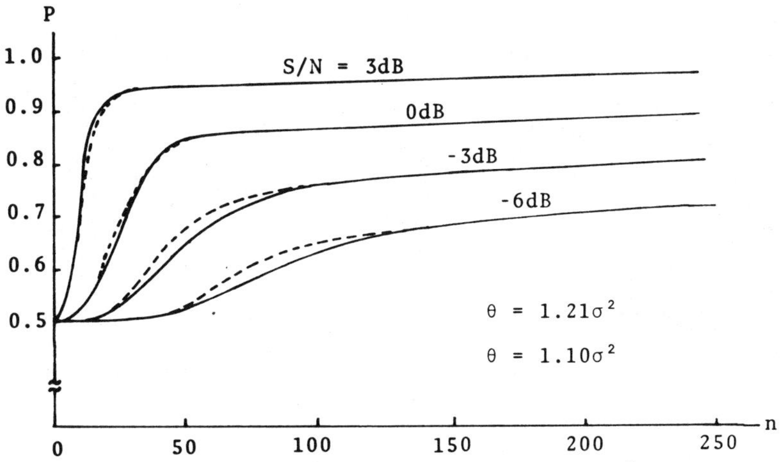

Figure 7. Learning Process when $\theta = 1.21\sigma^2$ and $1.10\sigma^2$.

As mentioned before, the threshold value of the detector affects to the values of p and q and thus the learning performance. Fugure 7 shows the learning performance when the threshold θ equals $1.21\sigma^2$ and $1.10\sigma^2$. In Figure 8, the learning rate when the dimension m of the machine is 10, 20 or 40 is indicated by a solid line, a dotted line or a broken line, respectively.

It is well known that the absolute correction rule in learning process generally leads to a more rapid learning rate than the fixed-increment rule and gives the same result in a supervised machine. If the absolute correct rule is employed in the machine

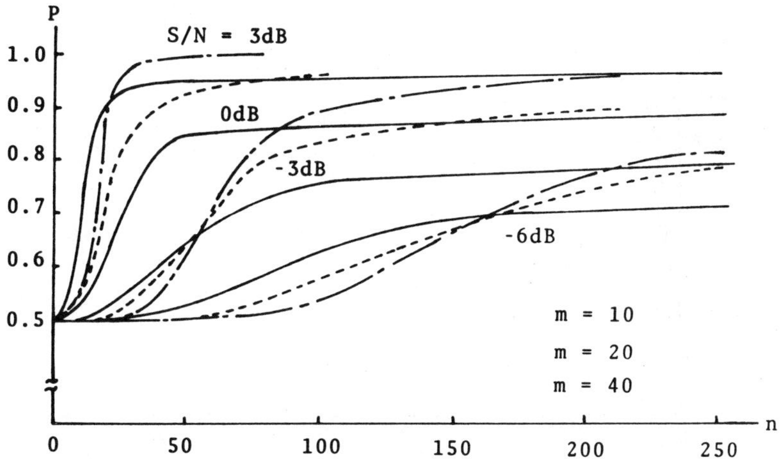

Figure 8. Learning Process when m = 10, 20 and 40.

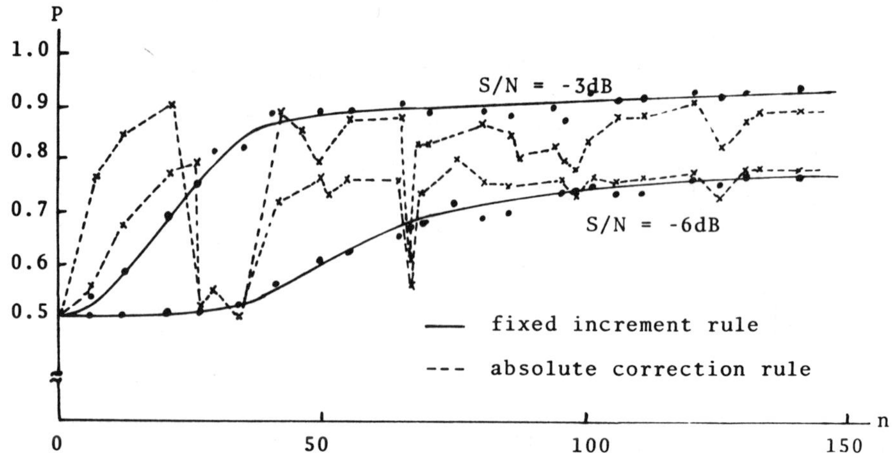

Figure 9. Learning Process by a Fixed Increment Rule and an Absolute Correction Rule.

shown in Figure 1, however, the probability of correct answers is oscillatory in time as indicated in Figure 9. This is because the machine is incorrectly taught when the acting teacher provides false information. Therefore, in the present machine the fixed-increment rule gives a better performance than the absolute correction rule.

Figure 10 shows the learning process of the machine in separating signals S_1 and S_2 into two classes. In this test, the power of signal S_1 is assumed to be twice that of signal S_2, and

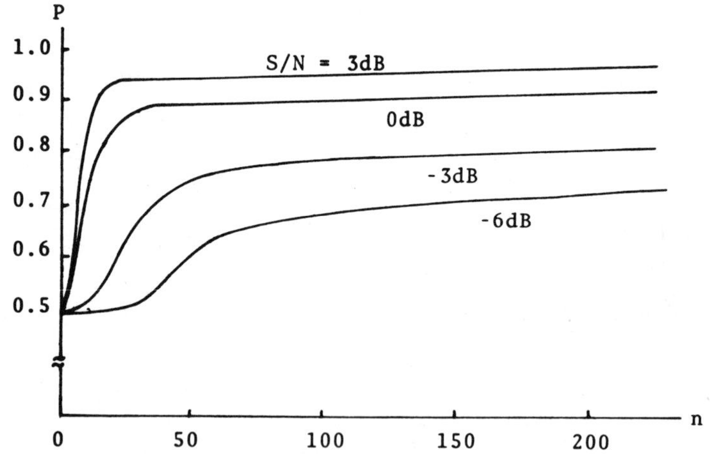

Figure 10. Learning Process in the Case There Exist Two Signals: S_1 and S_2.

the values of signal-to-noise ratio in this figure indicate the signal-to-noise ratio of S_2.

The above results are somewhat difficult to compare with those obtained in some parametric machines as they were obtained in a different manner. However, it can be concluded that we attained reasonably rapid convergence in this machine.

VII. CONCLUSIONS

In this paper, we have proposed a mathematical model for the class of problems in which non-parametric learning is performed without a teacher. We found that if the signal detector responds correctly with a probability more than 0.5, then the machine is able to learn by itself. Thus, the advantage of this machine is that the system is as simple as, while giving equal performance to the supervised one.

ACKNOWLEDGEMENT

The author would like to thank Dr. K. Tanaka, Professor of Osaka University, for the interesting and stimulating discussions on subjects pertaining to this paper.

REFERENCES

1. C. V. Jakowatz, et al.; "Adaptive Waveform Recognition," The 4th International Symposium on Information Theory, London, 1960.
2. N. J. Nilsson, Learning Machines, McGraw-Hill Book Co., 1965.

RECOGNITION SYSTEM FOR HANDWRITTEN LETTERS

SIMULATING VISUAL NERVOUS SYSTEM

Katsuhiko Fujii and Tatsuya Morita

Osaka University

Osaka, Japan

1. INTRODUCTION

Human beings and animals have excellent pattern recognition faculties. However, there are many difficulties when this task is to be performed by information processing equipment. Therefore, a recognition system constructed to have characteristics of the visual nervous system could possibly be applied to the recognition of letters and figures. In this paper, composition of a recognition system by simulating the visual nervous system is discussed.

First, the nervous structure of the visual system will be simulated by so-called lateral inhibition structure. This structure has the function of filtering characteristic portions which are important information for recognition. We will propose a design method of the lateral inhibition structure for extracting characteristic properties such as end portions, connecting portions and line orientations.

Since property patterns extracted from one kind of handwritten letters are still diverse, the determination of the identification logic is very difficult. Also, a classifier having learning function better expresses the characteristics of information processing in the human brain. Therefore, we will propose an application of a cascade connection of the lateral inhibition structure and Adaline learning system to handwritten character recognition system.

2. INFORMATION PROCESSING IN THE VISUAL SYSTEM

The visual system of the higher animal is constructed as shown in Figure 1. Visual patterns are transformed to the nervous impulse

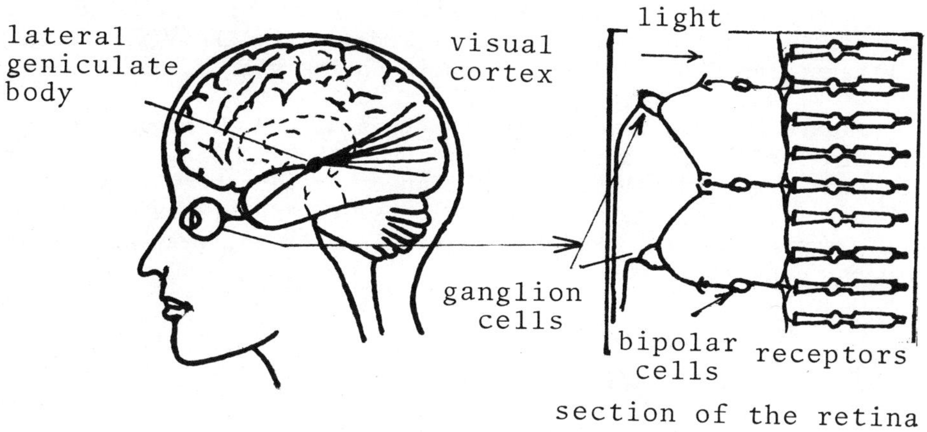

Figure 1. Schematic Illustration of the Visual System in Higher Animal.

frequencies by the receptors. The information transmits through the lateral geniculate body and is recognized in the visual cortex. Each cell receives information directly or indirectly from its own region on the retina termed 'receptive field'. S. W. Kuffler discovered two types of neurons in the cat's retina, one having 'on-center' receptive field and the other having 'off-center' field. The on-center cell excites when the central region of the field is stimulated by a spot of light, and it is inhibited when the peripheral region is stimulated. The off-center cell, vice versa.

While neurons in the cat's visual cortex respond to a white or a black line or an edge separating white from black as discovered by D. H. Hubel and T. N. Wiesel. It is supposed that these cells detect the orientation of the line or the edge. When the line shifts to a new position without changing its orientation, the response of cells termed 'simple' decreases. Other cells called 'complex' respond to the line, regardless of its position. Cells which are more organized than complex cells are discovered and termed 'hyper-complex'.

Though, the mechanism of the human brain has not been revealed, the excellent recognition faculties are possibly achieved by the information processing mechanism as above mentioned.

3. A MATHEMATICAL MODEL OF THE VISUAL SYSTEM

It is supposed that the visual system is essentially constructed by the lateral inhibition (LI) structure which is found in the eye of the horse-shoe crab. That is, nervous impulses evoked from a receptor by a spot of light are reduced by stimulating neighbouring receptors. In this chapter, the LI structure will be simulated by a mathematical model and its characteristics will be analyzed.

Figure 2 shows two elementary types of the LI structure: 'forward inhibition' and 'backward inhibition'. Circles and lines represent cell bodies and neural connections, respectively. The response (q(y)) of the forward and backward inhibition structures to a visual pattern (p(x)) can be written respectively as following integral equations,

$$q(y) = \int_{-\infty}^{\infty} w(y-x)p(x)dx \tag{1}$$

$$q(y) = p(y)-\int_{-\infty}^{\infty} w(y-x)q(x)dx \tag{2}$$

where the weighting function w(y-x) represents an interaction coefficient from a cell (x) to another cell (y). These equations mean that the LI structure is considered as a spatial filter. The backward inhibition structure corresponds to the feedback control system, and can be transformed equivalently to a forward inhibition structure.

The LI structure is also considered as a model of the receptive field. For example, the on-center receptive field can be simulated by the following forward inhibition structure with the weighting function (w(r)).

$$q(x,y) = \int_{-\infty}^{\infty} \int_{-\infty}^{\infty} w(x-\xi,\ y-\eta)p(\xi,\eta)d\xi d\eta \tag{3}$$

$$w(r) = \frac{K_1}{2\pi\sigma_1{}^2}\exp(-\frac{r^2}{2\sigma_1{}^2}) - \frac{K_2}{2\pi\sigma_2{}^2}\exp(-\frac{r^2}{2\sigma_2{}^2}),\ r^2=x^2+y^2 \tag{4}$$

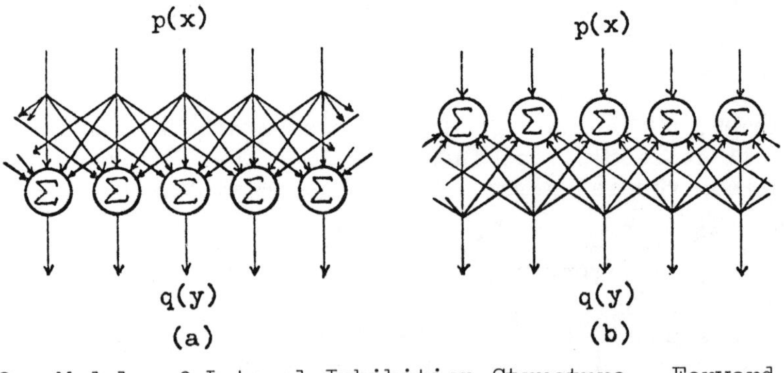

Figure 2. Models of Lateral Inhibition Structure. Forward Inhibition (a) and Backward Inhibition (b).

where K_1 and K_2 are coefficients of the excitatory and inhibitory connections, respectively. Also, σ_1 and σ_2 are parameters representing the spread of these neural connections.

Figure 3 (a) and (b) show the theoretical responses $(q(x,y))$ of the on-center neurons to a square and T-form figure, respectively. Neurons excite on the sides and the corners of the square, and are inhibited outside. This effect of outline emphasis accords with the Mach's phenomena known in psychology. The response to the T-form figure increases on such characteristic portions as intersecting portion and end portions of lines. Therefore, it is concluded that the cell having on-center receptive field extracts the important information for visual pattern recognition.

4. DESIGN METHOD FOR PROPERTY FILTERS

Since the nerves connect uniformly in the LI structure, the property can be filtered regardless of position and size of the present pattern, if the size is above the resolving power of the LI structure. Also, since the LI structure conducts spatial integration, it is not apt to be affected by the distortion of the patterns.

Two methods can be considered to utilize the LI structure in a property filter. (1) One method is simulating the visual nervous system in a multi-layer LI structure. This method is advantageous

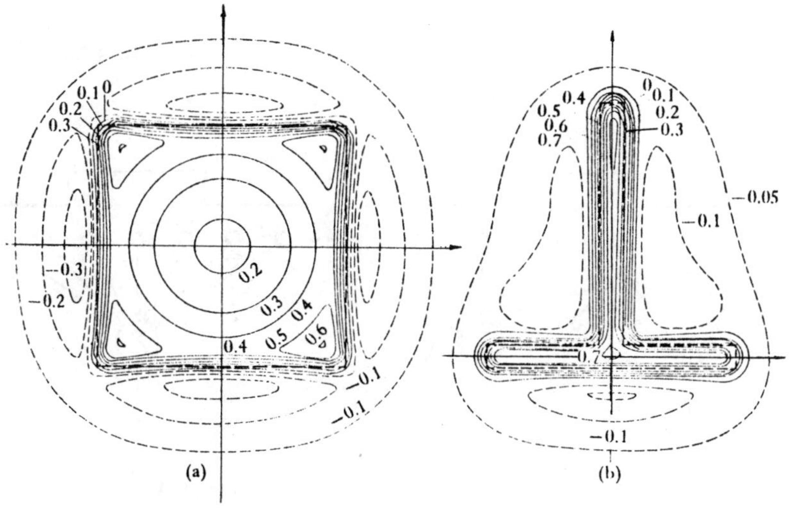

Figure 3. Responses of the Lateral Inhibition Model to Two Figures, Square (a) and T-form (b).

to consider nonlinearity of each neurons in the visual system. However, it is difficult to regulate the connections of successive nerve layers. (2) If nonlinearity of each nerve layer is disregarded, it is possible to express the entire visual system equivalently by a 1-layer LI structure. Though the anatomical relationship to the visual system is lost, the equivalent 1-layer LI structure is more convenient to regulate the coefficients of connections.

In this paper, we adopt method (2) and will design the LI structure for property filtering. When m kinds of properties (A, B, ... M) are given in the LI structure shown in Figure 4 (a), the parameters of the LI structure which filters any one property can be obtained as follows. If the interval between properties is greater than the spread of the nerve connection, Figure 4 (a) becomes approximately Figure 4 (b). Therefore, outputs q_A, and q_B, ..., q_M responding to the properties A, B, ... M become

$$q_K = \sum_{i=-n}^{n} p_{K_i} w_i \qquad K=A,B,\ldots M \tag{5}$$

where p_{K_i} is 1 or 0 according to whether the i-th receptor is stimulated when property K is given to the LI structure. Also, w_i is the coefficient of the nerve connection to be determined, and n expresses the spread of the connection. When the spread of the connection is chosen large, there are instances when the experimental results on property extracting do not accord with the specification of the filter design. Because it was assumed that the spread of the connection must be smaller than the interval between properties. However, when the spread is made small, freedom of nerve connection declines and the property filtering function deteriorates.

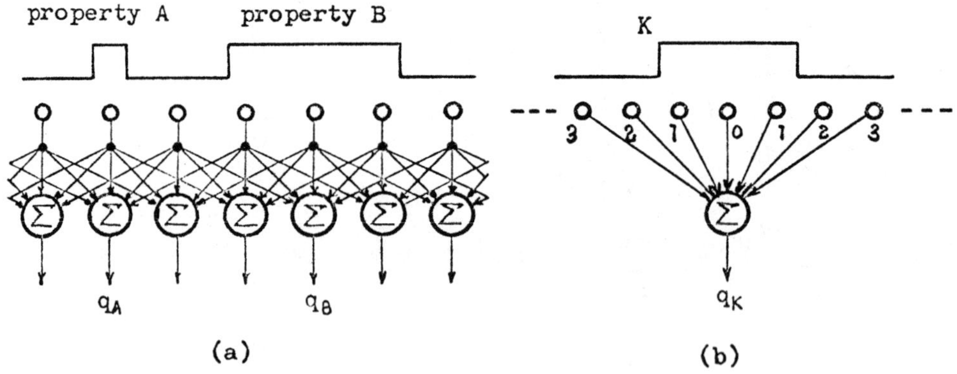

Figure 4. Property Filter Using Lateral Inhibition Structure (a) and its Elemental Unit (b).

When the nerve connection has symmetry, Eq. (5) can be expressed as

$$q_K = \sum_{i=0}^{n} N_i p_{K_i} w_i = \sum_{i=0}^{n} K_i \bar{W}_i$$
$$K_i = \sqrt{N_i} p_{K_i}, \quad \bar{W}_i = \sqrt{N_i} w_i \tag{6}$$

where N_i is the number of receptors in symmetrical positions. Since any coefficient of the nerve connection must be a finite value, the next restrictive condition is assumed.

$$\sum_{i=-n}^{n} w_i^2 = \sum_{i=0}^{n} \bar{W}_i^2 = 1 \tag{7}$$

Satisfying the condition and maintaining q_B, q_C, ..., q_M as small values (for example 0), the coefficient w_i which filters only one property A is obtained as,

$$\bar{W}_i = \sum_{n+1^C m-1} \begin{vmatrix} q_A, A_j, \ldots A_k \\ \\ \\ q_M, M_j, \ldots M_k \end{vmatrix} \begin{vmatrix} A_i, A_j \ldots A_k \\ \cdot \quad \cdot \\ \cdot \quad \cdot \\ \cdot \quad \cdot \\ \cdot \quad \cdot \\ M_i, M_j \ldots M_k \end{vmatrix} \Bigg/ \sum_{n+1^C m} \begin{vmatrix} A_u \ldots A_v \\ \cdot \quad \cdot \\ \cdot \quad \cdot \\ \cdot \quad \cdot \\ \cdot \quad \cdot \\ M_u \ldots M_v \end{vmatrix}^2 \tag{8}$$

where $\sum$ expresses the sum of ${}_{n+1}C_{m-1}$ or ${}_{n+1}C_m$ determinants made by taking m-1 or m inputs without duplication from n+1 inputs corresponding to i=0,1,...n.

Also, the relationship between q_A, q_B, ...q_M is given by the ellipsoid in m-dimensional space as,

$$\sum_{n+1^C m-1} \begin{vmatrix} q_A, A_j, \ldots, A_k \\ \cdot \quad \cdot \\ \cdot \quad \cdot \\ \cdot \quad \cdot \\ \cdot \quad \cdot \\ q_M, M_j, \ldots M_k \end{vmatrix} = \sum_{n+1^C m} \begin{vmatrix} A_u \ldots A_v \\ \cdot \quad \cdot \\ \cdot \quad \cdot \\ \cdot \quad \cdot \\ \cdot \quad \cdot \\ M_u \ldots M_v \end{vmatrix}^2 \tag{9}$$

Therefore, it is possible to obtain the maximum value of q_A from this equation.

When the properties are linearly dependent together regardless of i=0, 1, ... n as,

$$\alpha A_i + \beta B_i + \ldots\ldots + \mu M_i = 0 \tag{10}$$

the denominator of Eq. (8) becomes 0. In this case, outputs q_A, q_B,...q_M are also linearly dependent as

$$\alpha q_A + \beta q_B + \ldots\ldots + \mu q_M = 0 \tag{11}$$

and each output cannot be freely determined.

Figure 5 is an example where the horizontal segment of the Japanese letter nu (ヌ) is filtered by the LI structure. The coefficients of the nerve connections are determined so that the outputs of cells on the horizontal segments can become maximum value, preventing extraction of vertical segment and X-type intersecting portion. From the figure, it is appropriate to select the threshold value at 4. We will next give an example of application of the above mentioned design method to a handwritten letter recognition system.

5. A RECOGNITION MODEL FOR HANDWRITTEN LETTERS

A recognition system using LI structure as the property filters is shown in Figure 6. This recognition system may be applied to

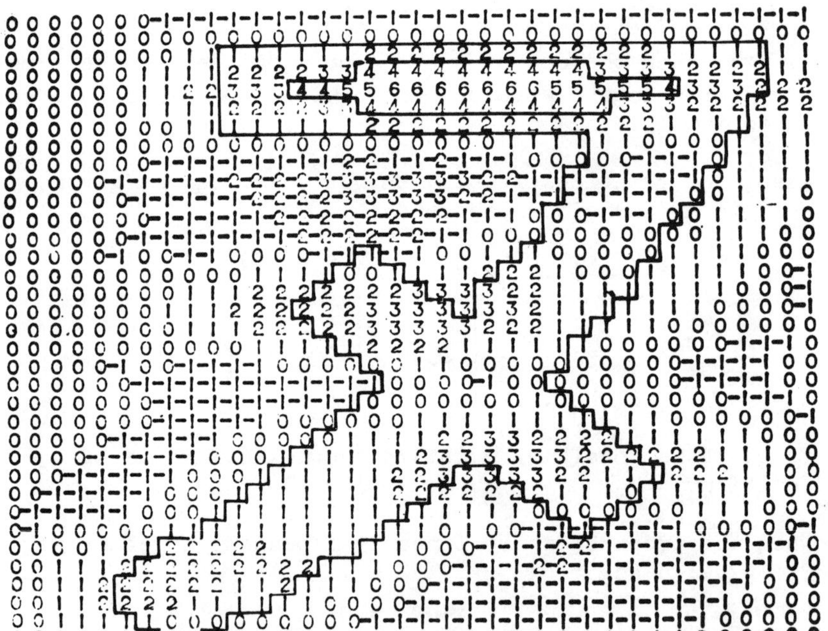

Figure 5. Horizontal Segment of Japanese Letter "nu" Extracted by Lateral Inhibition Structure.

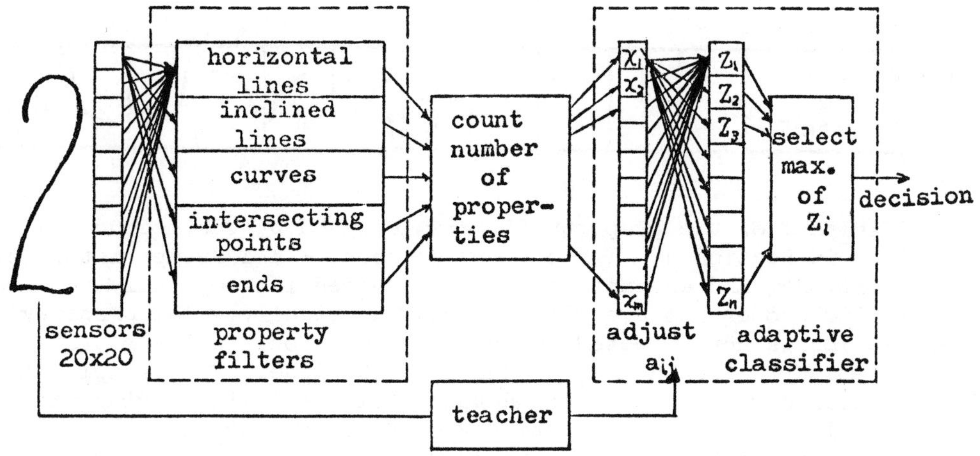

Figure 6. Pattern Recognition System Using Lateral Inhibition Structure.

various kinds of visual pattern recognition by altering the property filters. In this paper, the recognition of handwritten numerals is discussed.

The character is quantized by an artificial retina consisting of 20X20 receptors. This number corresponds to the one of the receptors on the retina stimulated when the smallest letters of a pocket type dictionary is looked at about 30 cm distance. The signals enter the property filters, where each property is extracted. A classifier is also designed to compute the linear discriminant function Z_i in relation to the position and number of the extracted properties. The character is identified as the one corresponding to the maximum in the functions Z_i. After identification, the coefficients of the functions are adjusted on the basis of the correct answer by a teacher. This recognition system is programed in a digital computer.

5.1. Property Filters

Generally, if the patterns are linear-separable, it is sible to classify these patterns by limited times of adjustments of the coefficients in the classifier. The purpose of property filtering is transforming the character information to linear-separable patterns. However, properties necessary for identification differ according to the character being identified, and their selection is difficult. In this paper, we have chosen the ones given in Table 1 for recognition of numerals. Figure 7 shows the

Table 1. Properties Which Become Information for Character Recognition

properties to be extracted	—	|	/	\	+	×	╮	╭	╰	╯	End

coefficients of the property filters designed by the method described in Section 4. The properties about + and X-type intersecting portions, and end portions are extracted isotropically and the remaining properties are filtered anisotropically. The property extraction is not affected by property position, but is restricted in relation to length and thickness of the line. For example, a horizontal segment filter can detect horizontal line having thickness of 1 or 2 cells and partially 3 cells, but not possible in case of greater thickness. Figure 8 shows property position extracted by each filter.

-1	-2	-2	-2	-1
0	-1	-2	-1	0
4	1	0	1	4
0	-1	-2	-1	0
-1	-2	-2	-2	-1

Horiz. segment
$\theta = 5$

-2	0	4	0	-2
-2	-1	1	-1	-2
-2	-1	0	-1	-2
-2	-1	1	-1	-2
-2	0	4	0	-2

Vertical segment
$\theta = 5$

-1	-1	-1	1	1
-1	-1	0	1	1
-1	0	0	0	-1
1	1	0	-1	-1
1	1	-1	-1	-1

Left downward oblique line
$\theta = 5$

1	1	-1	-1	-1
1	1	0	-1	-1
-1	0	0	0	-1
-1	-1	0	1	1
-1	-1	-1	1	1

Right downward oblique line
$\theta = 5$

4	-1	8	-1	-4
-1	1	2	1	-1
8	2	0	2	8
-1	1	2	1	-1
-4	-1	8	-1	-4

+-type intersecting point
$\theta = 38$

6	1	-3	1	6
1	2	0	2	1
-3	0	0	0	-3
1	2	0	2	1
6	1	-3	1	6

x-type intersecting point
$\theta = 26$

-4	-5	-1	-3	-1	0	0
1	1	-1	0	-1	0	0
4	1	0	0	0	-1	-1
0	1	-1	0	0	0	-3
-1	-1	0	-1	0	-1	-1
0	0	-1	1	1	1	-5
0	0	-1	0	4	1	-4

-type curve
$\theta = 10$

0	-2	-1	-2	0
-2	-1	-1	-1	-2
-1	-1	0	-1	-1
-2	-1	-1	-1	-2
0	-2	-1	-2	0

End point
$\theta = -8$

Figure 7. Nerve Connection Coefficients of Property Filters (θ is Threshold Value).

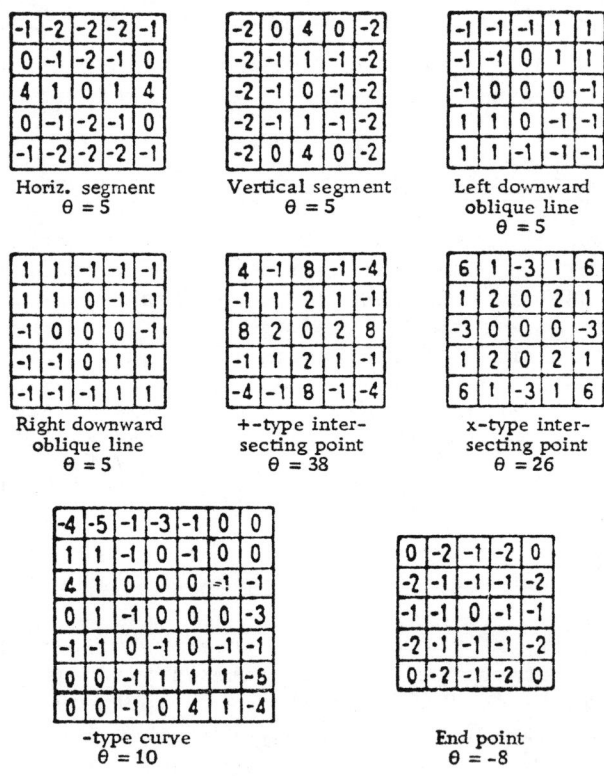

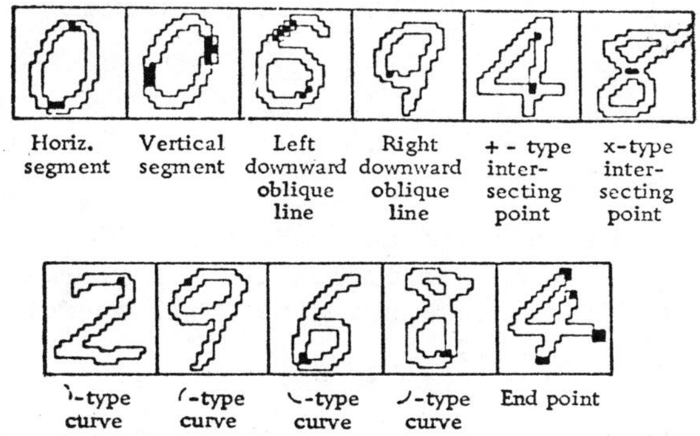

Figure 8. Examples of Property Filtering. Black Squares Show the Points Where the Properties are Extructed.

5.2. Property Discriminator

The cell outputs of each filter are scanned and when a cell emitting out-put is found, a counter corresponding to the property is added 1. Then, scanning procedure is resumed skipping over the vicinity of the discovered property position. The size of the vicinity is chosen suitably for each property and in the case of a horizontal segment, about 3 mesh lines is suitable. Each extracted property is treated as follows. The straight line group (horizontal, vertical, and oblique lines) is classified according to whether each number corresponds to 0, 1, 2, 3 or more. Curves are classified according to 0, 1, 2 or more, and + and X-type intersecting points are classified according to their presence or absence. In the case of end points, the center of gravity of the letter is computed and image plane is divided into 4 quadrants. The properties about ends are classified as to whether the number appearing in each quadrant is 0, 1, 2 or more. Thus, the property vector which consists of 44 bits information is obtained.

5.3. Pattern Classifier

Character information has been concentrated into the property vector, but properties obtained from one kind of letters are still very diverse. Therefore, it is difficult to predict the identification logic. Then, assuming that the property vector X is a linear-separable pattern, the classifier is constructed by the learning theory as follows. Linear discriminant function Z is computed in relation to the property vector $X=(x_1, x_2, \ldots x_n)$ as,

$$Z_i = \sum_{j=1}^{n} a_{ij} X_j, \qquad i=0,\ldots 9, \qquad n=44 \qquad (12)$$

and the letter is identified based on the maximum among the functions Z_i.

Coefficients a_{ij} are adjusted as,

$$a_{ij} = \log P(C_i/x_j) \tag{13}$$

on the basis of the conditional probability $P(C_i/x_j)$ that the input character is C_i when a property x_j is filtered. However, assuming that the characters are given with uniform probability,

$$a_{ij} = \log P(x_j/C_i) \tag{14}$$

may be used instead of Eq. (13) according to the Bayes' rule. Here, $P(x_j/C_i)$ is the conditional probability that the property x_j will be filtered when the character C_i is given. If the property vector X is the linear-separable pattern, it will ultimately be possible to construct the classifier which will make correct identification.

6. IDENTIFICATION EXPERIMENT

The performance of the recognition system was examined for handwritten letters. Material for the identification experiment was obtained using a pen of about 2 mm thickness and writing numerals 0 to 9 in a square frame 4 cm on each side. Even though it was requested that numerals were written carefully, there were notable individual differences in handwriting. These numerals were used with the recognition system by the following procedure.

Each sample was assigned a number according to the person who wrote it. The experiment proceeded with the numerals 0 to 9 of the first person, then 0 to 9 of the second person, etc. The teacher also taught the correct answer, after the recognition system gave its decision.

Figure 9 shows the results of the experiment; here abscissa is the number assigned to the person and ordinate is the number of correct answers, incorrect answers and rejections among the numerals 0 to 9. The numerals of the first person were all rejected since the recognition system had not yet learned any numerals. As coefficient of the classifier were adjusted, correct answers increased, and final rate of correct answers was about 90%. After learning 400 numerals, the recognition system was made to identify various numerals as follows. (1) 100 suitably selected characters from 400 learned characters were again subjected to identification. In this case, all numerals could be identified. (2) When the numerals were small as compared with the surface of the artificial retina, it was not possible to utilize effectively the receptors.

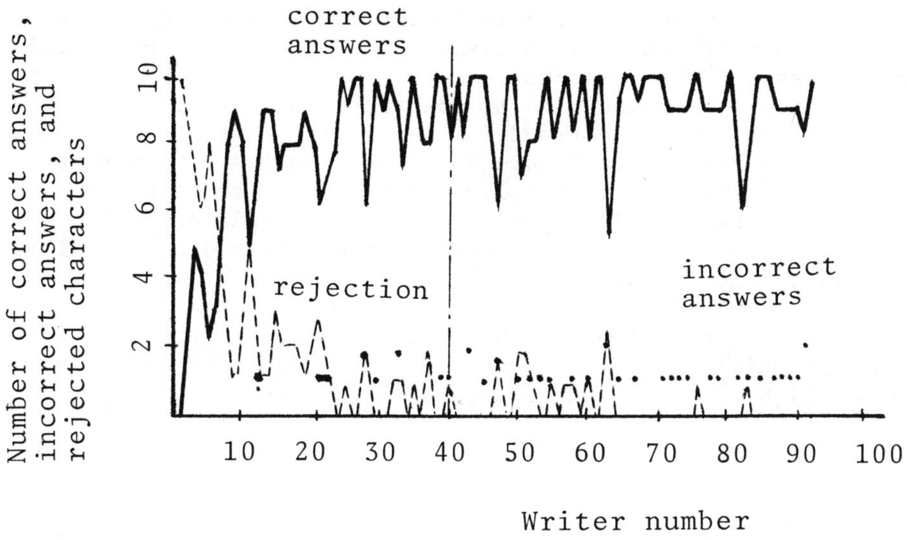

Figure 9. Learning Process of the Recognition System for Handwritten Numerals.

When the size of numerals is reduced to 2/3, the number of effective receptors declines to about half of the total. However, as is shown in Table 2 identification results were about 90% for reduction to 4/5 and 2/3. When compared with the correct answer rate of the original size, the identification results did not decline because of the size reduction. (3) When numerals were written with a slant, property filtering becomes difficult, since the orientation of lines is changed. When slant 15° clockwise, identification results were little influenced, but when slant 15° counterclockwise, the

Table 2. Identification Results After Learning 400 Characters.

	No. of characters	Correct answers	Rejects	Incorrect answers
Re-identified	100	100	0	0
Size reduced to 4/5	100	92	3	5
Size reduced to 2/3	100	89	3	8
Slant -15	100	93	3	4
Slant +15	100	78	5	17

identification results declined to about 80%. The reason of the difference in these results is considered as the handwritten numerals have a tendency to slant clockwise and the recognition system learned this orientation of numeral axis. (4) When the recognition system was made to identify somewhat wild written letters, the rate of correct answer for 520 numerals was still 90% as shown in Figure 9. Figure 10 shows examples of somewhat wild written letters used in the identification experiment. The letters which could be identified and others are classified by the marks C and E, respectively. The identification result did not influence notably by the handwritten.

In these experiments, the incorrect answers possibly come from the contradiction of the assumption that the filtered property vectors will become the linear separable patterns.

7. CONCLUSION

Refering to the structure of the visual nervous systems, the authors composed a recognition system having property filters constructed by the lateral inhibition structure. Since the recognition system has a learning function, it is not apt to be affected by character position, size, slant, etc., and it is not necessary to establish complex identification logic. Also, the information is processed in parallel in the recognition system, it will be applied to future recognition equipment of rapid information processing speed.

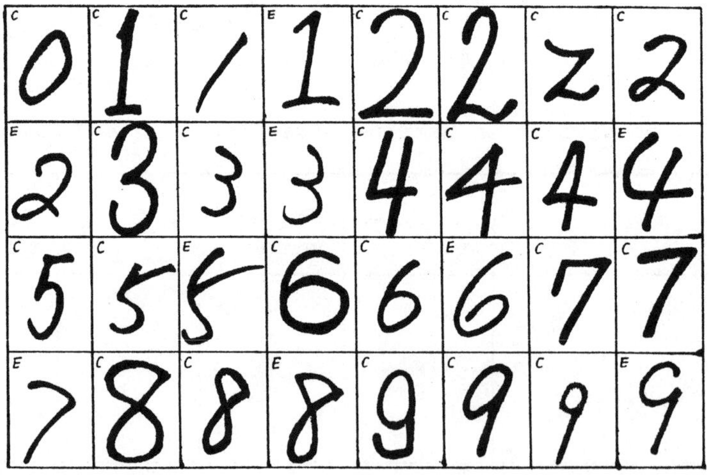

Figure 10. Examples of Handwritten Numerals used for Identification Experiment.

REFERENCES

1. H. K. Hartline and F. Ratliff, "Spatial Summation of Inhibitory Influence in the Eye of Limulus, and the Mutual Interaction of Receptor Units", J. Gen. Physiol., 41, p. 1049 (1958).

2. S. W. Kuffler, "Discharge Patterns and Functional Organization of Mammalian Retina", J. Neurophysiol., 16, p. 37 (1953).

3. D. H. Hubel and T. N. Wiesel, "Receptive Field and Functional Archetecture in the Non-striate Visual Areas (18 and 19) of the Cat", J. Neurophysiol., 28, 2, p. 229 (1965).

4. T. Morita and K. Fujii, "Function of Lateral Inhibition in Pattern Recognition Process of Biological Systems", C.E.J., 49, 10, p. 45 (1966).

5. K. Fujii, A. Matsuoka and T. Morita, "Analysis of the Optical Illusion by Lateral Inhibition", Tech. Report of Osaka Univ., 17, p. 445 (1967).

6. T. Morita and K. Fujii, "Recognition System for Handwritten Letters Simulating Visual Nervous System", C.E.J., 52, 7, p. 100 (1969).

7. N. J. Nilsson, Learning Machines: Function of Trainable Pattern Classifying Systems, McGraw Hill, N.Y. (1965).

SEQUENTIAL IDENTIFICATION BY MEANS OF GRADIENT LEARNING ALGORITHMS

J. M. Mendel

McDonnell Douglas Astronautics Company

Huntington Beach, California, U.S.A.

I. INTRODUCTION

The gradient identification algorithms considered herein are sequential algorithms in which estimates of a parameter vector, $\underline{b}_k$, (an n-vector) are obtained at time t_{k+1} from the preceding estimate at time t_k, from the gradient of a performance function that provides a measure of identification error, and from any additional information available at time t_{k+1}. During the past few years, numerous aspects of these identification algorithms have been studied ([1] and [2], for example). Analytical and experimental results have been obtained for the following classes of parameter identification problems:

Class 1 - Identification of constant parameters from perfect measurements.
Class 2 - Identification of time-varying parameters from perfect measurements.
Class 3 - Identification of constant parameters from noisy measurements.
Class 4 - Identification of time-varying parameters from noisy measurements.

II. IDENTIFICATION SYSTEM

Before presenting the various identification algorithms, some important notational definitions will be given. It has been shown [1,2] that many familiar parameter identification problems can be formulated as in Figure 1. The various quantities in the identification system are: actual input, $\underline{x}_k(n\times 1)$, which is usually

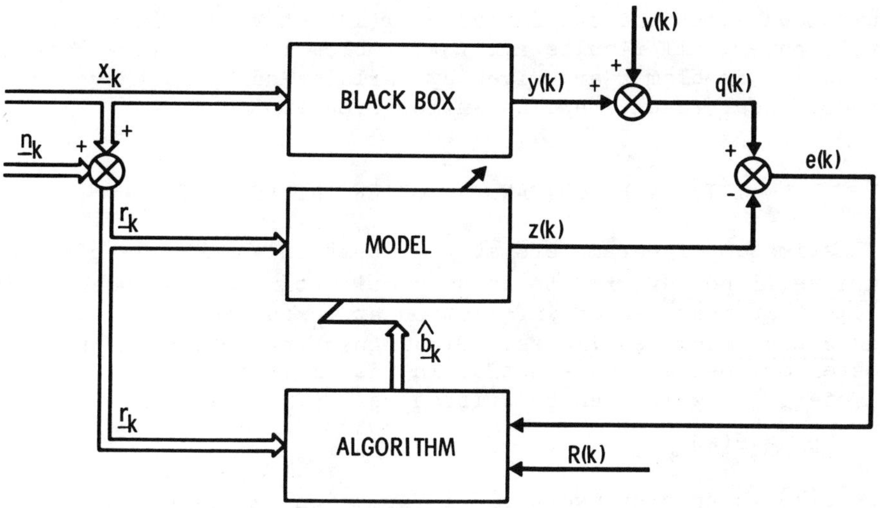

Figure 1. Identification System

inaccessible; input noise, $\underline{n}_k(n\times 1)$ with the following properties:

$$E\{\underline{n}_k\} = \underline{0} \text{ and } E\{\underline{n}_k\underline{n}_k'\} = \Sigma_n = \text{diag }(\sigma_1^2,\ldots,\sigma_n^2) \;; \qquad (1)$$

measured input, $\underline{r}_k(n\times 1)$, where

$$\underline{r}_k = \underline{x}_k + \underline{n}_k \qquad (2)$$

actual output, y(k) (scalar), which is usually inaccessible and is related to $\underline{x}_k$ in the following manner

$$y(k) = \sum_{i=1}^{n} b_i(k)x_i(k) = (\underline{b}_k,\underline{x}_k) \;; \qquad (3)$$

output noise, v(k) (scalar) with the following properties,

$$E\{v(k)\} = 0 \text{ and } E\{v^2(k)\} \leq V \;; \qquad (4)$$

observed output, q(k) (scalar), where

$$q(k) = y(k) + v(k) \;; \qquad (5)$$

model output, z(k) (scalar), where

$$z(k),\ (\hat{\underline{b}}_k,\underline{r}_k) \qquad (6)$$

in which $\hat{\underline{b}}_k$ is the estimate of $\underline{b}_k$; and error signal (scalar), e(k), where

$$e(k) = q(k) - z(k) \;. \qquad (7)$$

Three stochastic identification problems are described in the Appendix. These problems are distinguished by their different assumptions about the statistics of $\underline{x}(k)$ and $v(k)$. In the remainder of this paper, all results are for Problem 1. Some very useful results for Problem 2 are given by Saridis and Stein [3]. To date, no theoretical results are known for Problem 3.

III. A PRIORI AND A POSTERIORI ESTIMATES

Estimates of parameters at t_{k+1} that only make use of information t_k will be referred to as a priori estimates, whereas estimates at t_{k+1} that make use of information at t_k and t_{k+1} will be referred to as a posteriori estimates. Distinguishing between a priori and a posteriori estimates is useful in identification of a time-varying parameter, $\underline{b}_k$, which can be written as

$$\underline{b}_k = P(k)\underline{\beta}_k \tag{8}$$

where $P(k)$ is an $n\times n$ invertible information matrix whose elements are either measurable or specified ahead of time at t_k. In an a priori estimate of $\underline{b}_k$, the decomposition in eq. (8) is not utilized. On the other hand, in an a posteriori estimate of $\underline{b}_k$, $\underline{\beta}_k$ is identified first and then the estimate of $\underline{b}_k$ is formed from it and eq. (8).

To further clarify a priori and a posteriori estimates and the decomposition in eq. (8), the following example is offered. It is well known that the aerodynamic stability derivative, $M_\alpha(t)$ and control-surface effectiveness, $M_\delta(t)$, can be written as

$$M_\alpha(t) = \tilde{q}(t)M_{\alpha q}(t) \tag{9}$$

and

$$M_\delta(t) = \tilde{q}(t)M_{\delta q}(t). \tag{10}$$

In these equations, $\tilde{q}(t)$ is dynamic pressure; it varies by large amounts from one sampling instant to the next, but may either be known ahead of time or can be measured online by means of a pressure probe. $M_{\alpha q}(t)$ (shorthand for $(\partial C_m/\partial\alpha)S\ell(57.3/I_p)$) and $M_{\delta q}(t)$ (shorthand for $(\partial C_m/\partial\delta)S\ell(57.3/I_p)$) vary over smaller ranges of values and usually at slower rates than do $M_\alpha(t)$ and $M_\delta(t)$. A posteriori estimates of $M_\alpha(t)$ and $M_\delta(t)$ make use of the decompositions in eqs. (9) and (10), whereas a priori estimates do not. A posteriori estimates have been shown to be vastly superior to their a priori counterparts (as one might expect because they utilize all available information).

IV. IDENTIFICATION ALGORITHMS

A priori and a posteriori identification algorithms are summarized in Table I. A priori algorithms are applicable to all four

Table I
IDENTIFICATION ALGORITHMS

	A Priori Algorithms	A Posteriori Algorithms
Perfect Measurements	$\hat{\underline{b}}_{k+1} = \hat{\underline{b}}_k + e(k)R_{bP}(k)\underline{x}_k$	$\hat{\underline{\beta}}_{k+1} = \hat{\underline{\beta}}_k + e(k)R_{\beta P}(k)P'(k)\underline{x}_k$ $\hat{\underline{b}}_{k+1} = P(k+1)\hat{\underline{\beta}}_{k+1}$
Noisy Measurements	$\hat{\underline{b}}_{k+1} = \left[I + R_{bN}(k)\sum_n\right]\hat{\underline{b}}_k + e(k)R_{bN}(k)\underline{r}_k$	$\hat{\underline{\beta}}_{k+1} = \left[I + R_{\beta N}(k)P'(k)\sum_n P(k)\right]\hat{\underline{\beta}}_k + e(k)R_{\beta N}(k)P'(k)r_k$ $\hat{\underline{b}}_{k+1} = P(k+1)\hat{\underline{\beta}}_{k+1}$
Weighting Matrices	$R_{bP}(k) = \dfrac{\text{diag}(h_1,\ldots,h_n)}{\underline{x}_k'H\underline{x}_k}$ $R_{bN}(k) = \dfrac{\text{diag}(h_1,\ldots,h_n)}{\underline{r}_k'H\underline{r}_k}$	$R_{\beta P}(k) = \dfrac{\text{diag}(h_1,\ldots,h_n)}{\underline{x}_k'P'(k)HP(k)\underline{x}_k}$ $R_{\beta N}(k) = \dfrac{\text{diag}(h_1,\ldots,h_n)}{\underline{r}_k'P'(k)HP(k)\underline{r}_k}$

classes of identification problems, whereas a posteriori algorithms are applicable only to Class 2 and Class 4 identification problems. The weighting matrix, $R_{bP}(k)$, was derived from an error-correction principle. R_{bN} and $R_{\beta N}$ are heuristic adaptations of R_{bP}

The situation where

$$P(k) = p(k)I \tag{11}$$

is important in aerospace applications (see eqs. (9) and (10), for example). In the case of eq. (11), the a posteriori algorithms in Table I reduce to those summarized in Table II.

Applications of all the preceding gradient identification algorithms are discussed in references [1], [2], and [4].

V. COMPUTATIONAL REQUIREMENTS

Computational requirements have been obtained for Class 3 and Class 4 (a posteriori) algorithms. These requirements have been obtained for a representative airborne digital computer, the Honeywell SIGN-III. The SIGN-III is a general-purpose, high-speed, stored-program computer [5]; it is based on extensive use of integrated circuits and designed for airborne applications, such as guidance and navigation. In a recent survey of aerospace digital computers [4], the SIGN-III was classified as a late-1966 computer (date introduced). Some of its computational features are:

Table II
A POSTERIORI IDENTIFICATION ALGORITHMS: $P(k) = p(k)1$

	Perfect Measurements	Noisy Measurements
Vector Equations	$\hat{\underline{b}}_{k+1} = \frac{p(k+1)}{p(k)} [\hat{\underline{b}}_k + e(k)R_{bP}(k)\underline{x}_k]$	$\hat{\underline{b}}_{k+1} = \frac{p(k+1)}{p(k)} \{[I + R_{bN}(k)\Sigma_n] \hat{\underline{b}}_k + e(k)R_{bN}(k)\underline{r}_k\}$
Component Equations	$\hat{b}_i(k+1) = \frac{p(k+1)}{p(k)} [\hat{b}_i(k) + \frac{h_i e(k) x_i(k)}{\sum_{j=1}^{n} h_j x_j^2(k)}]$	$\hat{b}_i(k+1) = \frac{p(k+1)}{p(k)} \{[1 + \frac{h_i \sigma_i^2}{\sum_{j=1}^{n} h_j r_j^2(k)}] \hat{b}_i(k) + \frac{h_i e(k) r_i(k)}{\sum_{j=1}^{n} h_j r_j^2(k)}\}$
	$i = 1, 2, \ldots, n$	$i = 1, 2, \ldots, n$

4 μsec add time, 24 μsec multiply and divide times, 4,096 word memory capacity, and 20 bit word length.

Computational requirements--computational time and program size--are summarized in Table III for the a priori and a posteriori noisy-measurement algorithms. Computational time and program size were obtained from the SIGN-III Instruction Repertoire for the various equations, including add, multiply, divide, storage, index, cycle, and input and output requirements. The computational requirements are for one iteration and are seen to be quite small. For example, estimating two parameters by means of a posteriori algorithms would require 0.725 msec computational time and 54 memory cells.

Based on predictions of 1971 computational times [6], it is estimated that by 1971 these times will be reduced to 40 percent of their present values.

VI. CONCLUSIONS

The gradient identification algorithms which have been summarized in this paper are accurate (good convergence properties) and simple to implement. These properties make them very attractive for use in certain practical applications.

Table III
COMPUTATIONAL REQUIREMENTS (NOISY MEASUREMENTS)

	A Posteriori Algorithm	A Priori Algorithm
Computational Time (μsec)	131 + 297 n*	47 + 303 n
Program Size (Cells)	46 + 4 n	44 + 4 n

*n = number of parameters being identified

APPENDIX. THREE TYPES OF STOCHASTIC IDENTIFICATION PROBLEMS

By making different assumptions about the statistics of $\underline{x}(k)$ and $v(k)$, one can obtain three quite different and important identification problems. These problems are presented in increasing order of difficulty.

A. Problem 1

In this problem, the identification system is as depicted in Figure 1. The following assumptions are made concerning the statistics of $\underline{x}(k)$ and $v(k)$:

(1) $E\{\underline{x}(k)\underline{x}'(k)|\hat{\underline{b}}(k)\} = \Omega$, an unknown matrix that is not a function of $\hat{\underline{b}}(k)$; and

(2) $\underline{x}(k)$ and $v(k)$ are uncorrelated, so that

$$E\{\underline{x}(k)v(k)\} = E\{\underline{x}(k)\} \quad E\{v(k)\} .$$

Assumptions (1) and (2) mean that there is no coupling between $\underline{x}(k)$ and $\hat{\underline{b}}(k)$,and $\underline{x}(k)$ and $v(k)$.

Problem 1 is applicable to:

(a) Identification of coefficients in algebraic equations when some or all of the signals cannot be accurately measured.

(b) Identification of coefficients of difference equations when plant disturbances are not present and when noise distorts the measurements.

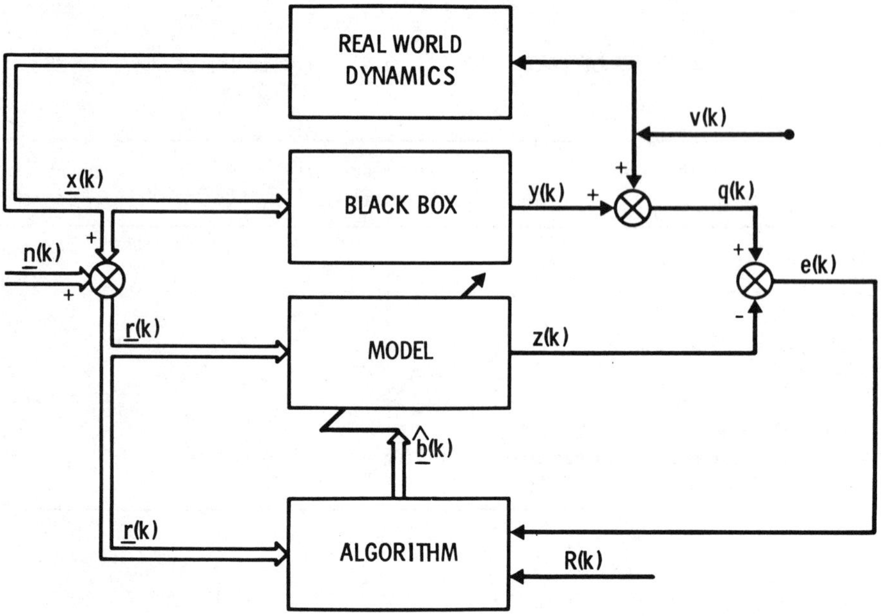

Figure 2. Identification System for Problem 2

B. Problem 2

The identification system in Problem 2 is as depicted in Figure 2. The following assumptions are made concerning the statistics of $\underline{x}(k)$ and $v(k)$:

(1) $E\{\underline{x}(k)\underline{x}'(k)|\hat{\underline{b}}(k)\} = \Omega$, an unknown matrix that is not a function of $\hat{\underline{b}}(k)$; and

(2) $\underline{x}(k)$ and $v(k)$ are correlated.

Now, there is coupling between $\underline{x}(k)$ and $v(k)$.

Problem 2 is applicable to identification of coefficients of difference equations when plant disturbances are present and when measurements are distorted by noise.

C. Problem 3

In this problem, the most difficult of the three, the identification system is as depicted in Figure 3. Now, not only is there coupling between $\underline{x}(k)$ and $v(k)$, but there is also coupling between $\underline{x}(k)$ and $\hat{\underline{b}}(k)$. In mathematical terms,

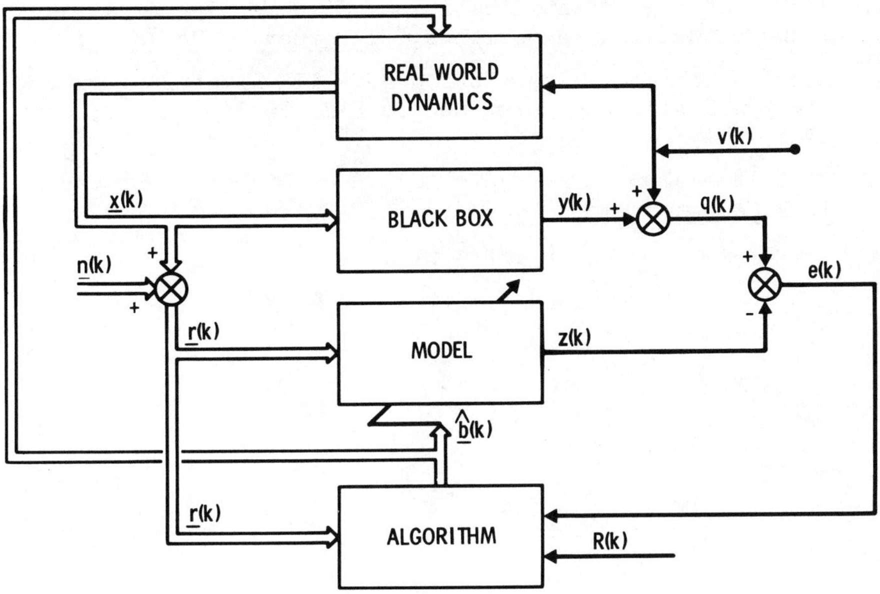

Figure 3. Identification System for Problem 3.

(1) $E\{\underline{x}(k)\underline{x}'(k)|\hat{\underline{b}}(k)\} = \Omega(\hat{\underline{b}}(k))$; and

(2) $\underline{x}(k)$ and $v(k)$ are correlated.

This problem is applicable to identification of coefficients of difference equations when plant disturbances are present, when measurements are distorted by noise, and when control gains are adjusted by expressions which are functions of $\hat{\underline{\theta}}(k)$, as in adaptive control.

REFERENCES

1. J. M. Mendel, "Gradient, Error-Correction Identification Algorithms," Information Sciences, Vol. 1, No. 1, pp. 23-42, Dec. 1968.

2. J. M. Mendel, "Gradient Identification for Linear Systems," in Adaptive, Learning and Pattern Recognition Systems, J. M. Mendel and K. S. Fu (eds.), Academic Press, 1970.

3. G. N. Saridis and G. Stein, "Stochastic Approximation Algorithms for Linear Discrete-Time System Identification," IEEE Trans. on Automatic Control, Vol. 13, pp. 515-523, Oct. 1968.

4. J. M. Mendel, "Identification of Time-Varying Parameters by means of Gradient Algorithms," Information Sciences, 1971.

5. Anon., "Technical Description SIGN-III Digital Computer," Honeywell Aerospace Div. Rept., R-ED 24551, Rev. B, Sept. 30, 1967.

6. D. O. Baechler, "Trends in Aerospace Digital Computer Design," IEEE Computer Group News, pp. 18-23, 1969.

STOCHASTIC APPROXIMATION ALGORITHMS FOR SYSTEM IDENTIFICATION USING NORMAL OPERATING DATA

Yasuyuki Funahashi and Kahei Nakamura

Nagoya University

Nagoya, Japan

1. INTRODUCTION

This paper considers the identification problem of linear control systems by use of stochastic approximation. There are two approaches to constructing models based on stochastic approximation: dynamical model approach and finite memory model one. In the former approach Saridis and Stein [1] have obtained the most general results. And in the latter approach Holmes [2] has established an algorithm giving an unbiased estimate. But it is common with these two works that (i) only white noise sequence is allowed for the input sequence and (ii) the updating of the estimates occurs at every finite time interval.

When no a priori information about system dynamic characteristics is known at all, it is reasonable to identify the system by subjecting it to white noise sequence. But it should be permitted to use time sequences other than white noise. The pseudo-white noise such as M-sequence (maximum length binary sequence) [3],which is developed for the identification method based on the correlation theory, is a potential input sequence for identification. But this sequence is not permitted as input in [1] and [2], because this is a periodic sequence.

On the other hand, if some a priori information about system characteristics is available, it is unwise to apply white noise to the system for identification. It is rather desirable, from the optimal control point of view, to apply to the system the optimal control computed from the model and to make more accurate the identification by using the corresponding output. This is especially desired when the system changes during its operation. That is, the optimum adaptive control of systems requires identification from the normal operating data.

From the above considerations, an algorithm is desired which permits deterministic sequences as input, other than white noise or any other time sequences generated by stochastic processes. In this paper a stochastic approximation algorithm is proposed which does not require to subject the system to any special disturbances and only uses input-output data which are obtained during its normal operating conditions. The algorithm corrects estimates at every sampling time. Let us now show the condition for the algorithm to converge. Under this condition the algorithm is shown to converge to the true value of the parameters in the mean square sense.

2. IDENTIFICATION ALGORITHM

Let us consider the identification problem of linear finite memory (N parameters) time-invariant time-discrete systems. Then for such systems we can relate the input sequence $\{u(k)\}$ and output sequence $\{y(k)\}$ by

$$y(k) = \sum_{i=1}^{N} \phi^i u(k-i) \tag{1}$$

where ϕ^i denotes the ith value of the sampled impulse response. This class of systems is considered since most practical systems possess finite memory or can be approximated by the set of values $\{\phi^i\}_{i=1}^N$ by choosing N sufficiently large. It is assumed that the input sequence is exactly measurable and the output is observed through additive white noise; i.e.,

$$z(k) = y(k) + w(k) \tag{2}$$

where $\{w(k)\}$ is independent random sequence with zero mean and has finite variance σ_w^2. Also $\{w(k)\}$ is assumed independent of $\{u(k)\}$ sequence. See Figure 1 for a block diagram. From the input sequence $u(1), u(2), u(3), u(4), \ldots$, and the corresponding noisy output sequence $z(1), z(2), z(3), z(4), \ldots$, we will identify the system recursively. Let the estimated value of the system at time k be $\{\phi_k^i\}_{i=1}^N$. Then the identifier output m(k) is given by

$$m(k) = \sum_{i=1}^{N} \phi_k^i \, u(k-i) \tag{3}$$

Using vector notations, eqs. (2) and (3) are written as:

$$z(k) = \langle \phi, U_{k-1} \rangle + w(k) , \tag{4}$$

$$m(k) = \langle \phi_{k-1}, U_{k-1} \rangle , \tag{5}$$

where

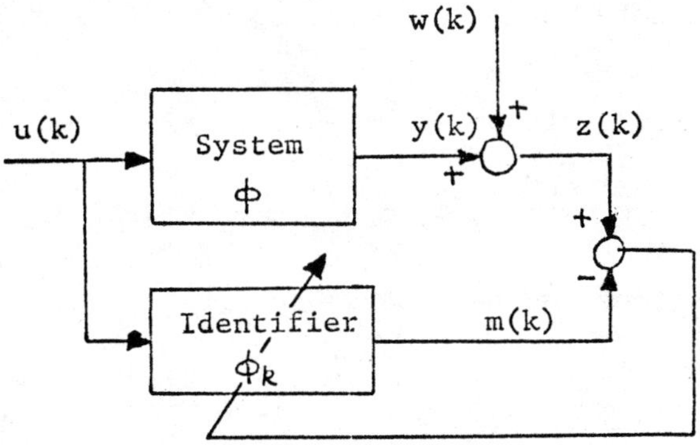

Figure 1. Block Diagram of the Identification System

$$U_k = \begin{bmatrix} u(k) \\ u(k-1) \\ \vdots \\ u(k-N+1) \end{bmatrix}, \phi = \begin{bmatrix} \phi^1 \\ \phi^2 \\ \vdots \\ \phi^N \end{bmatrix} \quad \phi_k = \begin{bmatrix} \phi_k^1 \\ \phi_k^2 \\ \vdots \\ \phi_k^N \end{bmatrix} \tag{6}$$

Define the estimate $\hat{\phi}$ of the parameter vector ϕ as the vector which minimizes the mean square error criterion $E\{(Z(k+1)-\langle\hat{\phi},U_k\rangle)^2\}$. This estimate $\hat{\phi}$ is to be computed recursively by means of Albert and Gardner type of stochastic approximation procedure [4]; i.e.,

$$\phi_{k+1} = \phi_k+\rho_k \frac{U_k}{\|U_k\|}[Z(k+1)-m(k+1)], \ k=N,N+1,N+2,... \tag{7}$$

Theorem 1. If the conditions (A), (B) and (C) are met, the estimated parameter sequence $\{\phi_k\}$ defined by (7), converges to the true parameter value ϕ in the mean square sense.

(A) Smoothing coefficients are a sequence of positive numbers such that

$$\sum_{k=N}^{\infty} \rho_k = +\infty \qquad \sum_{k=N}^{\infty} \rho_k^2 < +\infty .$$

(B) Input sequence $\{u(k)\}$ is such that

$$\lim\inf. \frac{1}{n} \lambda_{min} \left(\sum_{j=N}^{n} \frac{U_j U_j^T}{\|U_j\|} \right) = \alpha^2 > 0 . \tag{8}$$

where λ_{min} (Λ) denotes the minimum eigenvalue of the symmetric matrix Λ.

(C) Noise sequence $\{w(k)\}$ is white noise and is independent of input sequence $\{u(k)\}$.

Proof. Denote the estimate error at time k by $\tilde{\phi}_k$,

$$\tilde{\phi}_k = \phi_k - \phi \ . \tag{9}$$

Subtracting ϕ from both sides of eq. (7) yields

$$\tilde{\phi}_{k+1} = [I-\rho_k \frac{U_k U_k^T}{\|U_k\|}]\ \tilde{\phi}_k + \rho_k \frac{U_k}{\|U_k\|}\ w(k+1)\ . \tag{10}$$

Iterating this from k=n back to k=N gives,

$$\tilde{\phi}_{k+1} = \prod_{j=N}^{n} [I-\rho_j \frac{U_k U_k^T}{\|U_k\|}]\tilde{\phi}_k + \sum_{j=N}^{n} \prod_{i=j+1}^{n} [I-\rho_i \frac{U_i U_i^T}{\|U_i\|}]\rho_j \frac{U_j}{\|U_j\|}\ w(j+1), \tag{11}$$

where $\Pi_{j=m}^{n}$ means the matrix backward product; i.e., $\Pi_{j=m}^{n} A_j = A_n A_{n-1} \cdots A_m$.

Putting

$$P_m^n = \prod_{j=m}^{n} [I-\rho_j \frac{U_j U_j^T}{\|U_j\|}]\ , \tag{12}$$

we square both sides of eq. (11) and take expectations with respect to noise sequence w(N+1),w(N+2),...,w(n+1).

$$E\|\tilde{\phi}_{n+1}\|^2 = \|P_N^n\ \tilde{\phi}_N\|^2 + 2E\langle P_N^n \tilde{\phi}_N, \Sigma P_{j+1}^n\ {}_j \frac{U_j}{\|U_j\|}\ w(j+1)\rangle +$$

$$E\ \|\sum_{j=N}^{n} P_{j+1}^n\ \rho_j \frac{U_j}{\|U_j\|}\ w(j+1)\|^2 \tag{13}$$

As $\{w(k)\}_{k=N+1}^{n+1}$ is independent of each other and of $\{U_k\}_{k=N+1}^{n}$ and its mean is zero, the second term is zero. Therefore, eq. (13) becomes

$$E\|\tilde{\phi}_{n+1}\|^2 = \|P_N^n\ \tilde{\phi}_N\|^2 + \sum_{j=N}^{n} \rho_j^2\ \|P_{j+1}^n \frac{U_j}{\|U_j\|}\|^2\ \sigma_w^2\ .$$

We will now bound the $\|P_k^n\|$. As

$$P_k^n = I - \sum_{j=k}^{n} \rho_j \frac{U_j U_j^T}{\|U_j\|} + O(\rho^2)$$

where $O(\rho^2)$ denotes the uniformly bounded terms, it follows that

$$\|P_k^n\| = \lambda_{max}(P_k^n)$$

$$= \max_{\|x\|=1} \langle P_k^n\ x, x\rangle$$

$$= 1 - \min_{\|x\|=1} \sum_{j=k}^{n} \left\langle \rho_j \frac{U_j U_j^T}{\|U_j\|} x, x \right\rangle + O(\rho^2) \tag{14}$$

$$\leqq e^{-\lambda_{min}\left(\sum_{j=k}^{n} \rho_j \frac{U_j U_j^T}{\|U_j\|} \right)}$$

$$\therefore E\|\tilde{\phi}_{n+1}\|^2 \leqq \|\tilde{\phi}_N\|^2 e^{-2\lambda_{min}\left(\sum_{k=N}^{n} \rho_k \frac{U_k U_k^T}{\|U_k\|} \right)} + \sum_{j=N}^{n} \sigma_w^2 \rho_j^2 e^{-2\lambda_{min}\left(\sum_{k=j+1}^{n} \rho_k \frac{U_k U_k^T}{\|U_k\|} \right)} \tag{15}$$

From conditions (A) and (B), it can be shown that, for any fixed j,

$$\lambda_{min}\left(\sum_{k=j+1}^{n} \rho_k \frac{U_k U_k^T}{\|U_k\|} \right)$$

tends to infinite as n goes to infinity. From this it is concluded that the first term on the righthand side of eq. (15) tends to zero as $n \to \infty$. As for the second term, we can choose $\delta > 0$ such that

$$\sum_{k=N}^{\infty} \rho_k^{2-\delta} < \infty .$$

Let

$$x_n = \rho_n^{\delta} \text{ and } a_{n,k} = \rho_k^{2-\delta} e^{-\lambda_{min}\left(\sum_{j=k+1}^{n} \rho_j \frac{U_j U_j^T}{\|U_j\|} \right)} .$$

x_n tends to 0 as $n \to \infty$. For any fixed k, $a_{n,k} \to 0$ as $n \to \infty$. And

$$\Sigma a_{n,k} < \sum_{k=N}^{n} \rho_k^{2-\delta} < \sum_{k=N}^{\infty} \rho_k^{2-\delta} < +\infty .$$

So the Teoplitz lemma [5] implies

$$\sum_{k=N}^{n} a_{n,k} x_k = \sum_{k=N}^{n} \rho_k^2 e^{-\lambda_{min}\left(\sum_{j=k+1}^{n} \rho_j \frac{U_j U_j^T}{\|U_j\|} \right)} \to 0.$$

Thus it is shown that the second term on the righthand side of eq. (15) tends to zero; i.e., $E\|\phi_n - \phi\|^2 \to 0$. Q.E.D.

<u>Corollary 1</u>. If the sequence ρ_k merely tends to ρ for sufficiently small, the ϕ_k does not necessarily converge to ϕ and the limiting error is given by

$$E\|\phi_\infty - \phi\|^2 \leqq \frac{\sigma_w^2}{2\alpha^2(1-\alpha^2\rho)} \rho , \tag{16}$$

where α^2 is the positive constant given in eq. (8).

Proof. By eq. (15)

$$E\|\phi_n-\phi\|^2 \leqq \|\tilde{\phi}_N\|^2 e^{-2\rho\lambda_{min}\left(\sum_{k=N}^{n}\frac{U_kU_k^T}{\|U_k\|}\right)} + \rho^2\sigma_w^2 \sum_{k=N}^{n} e^{-2\alpha^2\rho(n-k)} .$$

The first term on the righthand side tends to zero exponentially. And the limiting square error is given by the second term; i.e.,

$$E\|\phi_\infty-\phi\|^2 \leqq \rho^2 \frac{\sigma_w^2}{1-e^{-2\alpha^2\rho}} \simeq \frac{\sigma_w^2}{2\alpha^2(1-\alpha^2\rho)} \rho . \qquad \text{Q.E.D.}$$

There may exist unmeasurable random disturbances in the system itself and/or from the surrounding environments. Then the output is given by

$$z(k) = \langle \phi, U_{k-1} \rangle + \sum_{i=1}^{n} \theta^i v(k-i) + w(k) , \tag{17}$$

where $\{v(k)\}$ denotes an independent random sequence. $\{\theta^i\}_{i=1}^N$ denotes the sampled impulse response to v. θ may be or may not be equal to ϕ. Let $\xi(k)$ denote the sum of the second and the third term of eq. (17). It is noted that $\xi(k)$ is not mutually independent, although $v(k)$ and $w(k)$ are independent random sequences. The $\xi(k)$ depends on the $\xi(k-1),\xi(k-2),\ldots,\xi(k-N)$ and independent of $\xi(k-j)$, $j \geqq N+1$. Even in this case, the algorithm (7) gives an unbiased estimate if the input sequence satisfies the condition (B).

Theorem 2. Let the output be given by eq. (17), where $\{v(k)\}$ and $\{w(k)\}$ are independent random sequences. If the conditions (A) and (B) are met, the estimates defined by eq. (7) converge to the true parameter value in the mean square sense.

Proof. Analogous to eq. (13), we obtain

$$E\|\tilde{\phi}_{n+1}\|^2 = \|P_N^n \tilde{\phi}_N\|^2 + 2E\langle P_N^n \tilde{\phi}_N, \Sigma P_{j+1}^n \rho_j \frac{U_j}{\|U_j\|} \xi(j+1) \rangle + E\|\sum_{j=N}^{n} P_{j+1}^n \rho_j \frac{U_j}{\|U_j\|} \xi(j+1)\|^2 . \tag{18}$$

where

$$\xi(j+1) = w(j+1)+\theta_1 v(j)+\theta_2 v(j-1)+\ldots+\theta_N v(j-N+1) .$$

Since $\{w(k)\}$ and $\{v(k)\}$ are independent, $E\{\xi(j+1)\} = 0$, and

$$\sigma(k) = E\{\xi(j+1)\xi(j+1+k)\}$$

$$= \begin{cases} \sigma_w^2+(\theta_1^2+\theta_2^2+\ldots+\theta_N^2)\,\sigma_v^2 & \text{when } k = 0 \\ (\theta_1\theta_{1+k}+\theta_2\theta_{2+k}+\ldots+\theta_{N-k}\theta_N)\,\sigma_v^2 & \text{when } 1 \leqq k \leqq N-1 \\ 0 & \text{when } k \geqq N\ . \end{cases}$$

The second term of eq. (18) is zero and the third term is as follows:

$$E\Big\|\sum_{j=N}^{n} P_{j+1}^{n}\rho_j \frac{U_j}{\|U_j\|}\ \xi(j+1)\Big\|^2 = \sum_{j=N}^{n}\ \sum_{\ell=N}^{n}\rho_j\rho_\ell\Big\langle P_{j+1}^{n}\frac{U_j}{\|U_j\|}\ ,$$

$$P_{\ell+1}^{n}\frac{U_\ell}{\|U_\ell\|}\Big\rangle\ E\{\xi(j+1)\xi(\ell+1)\}$$

$$= \sum_{j=N}^{n} \rho_j^2\sigma(0)\ \Big\|P_{j+1}^{n}\frac{U_j}{\|U_j\|}\Big\|^2 + 2\sum_{j=N}^{n}\ \sum_{k=1}^{N-1}\ \rho_j\rho_{j+k}\Big\langle P_{j+1}^{n}\frac{U_j}{\|U_j\|}\ ,$$

$$P_{j+k+1}^{n}\frac{U_{j+k}}{\|U_{n+k}\|}\Big\rangle\ \sigma(k)$$

$$\leqq \sigma(0)\sum_{j=N}^{n}\ \rho_j^2\ \|P_{j+1}^{n}\|^2 + 2\sum_{k+1}^{N-1}\sigma(k)\sum_{j=N}^{n}\ \rho_j\rho_{j+k}\ \|P_{j+1}^{n}\|\ \|P_{j+k+1}^{n}\|$$

$$\leqq \sum_{j=N}^{n}\ \sigma(0)\rho_j^2\ \|P_{j+1}^{n}\|^2+\sum_{k=1}^{N+1}\ \sigma(k)\Big[\sum_{j=N}^{n}\ \rho_j^2\ \|P_{j+1}^{n}\|^2+\sum_{j=N}^{n}\rho_{j+k}^2\ \|P_{j+1+k}\|^2\Big]$$

$$\to 0\ .$$

This proves $\lim_{n\to\infty}\ E\|\phi_n-\phi\|^2 = 0$. Q.E.D.

3. COMMENTS ON THE INPUT SEQUENCES

In almost all the identification methods, it is assumed that the input sequence for identification is restricted to independent random sequence; i.e., white noise sequence. In this paper the input sequence is broadened. Now let us illustrate some classes of input sequences by examining the condition (B). The first class is dependent random sequence.

Class (1). The input sequence whose spectral $\Phi(e^{j\omega})$ is strictly positive on a ω-set with finite positive measure on $(0,2\pi)$. α^2 in eq. (8) is, for this class, given by $\lambda_{min}\ E\{U_jU_j^T/\|U_j\|\}$.

Condition (B) is interpreted geometrically as follows: the input sequence spans the N-dimensional space infinitely often. For example, the special periodic sequence 011011··· satisfies condition (B) in the case of N=3 and will produce unbiased estimate. The maximum length binary sequence is another example. These examples can be stated more generally as the second class.

Class (2). Input sequence for which there exists a sequence of integers $N = v_1 < v_2 < \ldots$, such that, with $p_k = v_{k+1} - v_k$, $N \leqq p_k \leqq N_v < +\infty$, and

$$\liminf \frac{1}{p_k} \lambda_{min} \left(\Sigma_{j \in J_k} \frac{U_j U_j^T}{\|U_j\|} \right) = \beta^2 > 0$$

where

$$J_k = \{v_k, v_{k+1}, v_{k+2}, \ldots, v_{k+1} - 1\}.$$

α^2 is given by β^2 for this class.

4. CONCLUSION

A sequential identification procedure was presented for the identification of a time-invariant linear system whose memory was finite. The algorithm presented here is for on-line identification, and normal operating data are sufficient for the algorithm to converge to the true parameter value. Hence the presented procedure is closely related to the optimum adaptive control.

REFERENCES

1. G. N. Saridis and G. Stein, "A New Algorithm for Linear System Identification," IEEE Trans. on Automatic Control, Vol. AC-13, pp. 592-595, 1968.
2. J. K. Holmes, "Two Stochastic Approximation Procedures for Identifying Linear Systems," IEEE Trans. on Automatic Control, Vol. AC-14, pp. 292-295, 1969.
3. K. Izawa and K. Furuta, "Process Identification Using Correlation Coefficient," IFAC Symp. on Identification in Automatic Control Systems, 1967.
4. A. E. Albert and L. A. Gardner, Stochastic Approximation and Nonlinear Regression, MIT Press, 1967.
5. K. Knopp, Theory and Application of Infinite Series, Hafner, 1947.

ON UTILIZATION OF STRUCTURAL INFORMATION TO IMPROVE IDENTIFICATION ACCURACY

M. Aoki and P. C. Yue†

University of California

Los Angeles, California USA

INTRODUCTION

The so-called identification problem is that of estimating unknown constant system parameters of dynamic systems. The computation of maximum likelihood is known to be nontrivial and is full of various pitfalls [1]. The Koopmans-Levin algorithm for system identification, which was first discussed in the context of dynamic systems by Levin [2], has been found to be very efficient computationally as compared to the iterative type of algorithms since it only involves the solution of algebraic equations and a very simple eigenvalue problem [3][4]. This method can be motivated through the derivation of the maximum likelihood estimate but is not the true maximum likelihood estimate as claimed by Levin. It has been shown to yield (strongly) consistent estimates under very general conditions on the observation noise and simple restrictions on the systems properties as well as input characteristics. In particular, if the observation noise is Gaussian, then the estimates are known to be close approximations to those obtained via the method of maximum likelihood [3][4].

Identification of linear dynamic systems is usually treated in a general setting such that no priori knowledge about the system structure is utilized. Very often we do possess some a priori knowledge on the system structures. For example, it may be known a priori that certain submatrices of dynamic system matrices contain only zero elements. Such a priori knowledge should be incorporated in estimating system parameters since the computational

†Current address IBM Thomas J. Watson Research Center, Yorktown Heights, New York.

complexity may be reduced, sometimes dramatically, and since better estimates may be obtained by making full use of available information. The identification of dynamic system matrices utilizing a priori structural information was first considered by Aoki using the method of least squares in 1967. [5]

This paper shows in a brief outline that the Koopmans-Levin algorithm can be reformulated to utilize a priori structural knowledge about the system to gain significant computational savings; especially if the system has high dimension (e.g. a complex business organization). To illustrate this, the paper applies the developed algorithm to a system composed of four functional units, each possessing a set of known dynamics and operating under its own control policy, but at the same time, with unknown dynamic interactions and with unknown control intervention passing down from a higher level to the next level.

BASIC ALGORITHM

Consider estimating A and B matrices in

$$x(t+1) = Ax(t) + Bu(t) , \quad \dim x = n, \dim u = r$$

$$z(t) = x(t) + \eta(t)$$

$$v(t) = u(t) + \xi(t) \qquad (1)$$

from the observation record consisting of $v(t)$ and $z(t)$, $t = 0, .,$ $\ldots, N$.

The Koopmans-Levin algorithm can be reformulated as minimization with respect to A and B of

$$J = \operatorname{tr} [W_N M(M^T M)^{-1} M^T]$$

where

$$M^T = [-B, -A, I]$$

$$W_N = \frac{1}{N} \sum_{t=0}^{N-1} \begin{bmatrix} v(t) \\ z(t) \\ z(t+1) \end{bmatrix} [v^T(t), z^T(t), z^T(t+1)] \ : \quad (2\ n\times r)\times(2n+r)$$

Denote by g_j, $j = 1,2,\ldots,2n + r$, the (column) eigenvectors of W_N with non-decreasing eigenvalues. Then, it is seen that

$$\begin{bmatrix} g_1^T \\ \cdot \\ \cdot \\ \cdot \\ g_n^T \end{bmatrix} = [-\hat{B}, -\hat{A}, I]$$

In ref. [3] and [4] it has been established that the estimate converges almost surely to the true A and B matrices.

MODIFICATION TO ACCOMMODATE A PRIORI STRUCTURAL INFORMATION

We next show how the above algorithm can be applied to identify dynamic coupling terms A_{12} and A_{21} submatrices in A in (1) when the state and control vectors are

$$x(t) = [x_1^T(t), x_2^T(t)]^T , \qquad \dim x_1 = \ell, \dim x_2 = m, \ell + m = n$$

$$u(t) = [u_1^T(t), u_2^T(t)]^T$$

For simplicity of exposition, take B = I and u is observed exactly, i.e., $\xi(t) = 0$ for all t. This will be sufficient to indicate the needed modification.

Let the $(\ell\times\ell)$ symmetric positive definite matrix $(I + A_{11}A_{11}^T)$ have an L-U decomposition [6]

$$(I + A_{11}A_{11}^T) = RR^T$$

where R is a nonsingular $(\ell\times\ell)$ lower triangular matrix. Next, form the $(n\times n)$ matrix W_N by

$$W_N = \frac{1}{N}\sum_{t=0}^{N-1} y(t)y(t)^T$$

where

$$y(t) = \begin{bmatrix} z_2(t) \\ R^{-1}(z_1(t+1) - A_{11}z_1(t) - u_1(t)) \end{bmatrix} : \quad (m + \ell)\times 1$$

This matrix converges almost surely to EW_N as $N \to \infty$ which is

$$I + \lim_{N\to\infty} \frac{1}{N}\sum_{t=0}^{N-1} s(t)s(t)^T$$

where

$$s(t) = \begin{bmatrix} x_2(t) \\ R^{-1}(x_1(t+1) - A_{11}x_1(t) - u_1(t)) \end{bmatrix}$$

Since $[-A_{12},R]\, s(t) = 0$ for all t, we obtain $\hat{A}_{12}$ from

$$[-\hat{A}_{12}, R] = \begin{bmatrix} g_1^T \\ \vdots \\ g_\ell^T \end{bmatrix}$$

where $g_1, \ldots, g_\ell$ are the eigenvectors of W_N with the ℓ smallest eigenvalues.

The normalization involves the inversion of $(\ell \times \ell)$ matrix. Alternately $\hat{A}_{12}$ can be obtained from

$$[-A_{12}, R][g_{\ell+1}, \ldots, g_n] = 0$$

requiring the inversion of an $(m \times m)$ matrix. The procedure requiring less computational work may be chosen depending on $\ell > m$ or not. The estimate $\hat{A}_{21}$ may be similarly obtained by interchanging the subscript 1 with 2 in the above description.

EXAMPLE PROBLEM

We now give computational results of the problem mentioned in Introduction. See Figure 1. The system state and control vectors are composed of four subvectors

$$x(t) = (x_1^T(t), x_2^T(t), x_3^T(t), x_4^T(t))^T$$

and

$$u(t) = [u_1^T(t), \ldots, u_4^T(t)]$$

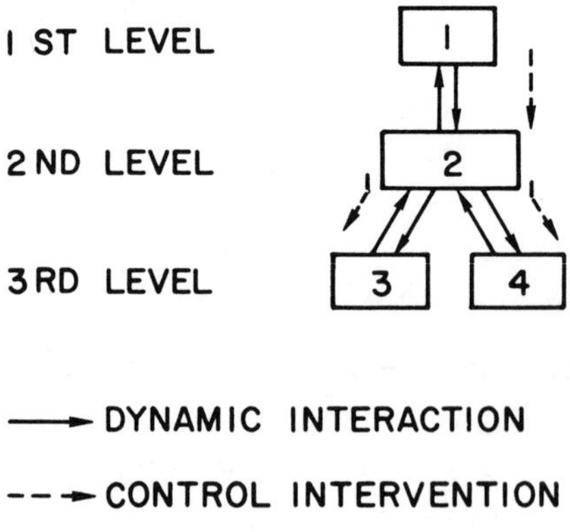

Figure 1.

with

$$A = \begin{bmatrix} A_{11} & A_{12} & 0 & 0 \\ A_{21} & A_{22} & A_{23} & A_{24} \\ 0 & A_{32} & A_{33} & 0 \\ 0 & A_{42} & 0 & A_{44} \end{bmatrix} \text{ and } B = \begin{bmatrix} B_{11} & 0 & 0 & 0 \\ B_{21} & B_{22} & 0 & 0 \\ 0 & B_{32} & B_{33} & 0 \\ 0 & B_{42} & 0 & B_{44} \end{bmatrix}$$

The vectors $x_i(t)$, $i=1,\dots,4$, and $u_i(t)$, $i=1,\dots,4$, denote the states of individual subsystems and their controls, respectively. It is assumed that the parameter matrices (A_{11}, B_{11}), ... (A_{44}, B_{44}) are known to individual subsystems; but the interaction characteristics (i.e., A_{12}, A_{21}, A_{23}, A_{24}, A_{32}, A_{42}) as well as the control intervention characteristics (i.e. B_{21}, B_{32}, B_{42}) are unknown. Noisy observations of the entire organization's state are available as

$$z(t) = x(t) + \eta(t),$$

where

$$E\eta(t) = 0, \quad E\eta(t)\eta^T(s) = \sigma^2\delta_{t,s}I$$

$$u^T(t) \triangleq (u_1^T(t),\ u_2^T(t), \dots u_4^T(t))$$

$$x^T(t) \triangleq (x_1^T(t),\ x_2^T(t), \dots x_4^T(t))$$

$$z^T(t) \triangleq (z_1^T(t),\ z_2^T(t), \dots z_4^T(t))$$

It is required to estimate the unknown parameters from $\{u(t), z(t)\}$, $0 \le t \le N$.

For simplicity, it is sufficient to describe the algorithm by assuming $x_i(t)$ to be (2×1) and $u_i(t)$ scalar such that $x(t)$ is (8×1) and $u(t)$ is (4×1). See Ref. [3] and [4] for detail of the algorithm. The estimates are next given. Then the result of numerical studies are given which show how the a priori knowledge help the identification accuracy.

$$\hat{A}_{12} = -G_2G_1^{-1}R_1 \quad : \quad 2\times2$$

where

$$\begin{bmatrix} G_1 \\ G_2 \end{bmatrix} = (g_1, g_2) \ : \ 4\times2 \ , \ R_1R_1^T = I + A_{11}A_{11}^T$$

and where g_i, $i=1,2$ are the eigenvectors of W_k corresponding to the smallest eigenvalues where

$$W_k \triangleq \sum_{t=0}^{N-1} z^t(z^t)^T, (z^t)^T \triangleq \left[[R_1^{-1}(z_1(t+1)-A_{11}z_1(t)-B_{11}u_1(t))]^T, z_2(t)^T\right].$$

Similarly $A_{j2}^T = -G_2G_1^{-1}R_j$, $R_jR_j^T = I+A_{jj}A_{jj}^T$, j=3,4 where G_1 and G_2 are defined similar to those with $\hat{A}_{12}$ except for the fact that W_k is now defined as $W_k=W_{11}-W_{12}W_{22}^{-1}W_{12}^T$ where

$$W_{11} = \sum_{t=0}^{N-1} z^t(z^t)^T,\ W_{12} = \sum_{0}^{N-1} z^t u_2(t),\ W_{22} = \sum_{t=0}^{N-1} u_2^2(t),$$

with

$$(z^t)^T = [R_j^{-1}(z_j(t+1)-A_{jj}z_j(t)-B_{jj}u_j(t))]^T, z_2(t)^T]$$

$$\hat{B}_{j2} = (R_j, -\hat{A}_{j2})W_{12} \quad : \quad 2\times1$$

$$\begin{bmatrix}\hat{A}_{21}^T\\ \hat{A}_{23}^T\\ \hat{A}_{24}^T\end{bmatrix} = -G_2G_1^{-1}R^2,\ R_2R_2^T = I+A_{22}A_{22}^T,\ \hat{B}_{21}=(R_{21},-\hat{A}_{21},-\hat{A}_{23},-\hat{A}_{24})W_{12}$$

with $W_k = W_{11} - W_{12}W_{22}^{-1}W_{12}^T$

where

$$W_{11} = \sum_{0}^{N-1} z^t(z^t)^T,\ W_{12} = \sum_{0}^{N-1} z^t u_1(t),\ W_{22} = \sum_{0}^{N-1} u_1^2(t)$$

where

$$(z^t)^T = [R_2^{-1}(z_2(t+1)-A_{22}z_2(t)-B_{22}u_2(t))]^T, z_1(t)^T, z_3(t)^T, z_4(t)^T$$

When no a priori structural information is available, the parameter matrices (A, B) in $x(t+1) = Ax(t)+Bu(t)$ are identified by the Koopmans-Levin method as

$$\hat{A}^T=-G_2G_1^{-1} \ :\ 8\times8,\quad \hat{B}=(I,-\hat{A})W_{12}W_{22}^{-1} \ :\ 8\times4,$$

where

$$\begin{bmatrix}G_1\\ G_2\end{bmatrix} = (g_1,\ldots,g_8),\quad W_k = W_{11}-W_{12}W_{22}^{-1}W_{12}^T$$

and

$$W_{11}=\sum_{t=0}^{N-1} \begin{bmatrix}z(t+1)\\ z(t)\end{bmatrix} [z(t+1)^T,\ z(t)^T],\quad W_{12}=\sum_{t=0}^{N-1}\begin{bmatrix}z(t+1)\\ z(t)\end{bmatrix}u^T(t)$$

$$W^{22} = \sum_{t=0}^{N-1} u(t)u(t)^T$$

where the symbols have similar meanings as above.

NUMERICAL STUDIES

This problem has been simulated on the computer and identification carried out with and without incorporating the structural information. The estimation error in percentage, e.g.

$$\| A_{i,j} - A_{i,j} \| / \| A_{i,j} \| \times 100\%$$

for one run is given in Tables 1 and 2, versus the number of observations. The true values of the parameter matrices are:

$$A_{11} = \begin{pmatrix} .9 & 0 \\ 0 & .8 \end{pmatrix},\ A_{12} = \begin{pmatrix} 0 & 0 \\ 0 & 0.2 \end{pmatrix},\ A_{21} = \begin{pmatrix} 0 & .3 \\ .1 & 0 \end{pmatrix},\ A_{22} = \begin{pmatrix} .6 & 0 \\ 0 & .5 \end{pmatrix}$$

$$A_{23} = \begin{pmatrix} .5 & 0 \\ 0 & 0 \end{pmatrix},\ A_{24} = \begin{pmatrix} .2 & 0 \\ 0 & .1 \end{pmatrix},\ A_{32} = \begin{pmatrix} .12 & 0 \\ 0 & 0 \end{pmatrix},\ A_{33} = \begin{pmatrix} .2 & 0 \\ 0 & .15 \end{pmatrix}$$

$$A_{42} = \begin{pmatrix} .08 & 0 \\ 0 & 0 \end{pmatrix},\ A_{44} = \begin{pmatrix} .1 & 0 \\ 0 & .05 \end{pmatrix},\ B_{11} = \begin{pmatrix} 1.0 \\ 1.0 \end{pmatrix},\ B_{21} = \begin{pmatrix} .5 \\ .6 \end{pmatrix}$$

$$B_{22} = \begin{pmatrix} 0 \\ 1.0 \end{pmatrix},\ B_{32} = \begin{pmatrix} .1 \\ 0 \end{pmatrix},\ B_{42} = \begin{pmatrix} 0 \\ .2 \end{pmatrix},\ B_{33} = \begin{pmatrix} .4 \\ .3 \end{pmatrix},\ B_{44} = \begin{pmatrix} .2 \\ .1 \end{pmatrix}$$

all components of x(0) are taken to be 1.0.

Table 1. Percentage Error of Subsystem Interaction and Control Intervention Parameters Using Different Numbers of Observations

No. of Obser.	Parameter Estimation Error %						
	I	III		IV		II	
	e_A	e_A	e_B	e_A	e_B	e_A	e_B
100	8.8	16.8	33.3	32.6	12.4	78.1	4.5
200	5.6	11.5	15.3	19.3	8.1	44.1	2.5
300	5.4	9.5	11.3	15.3	5.7	34.5	2.0
500	4.6	6.2	9.4	10.1	4.6	6.9	1.5
900	3.5	5.4	8.2	8.0	3.9	24.7	1.2

(Average of 14 samples)

e_A = interaction estimation error

e_B = intervention estimation error

Table 2. Percentage Error of Unknown Parameter Estimates Without Incorporating Structural Information

No. of Obser.	Parameter Estimation Error %						
	I	III		IV		II	
	e_A	e_A	e_B	e_A	e_B	e_A	e_B
100	411.1	315.3	58.6	586.5	30.7	645.1	9.0
200	158.4	135.1	23.3	226.2	13.1	264.9	4.4
300	151.7	196.7	26.4	212.6	12.3	444.2	5.7
500	112.4	106.7	28.7	130.5	8.6	200.9	2.5
900	129.7	133.6	10.8	171.5	4.9	213.4	1.3

(Average of 8 samples)

The input sequence was chosen to be mutually and serially independent and uniformly distributed over 0 and 1.0. The observation noise is also uniformly distributed over the interval between -.10 and +.10.

It is obvious upon comparison of Tables 1 and 2 that we obtain much better estimates by explicitly incorporating the structural information. The estimation error is also seen to decrease as the number of observations taken increases in both cases. The significant reduction in the order of eigenvalue problem and the size of matrix inversion is also evident.

DISCUSSIONS

When A matrix is expressed as $A = A_0 + A_u$, where A_0 contains only known elements of A and where A_u contains only unknown elements of A, the matrix A_u becomes sparse in many cases. Then A_u may satisfy certain linear constraint relation such as $PA_u = 0$ or $A_uQ = 0$ where P and Q are known.

When we can write A as $A_0 + A_1A_2$, where A_0 is an $(n\times n)$ known matrix, A_1 is an unknown $(n\times \ell)$ matrix and where A_2 is known $(\ell\times n)$ matrix, then the technique similar to that discussed in this paper can be used to reduce the computational load in identifying A_1. See ref. [3]-[5] for further details and the connection with the aggregation concept.

Additional examples are found in ref. [4]. Some additional results are shown in Figures 2 and 3.

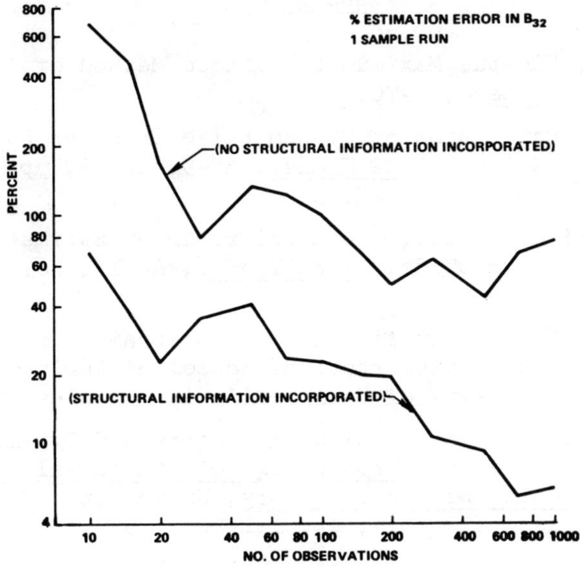

Figure 2.

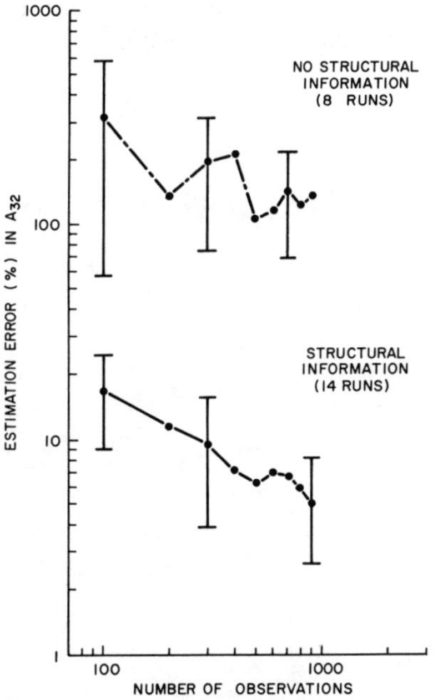

Figure 3.

REFERENCES

1. T. Bohlin, "On the Maximum Likelihood Method of Identification," IBM J. Res. & Dev., 1970.

2. M. Levin, "Estimation of System Pulse Transfer Function in the Presence of Noise," IEEE Trans., AC-9, No. 3, pp. 229-235, July 1964.

3. M. Aoki and P. C. Yue, "On A Priori Error Estimates of Some Identification Methods," IEEE Trans., AC-15, No. 5, pp.541-548 Oct. 1970.

4. P. C. Yue, "The Identification of Constrained Systems with High Dimension," Ph.D. Dissertation, School of Engineering and Applied Sciences, University of California, Los Angeles, March 1970.

5. M. Aoki, "The Identification of Constrained Dynamic Systems With High Dimension," Proceedings of 5th Annual Allerton Conference on Circuit and System Theory, University of Illinois, Allerton House, Illinois, pp. 191-200, 1967.

6. M. Aoki, Introduction to Optimization Techniques, Sec. 1.6, Macmillan Co., New York (1971).

ACKNOWLEDGMENT

The research reported in this paper was supported in part by National Science Foundation Grant GK-2032.

AN INCONSISTENCY BETWEEN THE RATE AND THE ACCURACY OF THE LEARNING METHOD FOR SYSTEM IDENTIFICATION AND ITS TRACKING CHARACTERISTICS

Atsuhiko Noda

Tokyo Institute of Technology

Tokyo, Japan

1. INTRODUCTION

A learning method for system identification has been proposed [1] which is based on the error-correcting training procedure in learning machines [2] and is an iteration method of identifying the dynamic characteristics of a linear system by use of a sampled weighting function, and detailed investigations have already been made on the fundamental characteristics of the method when an unknown system is a stationary linear one, the output of which is not corrupted by noise. A generalized method has also been proposed [3,4] which improves the rate of convergence using matrix weight.

This discussion concerns an inconsistency between the rate and the accuracy of a learning process in the case of the learning method for system identification, and the effect of the variation of unknown system parameters on the iterative identification of a linear system dynamics by means of the learning method of identification [5].

The process of the learning identification method may be considered to consist, roughly, of two stages: (1) the transient state (i.e., the initial stage of the approximation), and (2) the limit-accuracy state (i.e., the state in which the precision as the ensemble average has become almost invariable). In the former state, the influence of some initial error cannot yet be ignored, and the length of this state represents, roughly speaking, the rate of the learning process. In the latter state, the approximation has nearly converged and is fluctuating in the neighborhood of the true value of the unknown parameter according to the magnitude of the noise.

The amplitude of the fluctuation determines the achievable accuracy of the learning process as the ensemble average.

Now, there is an inconsistency between these two states of the learning process: shortening of the transient state is incompatible with an increase in accuracy. This incompatibility is examined qualitatively, and is utilized as a guiding principle in determining the error-correcting coefficient; that is to say, the step size of the learning method, in mediating between these two conflicting requirements of reducing the period of the transient state, and of increasing the limit-accuracy of the iteration method [5].

It is also shown that the learning method may be used to identify a slowly varying nonstationary system† which cannot be identified by the stochastic approximation method [6,7] and the influence of the variation in the nonstationary system parameters is investigated in some typical examples of nonstationary systems; i.e., random, ramp, and periodic varying system parameters. In particular, the greatest attainable limit-accuracy of the method is obtained in several cases of nonstationary systems.

Moreover, a discussion on the optimal value of the error-correcting coefficient, α, is given in relation to the incompatibility, and an adaptive step size method of learning identification is proposed as a tentative means of resolving the incompatibility so as to identify the variation in system parameters in greater detail [8].

2. CHARACTERISTICS OF THE LEARNING METHOD IN THE IDENTIFICATION OF A SYSTEM WITH NOISY OUTPUT

Denote by

$$w \triangleq (w_1, w_2, \ldots, w_N)' \qquad (1)^{\dagger\dagger}$$

an N-dimensional vector parameter to be estimated, which represents an approximated sampled weighting function of a linear, stable, and stationary system, and by a vector variable:

$$v_j \triangleq (v_1^{(j)}, v_2^{(j)}, \ldots, v_N^{(j)})' \qquad j=1,2,\ldots \qquad (2)$$

an approximation of w at the jth step of iterative approximation, which will be corrected successively by the fundamental method [1] of the learning identification:

†This means that the unknown parameter varies perpetually and slowly enough to be identified with the admissible tracking error, which may be determined at need.

††' indicates the transposition of a vector or a matrix.

$$\underline{v}_{j+1} = \underline{v}_j + \alpha \frac{\underline{x}_j}{|\underline{x}_j|^2} (y_j - z_j) \qquad j=1,2,\ldots \tag{3}$$

where

α: error-correcting coefficient, $0 < \alpha \ll 1$,

$x_j \triangleq (x_1^{(j)}, x_2^{(j)}, \ldots, x_N^{(j)})'$: input vector,

$|x_j|^2 \triangleq \sum_{i=1}^{N} (x_i^{(j)})^2$,

$y_j = x_j'w + n_j$: output of the unknown system with additive noise,

n_j: a sequence of random variables with $E[n_j]=0$ and $V[n_j]<\infty$,

$z_j = x_j'v_j$: output of the approximating linear system, v_j, at the jth step of iteration.

That is to say, the unknown parameter w is successively approximated by comparing the output of the unknown system with additive noise and the output of a hypothetical approximating system, v_j, and correcting the approximation vector according to the observed input vector of the unknown system, as shown in Figure 1.

The input vector, x_j, is obtained by grouping the sequence of sampled input signals, taken at the same sampling interval as the corresponding component, w_i, of the sampled weighting function w.

2.1 Limit-Accuracy of the Learning Method

Denote by e_{j+1}^2 the accuracy of the approximation by a mean-square error; i.e.,

$$e_{j+1}^2 \triangleq |v_{j+1} - w|^2 = \sum_{i=1}^{N} (v_i^{(j+1)} - w_i)^2 \tag{4}$$

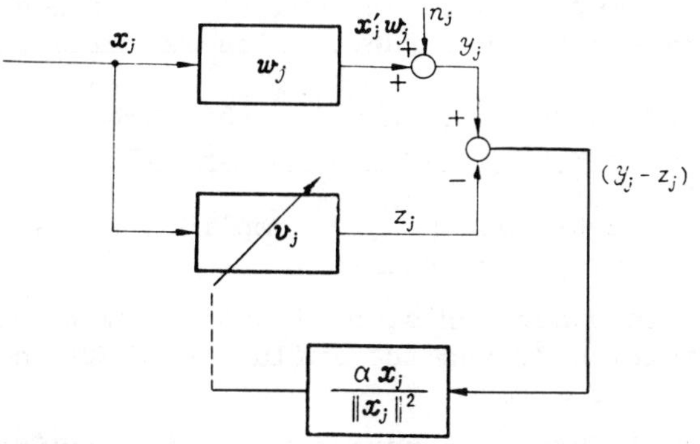

Figure 1. Fundamental Method of the Learning Identification.

which represents the accuracy of the identification immediately after completion of the jth step† of the learning identification, and by $\overline{e_{j+1}^2}$ the expectation of e_{j+1}^2 with respect to the sequences of stochsstic variables x_k and n_k (k=1,2,...,j), where $x_i^{(k)}$ and n_k are assumed to be independent stochastic variables and have the same symmetric probability density function.

Denote by ℓ the limit of the expectation of the mean-square error $\overline{e_{j+1}^2}$; i.e.,

$$\ell \triangleq \lim_{j\to\infty} \overline{e_j^2} = \frac{\alpha \overline{n^{*2}} N}{(2-\alpha)} \tag{5}$$

where

$$N \triangleq \dim(x_j), \quad n_j^* \triangleq \frac{n_j}{|\underline{x}_j|^2}, \quad \overline{n_j^{*2}} = \overline{n^{*2}} \text{ (const.)}$$

Eq. (5) implies that the expectation of the mean-square error $\overline{e_{j+1}^2}$ tends to ℓ (>0) with iterative approximation by (3) if the unknown parameter w is assumed to be stationary, and in this sense ℓ will be referred to as the limit-accuracy of the fundamental method of learning identification. The limit-accuracy depends on the error-correcting coefficient α, the dimension N of the unknown parameter w and on $\overline{n^{*2}}$ which is determined by the ensemble average of the noise magnitude. As $\overline{n^{*2}}$ and N are given constants, ℓ can be made as small as desired by reducing α.

2.2 Length of Transient-State of Learning Identification

The limit-accuracy state†† of the learning identification method may be defined as a situation in which the mean-square error $\overline{e_{j+1}^2}$ of the iterative approximation is in a circumscribed neighborhood of the limit-accuracy ℓ as the ensemble average, and the transient-state††† may also be defined as the period before attaining the approximate limit-accuracy, which is illustrated in Figure 2.

The length of iteration, that is, the number of iteration steps, may be called the length of the transient-state which is required

---

†The jth step is started at the jth sampling, and should be carried out within the jth sampling interval.

††This state is, in other words, a situation in which $|v_1-w|^2$ has been nearly extinguished, and the influence of the noise alone has remained to cause the error.

†††This state is, in short, a period which is required for the influence of the initial error $|v_1-w|^2$ to disappear.

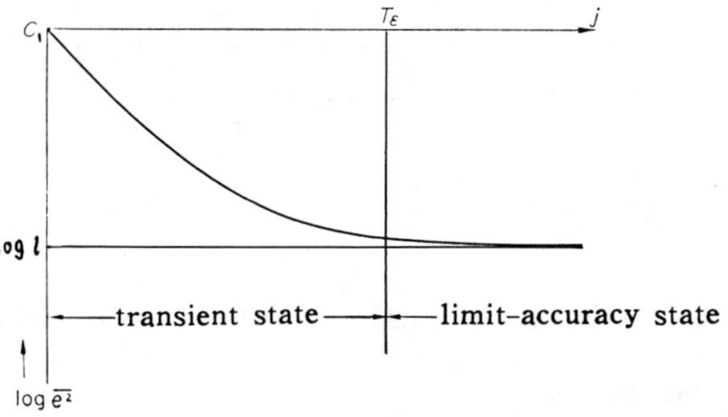

Figure 2. Limit-Accuracy State of the Learning Identification.

for the iterative approximation of the learning identification before attaining the state of limit-accuracy, and will be calculated as follows. Denote by T_ε the lowest iteration step j satisfying the condition

$$\beta_{j+1} < \varepsilon \ (\varepsilon: \text{const.},^\dagger \ 0 < \varepsilon << 1) \tag{6}$$

where

$$\beta_{j+1} = \frac{|\underline{b}_{j+1}|^2}{|\underline{b}_1|^2}, \quad b_j \triangleq \overline{v_j - w} \ ^{\dagger\dagger}$$

and T_ε will be referred to as the <u>length of ε-transient-state</u>, and is given by

$$T_\varepsilon = \left[\frac{\ln \varepsilon}{2 \ln(1-\alpha/N)}\right] + 1 \tag{7}$$

where [x] implies the integer part of x. For large N (e.g., N > 10), T_ε is given approximately as follows:

$$T_\varepsilon \simeq \frac{-N \ln \varepsilon}{2\alpha} \tag{8}$$

and T_ε is obviously monotonically decreasing with respect to both ε and α.

Incidentally, the remaining difference of the expectation $\overline{e_j^2}$ of the mean-square error from the limit-accuracy ℓ at $j=T_\varepsilon$, which is the last sampling instant in the ε-transient-state, will be as follows:

---

† If ε= 1/e, T_ε is practically equal to the time constant of the iterative approximation by the learning method.

†† β_j is the normalized quantity of the squared expectation of the bias b_j, which indicates an attenuation of b_j.

$$d_T^\varepsilon \triangleq \overline{e_j^2(j=T_\varepsilon)} - \ell = C_1\{1 - \frac{\alpha(2-\alpha)}{N}\}^{[\frac{\ln \varepsilon}{2\ln(1-\alpha/N)}] + 1} \simeq C_1\varepsilon$$

where

$$C_1 \triangleq e_1^2 - \frac{\alpha\, \overline{n^{*2}}\, N}{2-\alpha}$$

C_1 is determined by both the initial value of the mean-square error and the expectation of the magnitude of noise. In most cases, C_1 may be assumed as positive.†

2.3. A Relation Between the Limit-Accuracy and the Length of the Transient-State in the Learning Identification Method

The quantity of reduction of the mean-square error $\overline{e_j^2} > 0$ at the jth step of iteration by a single correction will be investigated below. After completion of the jth iterative correction, e_j^2 is a positive constant determined by a specific pair of series x_k and n_k $(k=1,2,\ldots,j-1)$ obtained out of stochastic series $\{x_k\}$ and $\{n_k\}$, and denoted by r_j the expectation of e_{j+1}^2/e_j^2 with respect to the stochastic variables x_j and n_j, i.e.,

$$r_j \triangleq \frac{\overline{e_{j+1}^2}}{e_j^2} \tag{9}$$

$$= \left(\frac{1}{N} + \frac{\overline{n_j^{*2}}}{e_j^2}\right)(\alpha-\alpha_j^{op})^2 + 1 - \frac{e_j^2}{N(e_j^2 + N\overline{n_j^{*2}})} \tag{10}$$

where

$$\alpha_j^{op} = \frac{e_j^2}{e_j^2 + N\overline{n_j^{*2}}} \tag{11}$$

Consequently, eq. (10) shows that the error-correcting coefficient α of procedure (3) should be α_j^{op} at the jth step of iteration to minimize r_j, and therefore, α_j^{op} may be referred to as the <u>optimal correcting coefficient</u> of the learning identification method. Nevertheless, α_j^{op} cannot be estimated at each step of iterative identification, because α_j^{op} depends on the unknown identification error e_j^2. However, (1) expresses qualitatively that α should be nearly equal to unity when the error e_j^2 is much greater than the noise magnitude $N\overline{n_j^{*2}}$. In other words, α should nearly equal unity

---

†C_1 may occasionally become non-positive. In such cases, the initial value v_1 has been unexpectedly chosen very close to the unknown parameter w.

to make an abrupt reduction of the identification error at a very early stage of transient-state of identification. On the other hand, the limit-accuracy is found to be the worst, as is discussed in 2.1., when $\alpha \simeq 1$.†

Accordingly, transient-state and limit-accuracy are inconsistent with each other and cannot be improved simultaneously; that is to say, shortening of the transient-state of the iteration is incompatible with increase of accuracy at the limit. In other words, the degrees of freedom in choosing between the rate and the accuracy (limit-accuracy) are confined to one, as follows:

$$\ell\, T_\varepsilon \simeq - \frac{\overline{n^{*2}}\, N^2 \ln \varepsilon}{4} \tag{12}$$

Eq. (12) represents practically the relation of inverse proportion between ℓ and T_ε; that is, an inconsistency between the rate and the accuracy (limit-accuracy).

Thus, eq. (12) will be made use of as a guiding principle in determining the error-correcting coefficient, α, of the iterative approximation by the learning method. In other words, the error-correcting coefficient should be chosen by giving priority to the more important demand, because the degrees of freedom are limited to one†† in the choice between rate and accuracy.

3. CHARACTERISTICS OF LEARNING METHOD IN NON-STATIONARY SYSTEM IDENTIFICATION

An iteration method is more effective in the case of estimating a non-stationary system parameter than a stationary one, because the parameter approximation can be carried out in real time so that it may reflect all of the data available concurrently. It should be mentioned that stochastic approximation is a kind of iterative approximation, but it must be essentially assumed that the unknown parameter is stationary, and it cannot successfully be employed for identification of a non-stationary parameter.††† On the other hand,

---

† α should satisfy $0<\alpha<1$, which is apparent from eq. (5).

†† The inconsistency between the rate and the accuracy originates in the fact that α is constant. This inconsistency will be settled by an adaptive adjustment of α in consideration of the process of approximation which is estimated by an adaptive stepsize approximation method which will be referred to later in this paper. Also, α varies with respect to j in the case of the stochastic approximation method, but α changes independently of the process of approximation, and the variation in that case does not increase the degrees of freedom in choosing the characteristic of iteration.

††† An estimate by stochastic approximation method is practically equivalent to taking the mean with respect to the total variation interval of the unknown parameter.

the learning identification method may be used in identifying a slowly-varying non-stationary parameter. In the following sections, characteristics of the method will be examined in the identification of some typical examples of non-stationary parameters.

Denote by w_j the unknown parameter at the jth iteration step, i.e.,

$$w_j \triangleq (w_1^{(j)}, w_2^{(j)}, \ldots, w_N^{(j)}) \quad (j=0,1,2,\ldots) \tag{13}$$

which is assumed to vary comparatively slowly with the iteration step j, and by v_j the estimate of the unknown parameter w_j which is corrected by the iterative approximation procedure of eq. (3) in the same manner as in the former section.†

Now, the estimate v_j follows up the variation of w_j.

3.1. Limit-Accuracy in the Estimation of a Non-Stationary Parameter with Random Variation

Denote by δw_j the variation of the unknown parameter, i.e.,

$$\delta w_j \triangleq w_{j+1} - w_j \tag{14}$$

Assuming that the variation δw_j $(j=0,1,2,\ldots)$ are stochastically independent vector parameters satisfying the condition $E[\delta w_j] = 0$ for any j, and that $\overline{n_j^{*2}}$ is stationary, i.e., $\overline{n_j^{*2}} = \overline{n^{*2}}$ (const.), and $E[|\delta w_{j-1}|^2] = \sigma^2$ yields

$$\tilde{\ell} \triangleq \lim_{j\to\infty} \overline{|v_{j+1} - w_j|^2} = \ell + \tilde{r} \tag{15}$$

where

$$\tilde{r} \simeq \frac{\sigma^2 N}{\alpha} \quad (N>10,\ 0<\alpha<<1) \tag{16}$$

This relation implies that the limit-accuracy $\tilde{\ell}$ in the non-stationary case is the sum of the limit-accuracy ℓ in the stationary case and the effect $\tilde{r}$ of the expected variation of the unknown parameter with respect to δw_k $(k=1,2,\ldots,j-1)$.

3.2. Ramp Response of the Learning Identification

Let the unknown parameter w_j be assumed to drift at a constant rate, i.e.,

$$w_{j+1} = w_n + \delta w \quad (\delta w\text{: a constant vector}) \tag{17}$$

---

†In the case of non-stationary parameter identification, the term 'convergence' may have no meaning but the characteristic of 'tracking' is essentially important.

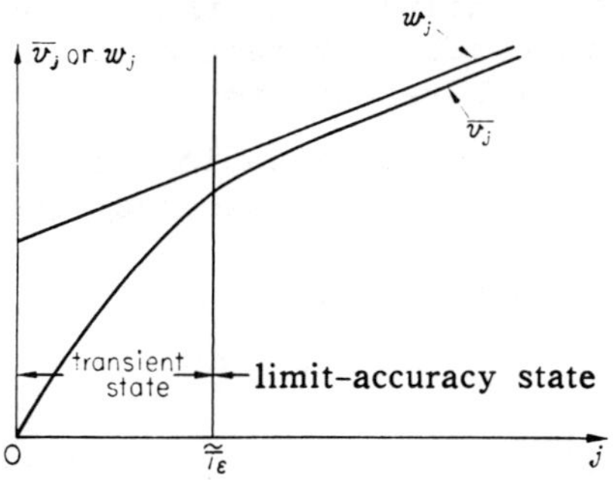

Figure 3. Limit-Accuracy State in the Case of Estimating a Non-Stationary Unknown Parameter by the Learning Method.

$$\ell = \ell + \tilde{r} \quad \text{where} \quad \tilde{r} \simeq \frac{N^2 |\underline{\delta w}|^2}{\alpha^2} \tag{18}$$

The effect of parameter variation $\tilde{r}$ does not disappear and thus influences the limit-accuracy $\tilde{\ell}$, which is distinguishable from the limit-accuracy of stationary parameter estimation, as shown in Figure 3.

3.3. Dynamic Characteristic of the Learning Method in the Identification of Periodic Varying Parameters

3.3.1. Assuming that each component of the unknown vector parameter w_j varies periodically with an independent period satisfying

$$w_j \triangleq w_0 + C \sin(j\omega + \tau) \tag{19}$$

where

$C \triangleq \text{diag}(c_1, c_2, \ldots, c_N)$: const. diagonal matrix
$\omega \triangleq (\omega_1, \omega_2, \ldots, \omega_N)$: constant vector
$\tau \triangleq (\tau_1, \tau_2, \ldots, \tau_N)$: phase vector

$$\sin(j\omega + \tau) \triangleq [\sin(j\omega_1 + \tau_1), \sin(j\omega_2 + \tau_2), \ldots, \sin(j\omega_N + \tau_N)] \tag{20}$$

The vector ω is assumed to be constant, each component of which represents one of the periodic variations of unknown parameters $w_i^{(j)}$ (i-1,2,...,N), and the vector τ is assumed to be a stochastic variable determining the boundary condition of the variation at random.

$$E[\tilde{\ell}] = \ell + E[\tilde{r}] \tag{21}$$

where

$$E[\tilde{r}] \simeq \frac{N}{\alpha} \sum_{i=1}^{N} c_i^2 \, \omega_i^2 \tag{22}$$

This is the quantity of influence on the components of the periodic variation of the unknown vector parameter on the limit-accuracy, the expectation taken with respect to τ.

Now, let $\overline{v_j^\infty}$ denote the expectation of v_j with respect to x_k and $n_k(k=1,2,\ldots,j-1)$ in the limit-accuracy state, and then,

$$\overline{v_j^\infty} \simeq w_0 + CA\sin(j\omega + \tau - \phi) \qquad (23)$$

where

$$A \triangleq \text{diag}\left(1-\frac{N^2\omega_1^2}{2\alpha^2},\ 1-\frac{N^2\omega_2^2}{2\alpha^2},\ \ldots,\ 1-\frac{N^2\omega_N^2}{2\alpha^2}\right)$$

$$\phi \triangleq \left(\frac{N\omega_1}{\alpha}, \frac{N\omega_2}{\alpha}, \ldots, \frac{N\omega_N}{\alpha}\right)$$

Therefore, by the expectation $\overline{v_j}$ of the approximation v_j, the period of each component of unknown parameter variation can be estimated by the rate of amplitude attenuation, A, and phase lag φ in the limit-accuracy state.

3.3.2. Assuming that all components of w_j vary synchronously with a period, i.e.,

$$w_j \triangleq w_0 + c\sin(j\omega + \tau) \quad \text{where} \quad c \triangleq (c_1, c_2, \ldots, c_N),\ \omega, \tau\text{:const.},$$

and letting $\overline{v_j^\infty}$ be the expectation of v_j with respect to x_k and n_k $(k=1,2,\ldots,j-1)$ in the limit-accuracy state, then from (23)

$$\overline{v_j^\infty} \simeq w_0 + ac\sin(j\omega + \tau - \phi)$$

where

$$a \triangleq 1 - \frac{N^2\omega^2}{2\alpha^2} \text{ : rate of amplitude attenuation,}$$

$$\phi \triangleq \frac{N\omega^2}{\alpha} \text{ : phase lag.}$$

Next, the minimum of limit-accuracy $\tilde{\ell}$ will be examined in the following section.

3.4. Extremum-Accuracy in the Identification of Non-Stationary Parameter

The highest attainable degree of limit-accuracy $\tilde{\ell}$ will be investigated in the non-stationary parameter estimation, which will be denoted by $\tilde{\ell}_{min}$ and will be referred to as <u>extremum-accuracy</u> of the learning identification method, and $\tilde{\ell}_{min}$ is attainable when $\alpha = \alpha_\ell$, which will be called the <u>extremum-accuracy coefficient</u>.

<u>3.4.1. The extremum-accuracy in the identification of a ramp varying parameter</u>. Assuming that the unknown parameter w_j varies at

a constant rate, then the extremum-accuracy of the learning method will be as follows:

$$\alpha_\ell \simeq \sqrt[3]{\frac{4pN^2}{q}}, \quad \tilde{\ell}_{min} \simeq \frac{3}{2}\sqrt[3]{\frac{p^2q\,N^2}{2}}$$

3.4.2. The extremum-accuracy in the identification of a parameter with random variation. The extremum-accuracy may be calculated in a similar manner as in the case of a ramp parameter.

$$\alpha_\ell \simeq \sqrt{\frac{2\,\sigma^2}{\overline{n^{*2}}}}, \quad \tilde{\ell}_{min} \simeq N\sqrt{2\sigma^2\overline{n^{*2}}}$$

4. AN ADAPTIVE STEPSIZE METHOD OF LEARNING IDENTIFICATION

4.1. Optimal Correcting Coefficient in the Approximation of a Non-Stationary Parameter

The optimal correcting coefficient α_j^{op} was obtained in the approximation of a stationary parameter in 2.3. A similar analysis will be made below in the case of non-stationary parameter estimation.

Denote by $e_{j,j}^2$ and $e_{j+1,j}^2$ the mean-square error immediately before and after the jth correction by procedure (3), respectively, i.e.,

$$e_{j,j}^2 \triangleq |v_j - w_j|^2, \quad e_{j+1,j}^2 \triangleq |v_{j+1} - w_j|^2 .$$

The jth correction by (3) is performed proportionally to the jth error vector: $(v_j - w_j)$; that is, not in a predictive manner considering the future variation: $\delta w_j = w_{j+1} - w_j$. Therefore, the performance evaluation of a single correction at the jth iterative step may be defined by an expectation $\tilde{R}_j$, i.e.,

$$\tilde{R}_j \triangleq \frac{\overline{e_{j+1,j}^2}}{e_{j,j}^2} = \{1 - \frac{\alpha(2-\alpha)}{N}\} + \frac{\alpha^2\overline{n_j^{*2}}}{e_{j,j}^2}$$

and $\tilde{R}_j$ should be minimized to make an optimal correction. α, which minimizes the expectation $\tilde{R}_j$ in the non-stationary parameter approximation, will be referred to as the optimal correcting coefficient in the non-stationary parameter approximation, and will be denoted by $\tilde{\alpha}_j^{op}$, i.e.,

$$\alpha_j^{op} = \frac{e_{j,j}^2}{e_{j,j}^2 + N\overline{n_j^{*2}}}$$

which is exactly similar to (11).

4.2. An Adaptive Correcting Coefficient

If the approximation by (3) were performed with the optimal sequence $\{\tilde{\alpha}_j^{op}\}$, the progress of iteration would undoubtedly be optimal in the approximation of a nonstationary parameter by the learning method. It is impossible, however, to define precisely $e_{j,j}^2 \triangleq |v_j - w_j|^2$ at any step j of $\tilde{\alpha}_j^{op}$. Thus, an estimate of $e_{j,j}^2$ should be adopted to determine an estimate of $\tilde{\alpha}_j^{op}$.

An estimate $\hat{e}_j^2$ of the mean-square error $e_{j,j}^2$ can be defined [8] as

$$\hat{e}_j^2 \triangleq N(Y_j - \overline{n_j^{*2}}) \tag{24}$$

where

$$Y_j \triangleq \frac{1}{m} \sum_{k=j}^{j-m+1} \hat{e}_{jk}^2 \quad (m\text{: positive integer}) \tag{25}$$

$$\hat{e}_{jk} \triangleq \frac{y_k - z_{jk}}{|\underline{x}_k|} \tag{26}$$

$$z_{jk} \triangleq (x_k, v_j), \quad k \leq j \tag{27}$$

If the quantity $\overline{Nn_j^{*2}}$ depending on the magnitude of noise is also unknown, $\hat{n}_j^2$ could be an estimate of $\overline{Nn_j^{*2}}$, even when $\overline{Nn_j^{*2}}$ varies slowly with step j, which will be calculated from the iterative equation of vector parameter t_j,

$$\underline{t}_{j+1} = \underline{t}_j + \gamma \frac{\underline{x}_j}{|\underline{x}_j|^2} (y_j - \underline{x}_j' \underline{t}_j) \tag{28}$$

where

$$y_j = x_j' w_j + n_j, \quad \gamma\text{: constant, } 0 < \gamma \ll 1.$$

Any j which is large enough (for example, $j > T_\varepsilon$, where T_ε is the length of the transient state of iterative eq. (28)), yields an approximation equation

$$\hat{n}_j^2 \simeq \frac{(2-\alpha)E[|\delta \underline{t}_j|^2]}{2\gamma^2} \tag{29}$$

where

$$\delta t_j \triangleq t_{j+1} - t_j$$

and E[ ] might be the time mean instead of the ensemble mean.

Denote by $\hat{\alpha}_j$ the optimal correcting coefficient $\tilde{\alpha}_j^{op}$ with substitutions of $\hat{e}_j^2$ for $e_{j,j}^2$ and $\hat{n}_j^2$ for $\overline{Nn_j^{*2}}$, i.e.,

$$\hat{\alpha}_j \triangleq \frac{\hat{e}_j^2}{\hat{e}_j^2 + \hat{n}_j^2} \tag{30}$$

$\hat{\alpha}_j$ may be used as an estimate for $\tilde{\alpha}_j^{op}$, and an iterative approximation with $\hat{\alpha}_j$ will be a quasi-optimal process by automatically adapting it to the variation of $e_{j,j}^2$ caused by the variation of the unknown parameter w_j. Thus, $\hat{\alpha}_j$ will be referred to as an adaptive correcting coefficient, and the iterative method with $\hat{\alpha}_j$ will be referred to as an adpative step size method, which will be effective, especially in the estimation of the variation of the unknown nonstationary parameter itself.

4.3 A Modification of the Adaptive Step Size Approximation Method Using a Hierarchical Learning

In the case of adaptive step size approximation method proposed in 4.2, an approximation $\hat{e}_j^2$, of the jth mean-square error e_j^2, was calculated by means of the moving average method, which may be substituted for an iterative moving average:

$$\alpha_{j+1}^* = \alpha_j^* + \varepsilon\left[\frac{(y_j - z_j)^2 - |\underline{x}_j|^2 E[n_j^{*2}]}{(y_j - z_j)^2} - \alpha_j^* \right]$$

where $0<\varepsilon\leqq 1$. This iteration attains an approximation of the adaptive correcting coefficient of adaptive step size approximation method. Therefore, this is a preparatory learning estimation of a parameter, which is available in a learning method, and will be referred to as a hierarchical learning, which will dispense with memory capacities required for an adaptive step size approximation method.

5. CONCLUSIONS

The effect of noise on estimation was investigated in the cases of the transient state and the limit-accuracy state in the identification of an unknown system parameter by the learning identification method. The transient state was prescribed by the length of ε-transient state and the stationary state by the limit-accuracy, and an inconsistency between the two was ascertained qualitatively.

In the identification of a nonstationary parameter, a limit-accuracy was obtained in a similar manner to the stationary case, and in addition, the characteristic of the method was examined for some typical examples of nonstationary parameters, and the extremum-accuracy was obtained, which was the minimum of the limit-accuracy in the identification of a nonstationary parameter.

These investigations revealed the characteristics of the learning identification method, especially the estimation accuracies in the nonstationary system identification. An adaptive step size approximation method was also proposed, which is an iterative approximation that automatically alters itself in an adaptive manner, and is now under detailed investigation. Further, a trial with a multi-model is now in progress to assure continual accuracy of the approximation which is estimated by the stochastic approximation method and the other models are by the learning method which are used to check the stationarity of the unknown system parameters.

REFERENCES

1. J. Nagumo and A. Noda, "A Learning Method for System Identification," IEEE Trans. on Automatic Control, Vol. AC-12, 1967, pp. 282-287.
2. B. Widrow, "Adaptive Sampled-Data Systems," 1959 IRE WESCON Conv. Rec., Pt. 4, pp. 74-85.
3. J. M. Mendel, "Gradient, Error-Correction Identification Algorithms," Information Sciences, 1, 1968, pp. 23-42.
4. J. M. Mendel and K. S. Fu, Adaptive, Learning and Pattern Recognition Systems: Theory and Applications, Academic Press, 1970.
5. A. Noda, "Effects of Noise and Parameter-Variation on the Learning Identification Method," Journal of SICE of Japan, 8-5, 1969, pp. 303-312, (in Japanese).
6. K. S. Fu, Sequential Methods in Pattern Recognition and Machine Learning, Academic Press, 1968.
7. A. E. Albert and L. A. Gardner, Jr., Stochastic Approximation and Nonlinear Regression, MIT Press, 1967.
8. A. Noda, "A System Identification by an Adaptive Approximation Method," 7th Preprint of SICE of Japan, 143, 1968, (in Japanese).

ACKNOWLEDGMENT

The author is grateful to Professor J. Nagumo and Assistant Professor S. Amari of Tokyo University for their suggestions and discussions.

WEIGHTING FUNCTION ESTIMATION IN DISTRIBUTED-PARAMETER SYSTEMS

Henry E. Lee
Westinghouse Electric Corp.
Annapolis, Maryland, U.S.A.

D. W. C. Shen
Univ. of Pennsylvania
Philadelphia, Pa., U.S.A.

ABSTRACT

An accelerated stochastic approximation algorithm is developed for the identification of weighting function associated with boundary control in one-dimensional linear distributed-parameter systems. The weighting function is assumed to be variable separable, and each variable is approximated by a finite number of orthonormal polynomials. In the absence of noise, this algorithm will converge in a finite number of steps. For adaptive control, on-line weighting function estimators are developed which use the optimal control function as input. These estimators are functional gradient algorithms based on least square approach. They can be used for estimating weighting function associated with either boundary or distributed control.

INTRODUCTION

The control of distributed-parameter systems requires a complete knowledge of the systems dynamic behavior. In practical situations, this complete knowledge is not always available and some identification procedures must be performed. When the form of partial differential equation describing the process dynamics is known, the determination of unknown coefficients is a parameter identification problem. On the other hand, if the order and exact form of the partial differential equation are not known a priori, the process can only be identified by estimating the system weighting functions.

Most of the useful techniques developed in the last few years are for determining the unknown parameters of an assumed partial differential equation model of the system [1,2]. Relatively fewer results exist on the estimation of weighting functions of the distributed-parameter systems. The objective of this paper is to develop computational algorithms for estimating the weighting functions of linear distributed-parameter systems. To ensure practical usefulness, these algorithms must have fast convergence property and be applicable to the situation where the measurements are subject to random noise.

FORMULATION OF THE OFF-LINE IDENTIFICATION PROBLEM

Consider one-dimensional, linear, stationary distributed-parameter systems controlled by boundary control functions which act at both ends of the one-dimensional space. The state of these systems can be expressed by:

$$q(x,t) = q_o(x,t) + \sum_{i=1}^{2} \int_0^t g_i(x,t-\tau)u_i(\tau)d\tau \tag{1}$$

where $q_o(x,t)$ is the state function due to initial condition. For off-line identification, $q_o(x,t)$ can be assumed zero. $u_1(t)$ and $u_2(t)$ are the two boundary control functions and $g_1(x,t)$ and $g_2(x,t)$ are the corresponding weighting functions.

For simplicity, assume only $u(t)[u_1(t)$ or $u_2(t)]$ is acting on the system. To identify weighting function $g(x,t)$ corresponding to $u(t)$, assume a record of noisy impulse response measurements $g_m(x,t)$ is available for every pair of (x,t) in the domain $0 \leq x \leq 1$, $0 \leq t \leq T$.

$$g_m(x,t) = g(x,t) + v \tag{2}$$

$g_m(x,t)$ is the measurement of $q(x,t)$ due to an instantaneous point source of unit strength introduced at the boundary control point. The noise v has zero mean, finite variance and statistically independent of x and t.

It is also assumed that the system weighting function $g(x,t)$ is piecewise continuous in the domain considered.

Assume that $g(x,t)$ is separable; i.e.,

$$g(x,t) = g_x(x) \cdot g_t(t). \tag{3}$$

Approximate $g_x(x)$ by a finite set of polynomials chosen from a complete system of orthonormal polynomials.

$$\hat{g}_x(x) = \sum_{j=1}^{M_x} c_{xj} P_j(x) \tag{4}$$

Likewise for $g_t(t)$,

$$\hat{g}_t(t) = \sum_{k=1}^{M_t} c_{tk} P_k(t) \tag{5}$$

From equations (3), (4), and (5), we see that g(x,t) can be approximated by:

$$\hat{g}(x,t) = \sum_{j=1}^{M_x} c_{xj} P_j(x) \sum_{k=1}^{M_t} c_{tk} P_k(t) = \sum_{i=1}^{N} c_i b_i(x,t) \tag{6}$$

where $N = M_x \cdot M_t$ and

$$b_{j+(k-1)M_x}(x,t) = P_j(x) P_k(t), \quad j=1,2,\ldots,M_x;\ k=1,2,\ldots,M_t \tag{7}$$

Using matrix notations, $\hat{g}(x,t)$ can be expressed by

$$\hat{g}(x,t) = \underline{c}^T \underline{b}(x,t) \tag{8}$$

where $\underline{c}$ and $\underline{b}(x,t)$ are both N x 1 vectors. Superscript T denotes the transpose of a vector or a matrix.

The identification problem is to determine $\underline{c}^*$ which minimizes the mean square error criterion.

$$J(\underline{c}) = E\{[g(x,t) - \hat{g}(x,t)]^2\} \tag{9}$$

Given the successive measurement pairs $\{g_m(m), (x,t)(m); m=1,2,\ldots\}$, define a sequence of estimates $\underline{c}(m)$ such that

$$\underline{c}(m) = \underline{c}(m-1) + r(m)\underline{b}(m)[g_m(m) - \underline{c}^T(m-1)\underline{b}(m)] \tag{10}$$

where

$$\lim_{m\to\infty} r(m)=0, \quad \lim_{M\to\infty} \sum_{m=1}^{M} r(m)= \infty, \quad \lim_{M\to\infty} \sum_{m=1}^{M} r^2(m)<\infty \tag{11}$$

Above is a stochastic approximation algorithm which is shown in [3] that the $\underline{c}$'s converge with probability 1 and to the value which minimizes the error criterion $J(\underline{c})$. This algorithm has a rather slow convergence rate.

ACCELERATED STOCHASTIC APPROXIMATION ALGORITHM

To improve the rate of convergence, let us consider an iterative procedure which will update the solution of m simultaneous linear equations when the additional input measurement $\underline{b}(m+1)$ and output measurement $g_m(m+1)$ are made available. The performance index to be minimized at the mth iteration is taken to be

$$I(m) = \frac{1}{2} \sum_{i=1}^{m} \{[g_m(i) - \underline{c}^T(m)\underline{b}(i)]^2\} \tag{12}$$

for all $m \leq N$, N being the dimension of the vector $\underline{c}$.

Assume the updating equation at (m+1)th iteration is

$$\underline{c}(m+1) = \underline{c}(m) + t(m+1)\underline{s}(m+1). \tag{13}$$

The step size $t(m+1)$ and direction vector $\underline{s}(m+1)$ can be determined by setting the derivative of $I(m+1)$ with respect to $\underline{c}(m+1)$ equal to zero vector.

$$\frac{\partial I(m+1)}{\partial \underline{c}(m+1)} = \sum_{i=1}^{m+1} [g_m(i) - \underline{c}^T(m+1)\underline{b}(i)]\ \underline{b}(i)\ . \tag{14}$$

So we obtain

$$[g_m(1)g_m(2)\ldots g_m(m+1)]-\underline{c}^T(m+1)\cdot[\underline{b}(1)\underline{b}(2)\ldots\underline{b}(m+1)] = \underline{0}. \tag{15}$$

Substituting $\underline{c}(m+1)$ from eq. (13) into eq. (15), we get

$$\begin{aligned}&[g_m(1)g_m(2)\ldots g_m(m+1)]-\underline{c}^T(m)[\underline{b}(1)\underline{b}(2)\ldots\underline{b}(m+1)]\\ &\quad -t(m+1)\underline{s}^T(m+1)[\underline{b}(1)\underline{b}(2)\ldots\underline{b}(m+1)] = \underline{0}\ .\end{aligned} \tag{16}$$

Since $\underline{c}(m)$ minimizes $I(m)$ at the mth iteration, so

$$[g_m(1)g_m(2)\ldots g_m(m)]-\underline{c}^T(m)\cdot[\underline{b}(1)\underline{b}(2)\ldots\underline{b}(m)] = \underline{0}\ . \tag{17}$$

Therefore, eq. (16) is reduced to

$$g_m(m+1)-\underline{c}^T(m)\underline{b}(m+1)-t(m+1)\underline{s}^T(m+1)\underline{b}(m+1) = 0 \tag{18}$$

and

$$\underline{s}^T(m+1)[\underline{b}(1)\underline{b}(2)\ldots\underline{b}(m)] = \underline{0}\ . \tag{19}$$

From eq. (18), with $\underline{s}^T(m+1)\underline{b}(m+1) = 1$, $t(m+1)$ is found to be

$$t(m+1) = g_m(m+1)-\underline{c}^T(m)\underline{b}(m+1)\ . \tag{20}$$

So the updating equation at (m+1)th iteration becomes

$$\underline{c}(m+1) = \underline{c}(m)+[g_m(m+1)-\underline{c}^T(m)\underline{b}(m+1)]\underline{s}(m+1) \tag{21}$$

where $\underline{s}(m+1)$ satisfies

$$\underline{s}^T(m+1)[\underline{b}(1)\underline{b}(2)\ldots\underline{b}(m+1)] = [00\ldots1]\ . \tag{22}$$

To compute the direction vector $\underline{s}(m)$, it is suggested (in [4]) that $\underline{s}(m+1)$ be equal to the first row of the inverse of $B(m+1)$ where

$$B(m+1) = [\underline{b}(m+1)\underline{b}(m)\ldots\underline{b}(1)\underline{w}_1\underline{w}_2\ldots\underline{w}_{N-m-1}] \tag{23}$$

$\underline{w}_i$ is an N-dimensional vector with all elements 0 except the ith element which is 1. The above choice of $\underline{s}(m+1)$ is made because the inverse of B(m+1) can be determined from the inverse of B(m) and a new measurement by the following relation

$$B^{-1}(m+1) = T^{-1}\left(B^{-1}(m)\ \frac{B^{-1}(m)\underline{b}(m+1)-\underline{w}_m}{(\underline{B}_n^{-1})^T(m)\underline{b}(m+1)}\ (\underline{B}_n^{-1})^T(m)\right) \tag{24}$$

where T^{-1} is the inverse of the shifting matrix and $\underline{B}_n^{-1}(m)$ is the nth column of the matrix $B^{-1}(m)$. The above relation can be verified by using the rank annihilation technique [5].

In the ideal cases where the N independent input and output measurements are not contaminated with noise, eq. (21) together with eq. (24) will determine $\underline{c}^*$ in N iterations. However, the measurement noise is usually present in most applications and the following accelerated stochastic approximation algorithm is formulated based on eq. (21).

$$\underline{c}(m+1) = \underline{c}(m)+r(m+1)[g_m(m+1)-\underline{c}^T(m)\underline{b}(m+1)]\cdot(\underline{B}_1^{-1})^T(m+1)\ . \tag{25}$$

$\underline{B}_1^{-1}(m)$ in the above equation is the 1st row of the matrix $B^{-1}(m)$ which is updated by eq. (24). r(m) is a real number such that for $m \geq 1$,

$$\lim_{m\to\infty} r(m) = 0,\quad \lim_{M\to\infty} \sum_{m=1}^{M} r(m) = \infty\ ,\quad \lim_{M\to\infty} \sum_{m=1}^{M} r^2(m) < \infty\ . \tag{26}$$

The proof of convergence of the accelerated stochastic approximation algorithm is based on the theorem of convergence by Dvoretzky [6]. It can be shown that all the conditions for convergence in Dvoretzky's theorem are satisfied by the accelerated stochastic approximation algorithm.

It must be said that, for noise-free case, both conjugate gradient and the method presented in this section have finite-step convergence property. However, it is difficult to update the conjugate gradient algorithm with new measurement if noise is present whereas the updating of the accelerated stochastic approximation algorithm is conveniently done by evaluating the inverse of B(m+1) in terms of the inverse of B(m) and a new measurement.

ON-LINE LEAST-SQUARE ESTIMATION

The recursive algorithm developed in the previous section is only applicable to off-line application since impulse response measurement is the required information for estimation. In the adaptive control of distributed-parameter systems, the estimation scheme has tc be performed under normal operation condition, no test signal can be injected. Here we consider on-line weighting function estimation procedure for either boundary or distributed control based on least-square estimation technique. The approach is first to formulate a minimization problem in the Hilbert space. Functional gradient algorithms are then used to minimize the quadratic error criterion.

One-dimensional, linear, stationary distributed-parameter systems subject to both boundary and distributed control functions can be represented by:

$$q(x,t) = \sum_{i=1}^{2} \int_0^t g_i(x,t-\tau)u_i(\tau)d\tau + \int_0^t \int_0^\ell g_d(x,x';t-\tau)[u_d(x';\tau)+q(x';0)\delta(\tau)]dx'd\tau \tag{27}$$

where $q(x,0)$ is the initial state function; $u_d(x,t)$ is the distributed control function, and $g_d(x,x',t)$ the corresponding weighting function. In the absence of distributed control, eq. (27) can be reduced to eq. (1).

Output measurements available for estimation are

$$q_m(x,t) = q(x,t) + n(x,t) \tag{28}$$

where $n(x,t)$ includes the effects of random disturbances entering the system and of the measurement errors.

First, consider the estimation of

$$\underline{g}(x,t) = \begin{pmatrix} g_1(x,t) \\ g_2(x,t) \end{pmatrix}$$

of boundary control function

$$\underline{u}(t) = \begin{pmatrix} u_1(t) \\ u_2(t) \end{pmatrix} .$$

The error criterion to be minimized can be formulated as follows:

$$I[\underline{g}(x,t)] = \int_0^T \int_0^\ell \{\underline{q}_m(x,t) - \int_0^t U(t-\tau)\underline{g}(x,\tau)d\tau\}^T \{\underline{q}_m(x,t) - \int_0^t U(t-\tau)\underline{g}(x,\tau)d\tau\}dxdt \tag{29}$$

where

$$U(t) = \begin{bmatrix} u_{11}(t) & u_{12}(t) \\ u_{21}(t) & u_{22}(t) \end{bmatrix} \quad , \; u_{1i}(t)$$

is the control function $u_i(t)$ from t=0 to t=T and $u_{2i}(t)$ is $u_i(t)$ from t=T to t=2T. The need for two sets of control inputs

$$\begin{pmatrix} u_{11}(t) \\ u_{12}(t) \end{pmatrix} \text{ and } \begin{pmatrix} u_{21}(t) \\ u_{22}(t) \end{pmatrix}$$

and two sets of output measurements $q_{m1}(x,t)$ and $q_{m2}(x,t)$ is because we wish to determine $g_1(x,t)$ and $g_2(x,t)$ simultaneously. Here $q_{m1}(x,t)$ is the output measurement $q_m(x,t)$ from t=0 to t=T and $q_{m2}(x,t)$ is $q_m(x,t)$ from t=T to t=2T.

Using functional analysis notations, the above error criterion can be expressed as:

$$I[\underline{g}(x,t)] = ||\underline{q}_m(x,t) - (L\underline{g})(x,t)||^2_{H_1(T)} \tag{30}$$

where the linear operator L is defined by

$$(L\underline{g})(x,t) = \int_0^t U(t-\tau)\underline{g}(x,\tau)d\tau \; . \tag{31}$$

$||\cdot||_{H_1(T)}$ in eq. (30) is a norm in the Hilbert space with Bochner integrable functions [7]. Element $\underline{g}(x,t)$ in space $H_1(T)$ is a function of two variables x and t which satisfies the following condition:

$$\int_0^T \int_0^\ell \underline{g}^T(x,t)\underline{g}(x,t)dxdt < \infty \; . \tag{32}$$

Inner product in space $H_1(T)$ is defined by

$$\langle \underline{c}(x,t), \underline{d}(x,t) \rangle_{H_1(T)} = \int_0^T \int_0^\ell \underline{c}^T(x,t)\underline{d}(x,t)dxdt \; . \tag{33}$$

Using the definition of adjoint operator,

$$\langle (L^*\underline{c})(x,t), \underline{d}(x,t) \rangle_{H_1(T)} = \langle \underline{c}(x,t), (L\underline{d})(x,t) \rangle_{H_1(T)} \tag{34}$$

we find

$$(L^*\underline{c})(x,t) = \int_t^T U^T(z-t)\underline{c}(x,z)dz \; . \tag{35}$$

Expanding eq. (30), the actual criterion to be minimized is found to be:

$$J[\underline{g}(x,t)] = \langle \underline{g}(x,t), (L^*L\underline{g})(x,t) \rangle_{H_1(T)} - 2\langle (L^*\underline{q}_m)(x,t), \underline{g}(x,t) \rangle_{H_1(T)} \cdot \tag{36}$$

Similarly, when distributed control function $u_d(x,t)$ is acting on the system, the identification of corresponding weighting function $g_d(x,x',t)$ can be formulated as minimizing the following error criterion:

$$J_d[g_d(x,x',t)] = \langle g_d(x,x',t),(L_d^*L_dg_d)(x,x',t)\rangle_{H_2(T)} - 2\langle (L_d^*q_m)(x,x',t),g_d(x,x',t)\rangle_{H_2(T)} \tag{37}$$

where

$$(L_dg_d)(x,t) = \int_0^t\int_0^\ell g_d(x,x',\tau)[u_d(x',t-\tau)+q(x',0)\delta(t-\tau)]dx'd\tau \tag{38}$$

and

$$(L_d^*q)(x,x',t) = \int_t^T [u_d(x',\tau-t)+q(x',0)\delta(\tau-t)]q(x,\tau)d\tau \tag{39}$$

Elements in space $H_2(T)$ are Bochner integrable functions with three variables x, x' and t. Inner product in $H_2(T)$ is defined by

$$\langle \underline{c}(x,x',t),\underline{d}(x,x',t)\rangle_{H_2(T)} = \int_0^T\int_0^\ell\int_0^\ell \underline{c}^T(x,x',t)\underline{d}(x,x',t)dx'dxdt. \tag{40}$$

FUNCTIONAL GRADIENT ALGORITHMS

To minimize $J[\underline{g}(x,t)]$ of eq. (36), we consider the following iterative procedure:

$$\underline{g}_{i+1}(x,t) = \underline{g}_i(x,t) + a_i\underline{z}_i(x,t) . \tag{41}$$

Taking the Gateau differential of $J[g_i(x,t)]$, we get

$$dJ[\underline{g}_i(x,t)] = \left.\frac{\partial J[\underline{g}_i(x,t) + a_i\underline{z}_i(x,t)]}{\partial a_i}\right|_{a_i=0}$$

$$= 2\langle (L^*L\underline{g}_i)(x,t)-(L^*\underline{q}_m)(x,t),\underline{z}_i(x,t)\rangle_{H_1(T)} \tag{42}$$

$$dJ[\underline{g}_i(x,t)] \text{ is maximum if } \underline{z}_i(x,t)=(L^*L\underline{g}_i)(x,t)-(L^*\underline{q}_m)(x,t) \tag{43}$$

After $\underline{z}_i(x,t)$ is chosen, take derivative of $J[g_i(x,t)+a_i\underline{z}_i(x,t)]$ with respect to a_i, and set it equal to 0. We obtain

$$a_i = -\frac{\langle (L^*L\underline{g}_i)(x,t)-(L^*\underline{q}_m)(x,t),\underline{z}_i(x,t)\rangle_{H_1(T)}}{\langle \underline{z}_i(x,t),(L^*L\underline{z}_i)(x,t)\rangle_{H_1(T)}} . \tag{44}$$

Equations (43) and (44) together with (41) form the functional steepest descent algorithm.

The proof of convergence and an estimate of the rate of convergence of steepest descent method of approximation can be found in Kantarovich's original work [8].

The above algorithm suffers from slow convergence, especially near the optimum. To improve the convergence rate, $\underline{z}_i(x,t)$'s are chosen such that they satisfy the following condition:

$$\langle \underline{z}_j(x,t), (L^*L\underline{z}_i)(x,t)\rangle_{H_1(T)} = 0 \quad \text{for } i \neq j . \tag{45}$$

$\underline{z}_i(x,t)$'s are called L*L-orthogonal or L*L-conjugate vectors. The derivation of this functional conjugate gradient algorithm is in reference [9]. It is formulated as follows.

With initial assumption $\underline{g}_o(x,t)$,

$$\left.\begin{aligned}
\underline{r}_i(x,t) &= \underline{r}_{i-1}(x,t) - a_{i-1}(L^*L\underline{z}_{i-1})(x,t) \\
b_{i-1} &= -\frac{\langle \underline{z}_{i-1}(x,t), (L^*L\underline{r}_i)(x,t)\rangle_{H_1(T)}}{\langle \underline{z}_{i-1}(x,t), (L^*L\underline{z}_{i-1})(x,t)\rangle_{H_1(T)}} \\
\underline{z}_i(x,t) &= \underline{r}_i(x,t) + b_{i-1}\underline{z}_{i-1}(x,t) \\
a_i &= \frac{\langle \underline{z}(x,t), \underline{r}_i(x,t)\rangle_{H_1(T)}}{\langle \underline{z}_i(x,t), (L^*L\underline{z}_i)(x,t)\rangle_{H_1(T)}} \\
\underline{g}_{i+1}(x,t) &= \underline{g}_i(x,t) + a_i\underline{z}_i(x,t)
\end{aligned}\right\} \quad i=0,1,\ldots,n-1 \tag{46}$$

with $\underline{z}_{-1}(x,t)=0$ and $\underline{r}_{-1}(x,t)=(L^*\underline{q}_m)(x,t)-(L^*L\underline{g}_o)(x,t)$.

Both algorithms shown above can be extended for identifying the weighting function $g_d(x,x',t)$ for distributed control $u_d(x,t)$, since the error criterion to be minimized in eq. (37) has the same form as eq. (36). It is only necessary to replace operators L and L* by L_d and L_d^*, and all inner products are redefined in the $H_2(T)$ space.

CONCLUSION

This paper has provided computational algorithms for effectively solving the estimation problem for a class of distributed-parameter systems. The inputs and measurements may be subject to random fluctuation which is a function of space as well as of time. In applying the on-line gradient algorithms, no statistical assumptions need be made concerning the nature of the input disturbance or of the measurement error. The absence of statistical data corresponds closely to the adaptive control situation where the controller is operating on a real time scale.

REFERENCES

1. F. J. Perdreauville and R. E. Goodson, "Identification of Systems Described by Partial Differential Equations," Trans. ASME, J. of Basic Engr., Ser. D, Vol. 88, No. 2, pp. 463-468, 1966.

2. P. L. Collins and H. C. Khatri, "Identification of Distributed Parameter Systems Using Finite Differences," Trans. ASME, J. of Basic Engr., Ser. D, Vol. 91, No. 2, pp. 239-245, 1969.

3. G. N. Saridis and P. C. Badavas, "Identifying Solutions of Distributed Parameter Systems by Stochastic Approximation," Proc. of 7th IEEE Symp. on Adaptive Processes, December 1968.

4. G. S. Levy, "Accelerated Gradient Pattern Recognition: An Application to System Identification," M.S. Thesis, Department of Electrical Engineering, McGill University, 1968.

5. H. S. Wilf and A. Ralston, Mathematical Methods for Digital Computers, Wiley and Sons, 1960.

6. A. Dvoretzky, "On Stochastic Approximation," Proc. Third Berkeley Symp. on Math. Stat. and Prob., Vol. 1, 1956.

7. E. I. Axelband, "The Structure of the Optimal Tracking Problem for Distributed-Parameter Systems," IEEE Trans. on Automatic Control, Vol. AC-13, pp. 50-56, 1968.

8. L. V. Kantarovich, "Functional Analysis and Applied Mathematics," Uspekhi Mat. Nauk, Vol. 3, p. 89, 1948.

9. H. E. Lee, "Identification and Optimal Control of Stochastic Distributed-Parameter Systems," Ph.D. Dissertation, Univ. of Pennsylvania, 1970.

SYSTEM IDENTIFICATIONS BY A NONLINEAR FILTER

Setsuzo Tsuji and Kousuke Kumamaru

Kyushu University

Fukuoka, Japan

1. INTRODUCTION

Most system identification problems may be reduced to the state estimation in a nonlinear system. Athans [4] synthesized a filter for nonlinear continuous systems with discrete observations through the Taylor series expansion of nonlinear functions up to the second order term about the estimated state and the minimization of the estimation error covariance under the criterion of unbiased estimate. It has been shown that the Athans' consideration about the second order term would give the better results compared with the existing methods such as the approximate linearization method. However, in this estimation process there is tedious task to solve the parallel nonlinear differential equations about the state and the estimation error covariance. In this paper the authors have derived a forward recursive algorithm for the nonlinear filter by applying the Taylor series expansion method to the nonlinear discrete systems, in which the computation of the error covariance is simplified due to the discrete form. It has been found that this estimation method is available effectively to the identification of unknown transfer function or the learning of the unknown function in a system. Moreover, it is possible to learn nonstationary functions by expanding into the independent function series with unknown varying coefficients.

2. SECOND ORDER FILTER FOR THE NONLINEAR DISCRETE SYSTEM

Consider the following n-dimensional nonlinear discrete system with m-dimensional observation:

$$X_{k+1} = f(X_k) + W_k \tag{1}$$
$$y_k = h(X_k) + V_k \tag{2}$$

where X_k is an n×1 state vector, y_k is an m×1 observation vector, $f(X_k)$ and $h(X_k)$ are n×1 and m×1 nonlinear function vectors, W_k is an n×1 white Gaussian noise vector with zero mean and covariance Q_k, and V_k is an m×1 white Gaussian noise vector with zero mean and covariance R_k. Control vector is not represented explicitly in the system equation because this vector is generally considered as the known variable in the estimation problem.

Definitions.

(1) $y^k = \{y_o y_1 \ldots y_k\}$: observed information up to the kth stage
(2) $\hat{X}_{k/k}$: estimate of X_k based on y^k
(3) $\hat{X}_{k+1/k}$: prediction of X_{k+1} based on y^k
(4) $\tilde{X}_{k/k} = X_k - \hat{X}_{k/k}$: estimation error of X_k
(5) $\tilde{X}_{k+1/k} = X_{k+1} - \hat{X}_{k+1/k}$: prediction error of X_{k+1}
(6) $C_{k/k} = E(\tilde{X}_{k/k}\tilde{X}_{k/k}/y^k)$, $C_{k+1/k} = E(\tilde{X}_{k+1/k}\tilde{X}'_{k+1/k}/y^k)$
$E(./y^k)$ is expected value conditioned on y^k.

Assumptions.

(1) Initial state X_0 is Gaussian distributed with mean $\hat{X}_{0/-1}$ and and covariance $C_{0/-1}$, which are given a priori.
(2) $f(X_k)$ and $h(X_k)$ are continuous functions which have at least second order derivatives.
(3) $\tilde{X}_{k/k}$ and $\tilde{X}_{k+L/k}$ are approximately Gaussian Distributed.
(4) $\hat{X}_{k/k}$ and $\hat{X}_{k+1/k}$ are unbiased estimate of X_k and unbiased prediction of X_{k+1}, i.e., $E(\tilde{X}_{k/k}/y^k)=0$, $E(\tilde{X}_{k+1/k}/y^k)=0$.†

Suppose that $f(X_k)$ and $h(X_{k+1})$ are approximated by the second order Taylor expansion about $\hat{X}_{k/k}$ and $\hat{X}_{k+1/k}$, respectively, as follows:

$$X_{k+1} = f(\hat{X}_{k/k}) + f_x(\hat{X}_{k/k})\,(X_k-\hat{X}_{k/k}) + \frac{1}{2}\sum_{i=1}^{n}\phi_i(X_k-\hat{X}_{k/k})'f^i_{xx}(\hat{X}_{k/k})\,(X_k-\hat{X}_{k/k}) + W_k \tag{3}$$

†Under the assumption (4), $C_{k/k}$ and $C_{k+1/k}$ given by the definition (6) are exactly covariance matrices.

$$y_{k+1} = h(\hat{X}_{k+1/k}) + h_x(\hat{X}_{k+1/k})(X_{k+1}-\hat{X}_{k+1/k}) + \frac{1}{2}\sum_{i=1}^{m} \psi_i (X_{k+1}-\hat{X}_{k+1/k})' h^i_{xx}(\hat{X}_{k+1/k})(X_{k+1}-\hat{X}_{k+1/k})+V_{k+1} \tag{4}$$

where

$$X_k = (X_k^1\ X_k^2\ \ldots\ X_k^n)', \quad f(X_k) = (f^1(X_k)f^2(X_k)\ \ldots\ f^n(X_k))',$$

$$h(X_k) = (h^1(x_k)h^2(X_k)\ \ldots\ h^m(X_k))',$$

ϕ_i = n×1 natural basis vector,

ψ_i = m×1 natural basis vector

$$(f_x(\hat{X}))^{ij} = \left.\frac{\partial f^i(X)}{\partial X^j}\right|_{X=\hat{X}}, \quad (f^i_{xx}(\hat{X}))^{jk} = \left.\frac{\partial^2 f^i(X)}{\partial X^j \partial X^k}\right|_{X=\hat{X}},$$

$$(h_x(\hat{X}))^{ij} = \left.\frac{\partial h^i(X)}{\partial X^j}\right|_{X=\hat{X}}, \quad (h^i_{xx}(\hat{X}))^{jk} = \left.\frac{\partial^2 h^i(X)}{\partial X^j \partial X^k}\right|_{X=\hat{X}}.$$

Derivation of $\hat{X}_{k+1/k}$. We assume that $\hat{X}_{k+1/k}$ is given by the following formula:

$$\hat{X}_{k+1/k} = f(\hat{X}_{k/k}) + \mu(X_{k/k}) \tag{5}$$

where $\mu(\hat{X}_{k/k})$ is a bias correction term which guarantees the unbiased prediction condition of X_{k+1} and is determined as below. From eqs. (3), (5) and the assumption (4), we obtain:

$$E(\tilde{X}_{k+1/k}/y^k) = E\{(f_x(\hat{X}_{k/k})\tilde{X}_{k/k} + \frac{1}{2}\sum_{i=1}^{n} \phi_i \tilde{X}'_{k/k} f^i_{xx}(\hat{X}_{k/k})\tilde{X}_{k/k} + W_k - \mu(\hat{X}_{k/k}))/y^k\} = 0 \tag{6}$$

Using the assumption (4) and the definition (6) in eq. (6) yields

$$\mu(\hat{X}_{k/k}) = \frac{1}{2}\sum_{i=1}^{n} \phi_i \mathrm{tr}(f^i_{xx}(\hat{X}_{k/k})C_{k/k}) \tag{7}$$

Therefore

$$\hat{X}_{k+1/k} = f(\hat{X}_{k/k}) + \frac{1}{2}\sum_{i=1}^{n} \phi_i \mathrm{tr}(f^i_{xx}(\hat{X}_{k/k})C_{k/k}) \tag{8}$$

Derivation of $\hat{X}_{k+1/k+1}$. We assume that $\hat{X}_{k+1/k+1}$ is given by the following formula:

$$\hat{X}_{k+1/k+1} = \hat{X}_{k+1/k} + K_{k+1}(y_{k+1} - h(\hat{X}_{k+1/k}) - \pi(\hat{X}_{k+1/k})) \tag{9}$$

where K_{k+1} is n×m filter gain matrix and $\pi(\hat{X}_{k+1/k})$ is another biased correction term. $\pi(\hat{X}_{k+1/k})$ is determined as below. Define

$$\hat{y}_{k+1/k} = h(\hat{X}_{k+1/k}) + \pi(\hat{X}_{k+1/k}) \tag{10}$$

$$\hat{y}_{k+1/k} = y_{k+1} - \hat{y}_{k+1/k} \tag{11}$$

Then from eq. (4) we obtain

$$\tilde{y}_{k+1/k} = h_x(\hat{X}_{k+1/k})\tilde{X}_{k+1/k} + \frac{1}{2}\sum_{i=1}^{m}\psi_i\tilde{X}_{k+1/k}h^i_{xx}(\hat{X}_{k+1/k})\tilde{X}_{k+1/k} + V_{k+1} - \pi(\hat{X}_{k+1/k}) \tag{12}$$

We determine $\pi(X_{k+1/k})$ as such value that makes the expectation of $\tilde{y}_{k+1/k}$ conditioned on y^k zero. Using the assumption (4) and the definition (6) in the expectation of eq. (12) yields

$$\pi(\hat{X}_{k+1/k}) = \frac{1}{2}\sum_{i=1}^{m}\psi_i tr(h^i_{xx}(\hat{X}_{k+1/k})C_{k+1/k}) \tag{13}$$

Therefore

$$\hat{X}_{k+1/k+1} = \hat{X}_{k+1/k} + K_{k+1}(y_{k+1} - h(\hat{X}_{k+1/k}) - \frac{1}{2}\sum_{i=1}^{m}\psi_i tr(h^i_{xx}(\hat{X}_{k+1/k})C_{k+1/k})) \tag{14}$$

For this bias correction term $\pi(\hat{X}_{k+1/k})$,

$$E(\tilde{X}_{k+1/k+1}/y^k) = 0 \tag{15}$$

<u>Derivation of the Recursive Equations for $C_{k/k}$ and $C_{k+1/k}$</u>. From the definition (6) and eqs. (3) and (8), $C_{k+1/k}$ is given by

$$C_{k+1/k} = E(\tilde{X}_{k+1/k}\tilde{X}'_{k+1/k}/y^k) = E\{(f_x(\hat{X}_{k/k})\tilde{X}_{k/k} + \frac{1}{2}\sum_{i=1}^{n}\phi_i \cdot tr(f^i_{xx}(\hat{X}_{k/k})(\tilde{X}_{k/k}\tilde{X}'_{k/k} - C_{k/k})) + W_k)(f_x(\hat{X}_{k/k})\tilde{X}_{k/k} + \frac{1}{2}\sum_{i=1}^{n}\phi_i tr(f^i_{xx}(\hat{X}_{k/k})(\tilde{X}_{k/k}\tilde{X}'_{k/k} - C_{k/k})) + W_k)'/y^k\} \tag{16}$$

The following theorem is used in the expectation of eq. (16).

<u>Theorem 1</u>. When X is an n-dimensional Gaussian variable with zero mean and covariance Q, the following formulas are satisfied for the arbitrary n×n matrices A and B,

$$E(X\ tr(AXX')) = 0 \tag{17}$$

$$E(tr(AXX'BXX')) = E(tr(AXX')tr(BXX'))$$
$$= 2\ tr(ACBC) + tr(AC)tr(BC) \tag{18}$$

Using eqs. (17), (18) and definition (6) yields

$$C_{k+1/k} = f_x(\hat{X}_{k/k})C_{k/k}f_x(\hat{X}_{k/k})' + D_k + Q_k \tag{19}$$

where D_k is given by

$$(D_k)^{ij} = \frac{1}{2}\ tr(f^i_{xx}(\hat{X}_{k/k})C_{k/k}f^j_{xx}(\hat{X}_{k/k})C_{k/k}) \tag{20}$$

On the other hand, from eq. (14) we obtain

$$E(\tilde{X}_{k+1/k+1}\tilde{X}'_{k+1/k+1}/y^k) = E\{(\tilde{X}_{k+1/k}-K_{k+1}(h_x(\hat{X}_{k+1/k})\tilde{X}_{k+1/k} + \frac{1}{2}\sum_{i=1}^{m}\psi_i tr(h^i_{xx}(\hat{X}_{k+1/k})(\tilde{X}_{k+1/k}\tilde{X}'_{k+1/k}-C_{k+1/k}))+V_{k+1}))\cdot (\tilde{X}_{k+1/k}-K_{k+1}(h^i_x(\hat{X}_{k+1/k})\tilde{X}_{k+1/k} + \frac{1}{2}\sum_{i=1}^{m}\psi_i tr(h^i_{xx}(\hat{X}_{k+1/k})\cdot (\tilde{X}_{k+1/k}\tilde{X}'_{k+1/k}-C_{k+1/k})) + V_{k+1}))'/y^k\} \tag{21}$$

Using eqs. (17) and (18) and the definition (6) in eq. (21) yields

$$E(\tilde{X}_{k+1/k+1}\tilde{X}_{k+1/k+1}/y^k) = (\Pi - K_{k+1}h_x(\hat{X}_{k+1/k}))C_{k+1/k} (\Pi - K_{k+1}h_x(\hat{X}_{k+1/k}))' + K_{k+1}(R_{k+1}+L_{k+1})K'_{k+1} \tag{22}$$

where Π is an n×n unit matrix and L_{k+1} is given by

$$(L_{k+1})^{ij} = \frac{1}{2}\ tr(h^i_{xx}(\hat{X}_{k+1/k})C_{k+1/k}h^j_x(\hat{X}_{k+1/k})C_{k+1/k}) \tag{23}$$

We determine the optimal filter gain K_{k+1} as such value that minimizes the weighted mean square error $\tilde{X}_{k+1/k+1}$ conditioned on y^k, i.e.,

$$E(\tilde{X}'_{k+1/k+1}M_{k+1}\tilde{X}_{k+1/k+1}/y^k) = tr(M_{k+1}E(\tilde{X}_{k+1/k+1}\tilde{X}'_{k+1/k+1}/y^k)),$$

where M_{k+1} is an n×n positive symmetric weighting matrix. Therefore, K_{k+1} is given by

$$\frac{\partial tr(M_{k+1}E(\tilde{X}_{k+1/k+1}\tilde{X}'_{k+1/k+1}/y^k))}{\partial K_{k+1}} = 0 \tag{24}$$

In the calculation of eq. (24), the following gradient matrix formulas are used.

Gradient Matrix Formulas.

$$\frac{\partial}{\partial K}\,\mathrm{tr}(AKB') = A'B \tag{25}$$

$$\frac{\partial}{\partial K}\,\mathrm{tr}(BK') = B \tag{26}$$

$$\frac{\partial}{\partial K}\,\mathrm{tr}(AKCK') = A'KC' + AKC \tag{27}$$

where A, B and C are arbitrary matrices for which the above matrix calculations are possible.

Consequently, the optimal filter gain is given by

$$K_{k+1} = C_{k+1/k}h'_x(\hat{X}_{k+1/k})(h_x(\hat{X}_{k+1/k})C_{k+1/k}h'_x(\hat{X}_{k+1/k}) + L_{k+1} + R_{k+1})^{-1} \tag{28}$$

Substituting eq. (28) into eq. (22) yields

$$E(\tilde{X}_{k+1/k+1}\tilde{X}'_{k+1/k+1}/y^k) = C_{k+1/k}-C_{k+1/k}h'_x(\hat{X}_{k+1/k})(h_x(\hat{X}_{k+1/k})\cdot C_{k+1/k}h'_x(\hat{X}_{k+1/k})+L_{k+1}+R_{k+1})^{-1}\,h_x(\hat{X}_{k+1/k})C_{k+1/k} \tag{29}$$

In order to derive the recursive relation between $C_{k+1/k+1}$ and $C_{k+1/k}$, the following theorem is used.

Theorem 2.† Under the optimal filter gain given by eq. (29), the following properties are satisfied to the estimation error $\hat{X}_{k+1/k+1}$,

$$E(\tilde{X}_{k+1/k+1}/y^{k+1}) = E(\tilde{X}_{k+1/k+1}/y^k) \tag{30}$$

$$E(\tilde{X}_{k+1/k+1}\tilde{X}'_{k+1/k+1}/y^{k+1}) = E(\tilde{X}_{k+1/k+1}\tilde{X}'_{k+1/k+1}/y^k) \tag{31}$$

Consequently, by eqs. (29) and (31), we obtain

$$C_{k+1/k+1} = C_{k+1/k}-C_{k+1/k}h'_x(\hat{X}_{k+1/k})(h_x(\hat{X}_{k+1/k})C_{k+1/k}h'_x(\hat{X}_{k+1/k}) + L_{k+1} + R_{k+1})^{-1}h_x(\hat{X}_{k+1/k})C_{k+1/k} \tag{32}$$

The minimum of the weighted mean square error $\tilde{X}_{k+1/k+1}$ conditioned on y^k is

$$\mathrm{Min}\ E(\tilde{X}'_{k+1/k+1}M_{k+1}\tilde{X}_{k+1/k+1}/y^k) = \mathrm{tr}(M_{k+1}C_{k+1/k+1}) \tag{33}$$

---

†This theorem is proved by using the orthogonality of $\tilde{X}_{k+1/k+1}$ and $\tilde{y}_{k+1/k}$ (see [9]).

Moreover, from eqs. (15) and (30) the following unbiased estimate condition is obtained:

$$E(\tilde{X}_{k+1/k+1}/y^{k+1}) = E(\tilde{X}_{k+1/k+1}/y^{k}) = 0 \tag{34}$$

Summary of the Filter Algorithm.

$$\hat{X}_{k+1/k} = f(\hat{X}_{k/k}) + \frac{1}{2}\sum_{i=1}^{n}\phi_i \ tr(f^i_{xx}(\hat{X}_{k/k})C_{k/k}) \tag{35}$$

$$\hat{X}_{k+1/k+1} = \hat{X}_{k+1/k} + K_{k+1}(y_{k+1} - h(\hat{X}_{k+1/k}) - \frac{1}{2}\sum_{i=1}^{m}\psi_i tr(h^i_{xx}(\hat{X}_{k+1/k}\ C_{k+1/k})) \tag{36}$$

$$K_{k+1} = C_{k+1/k}h'_x(\hat{X}_{k+1/k})(h_x(\hat{X}_{k+1/k})C_{k+1/k}h'_x(\hat{X}_{k+1/k}) + L_{k+1} + R_{k+1})^{-1} \tag{37}$$

$$(L_{k+1})^{ij} = \frac{1}{2}\ tr(h^i_{xx}(\hat{X}_{k+1/k})C_{k+1/k}h^j_{xx}(\hat{X}_{k+1/k})C_{k+1/k}) \tag{38}$$

$$C_{k+1/k+1} = C_{k+1/k} - C_{k+1/k}h'_x(\hat{X}_{k+1/k})(h_x(\hat{X}_{k+1/k})C_{k+1/k}h'_x(\hat{X}_{k+1/k}) + L_{k+1} + R_{k+1})^{-1}h_x(\hat{X}_{k+1/k})C_{k+1/k} \tag{39}$$

$$C_{k+1/k} = f_x(\hat{X}_{k/k})C_{k/k}f_x(\hat{X}_{k/k})' + D_k + Q_k \tag{40}$$

$$(D_k)^{ij} = \frac{1}{2}\ tr(f^i_{xx}(\hat{X}_{k/k})C_{k/k}f^j_{xx}(\hat{X}_{k/k})C_{k/k}) \tag{41}$$

With a priori information $\hat{X}_{0/-1}$, $C_{0/-1}$, the nonlinear state estimation can be sequentially processed through the eqs. (35)-(41).

3. APPLICATIONS

This nonlinear filter seems to be available to many learning-identification problems. We treat here the following three cases as the application examples of the nonlinear filter and show that the satisfactory learning-identification results have been obtained with the proper initial values.

3.1 Identification of Unknown Transfer Function By Single Input-Output Pair

Unknown linear dynamic systems of lumped constant with single input and output are expressible by the transfer function with unknown parameters;

$$F(s) = \frac{Y(s)}{U(s)} = \frac{b_n s^{n-1} + b_{n-1} s^{n-2} + \ldots + b_1}{(s-\lambda_1)(s-\lambda_2)\ \ldots\ (s-\lambda_n)}$$

The eigenvalues $\lambda_i (i=1,2,\ldots,n)$ are assumed to be all distinct.[†] This system is described by the following canonical system and linear observation equation

$$\dot{x} = \Gamma x + \Delta u \tag{42}$$

$$y = Mx \tag{43}$$

where

$$\Gamma = \begin{bmatrix} \lambda_1 & & & 0 \\ & \lambda_2 & & \\ & & \ddots & \\ 0 & & & \lambda_n \end{bmatrix}, \qquad \begin{aligned} \Delta &= (1\ 1\ \ldots\ 1)' \\ M &= (m^1\ m^2\ \ldots\ m^n) \\ m^i &= G(s)(s-\lambda_i)|_{s=\lambda_i} \end{aligned}$$

This continuous system is transformed to the n-dimensional discrete system corrupted by additive Gaussian noises with single control and observation output and the system transition, driving and observation matrices are unknown;

$$x_{k+1} = \Phi x_k + Gu_k + w_k \tag{44}$$

$$y_k = Mx_k + v_k \tag{45}$$

where

$$\Phi = \begin{bmatrix} e^{\lambda_1 T} & & & 0 \\ & e^{\lambda_2 T} & & \\ & & \ddots & \\ 0 & & & e^{\lambda_n T} \end{bmatrix} = |\phi^{ij}|, \qquad \begin{aligned} \Phi^d &= (\phi^{11}\ \phi^{22}\ \ldots\ \phi^{nn})' \\ T &= \text{sampling period} \\ G &= (g^1\ g^2\ \ldots\ g^n)' \\ g^i &= (\phi^{ii}-1)T/\log \phi^{ii}\ . \end{aligned}$$

This is considered as the nonlinear discrete system about the augmented state vector X_k composed of the original states and all of the unknown parameters, i.e.,

$$X_k = (x_k' \Phi^{d\prime} M)' = (X_k^1\ X_k^2\ \ldots\ X_k^{3n})'\ .$$

It should be noticed that the order of the augmented state vector is reduced to 3n from 4n by using the interrelation between Φ and G through eigenvalues. Consequently, parameter estimation, i.e., system identification, is performed by using the second order filter as the augmented state estimation problem in the nonlinear system described by the following dynamic and observation equations:

$$x_{k+1} = f(X_k, u_k, w_k)\ , \qquad y_k = h(X_k) + v_k\ ,$$

[†]It is also possible to identify the transfer function with any complex or multiple eigenvalues by the proper modifications.

$$f(X_k, u_k, w_k) = \begin{bmatrix} X_k^{n+i} \cdot X_k^i + ((X_k^{n+i} - 1)T/\log X_k^{n+i}) u_k + w_k \\ X_k^j \end{bmatrix}$$

$$i=1,2,\ldots,n, \quad j=n+1,n+2,\ldots,3n$$

$$h(X_k) = Mx_k = \sum_{i=1}^{n} X_k^{2n+i} X_k^i .$$

As an example, a 6th order unknown linear system is taken and the identification results are shown in Figures 1 and 2 and Table 1.

Table 1. Identification Results

	true value	estimate value
λ_1	-1.20	-1.22
λ_2	-1.00	-1.01
λ_3	-0.80	-0.81
λ_4	-0.50	-0.51
λ_5	-0.30	-0.30
λ_6	-0.20	-0.20

	true value	estimate value
b_1	0.10	0.11
b_2	0.10	0.11
b_3	0.50	0.58
b_4	0.20	0.38
b_5	0.10	0.30
b_6	0.10	0.18

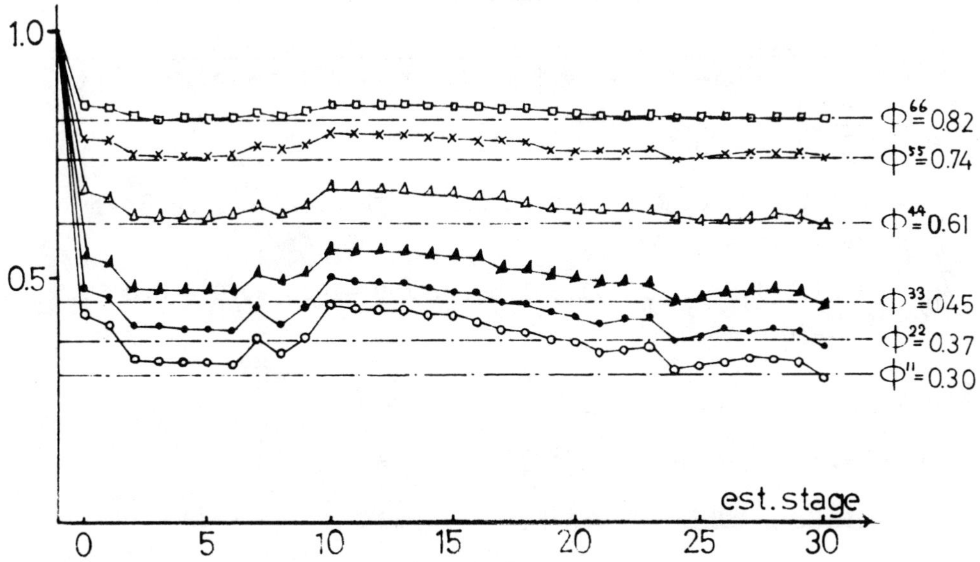

Figure 1. Estimation of Transition Matrix.

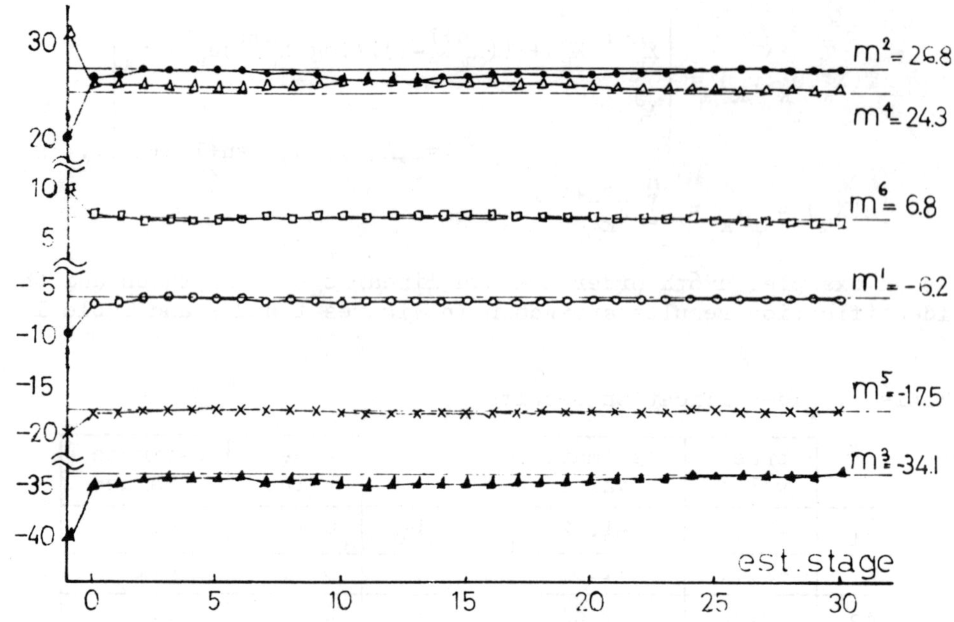

Figure 2. Estimation of Observation Matrix.

<u>Specifications for Figure 1, Figure 2 and Table 1.</u>

$$G(s) = \frac{0.1s^5 + 0.1s^4 + 0.2s^3 + 0.5s^2 + 0.1s + 0.1}{(s + 1.2)(s + 1.0)(s + 0.8)(s + 0.5)(s + 0.3)(s + 0.2)}$$

$$X_0 = (x_0^1\ x_0^2\ \ldots\ x_0^6\ \phi^{11}\ \phi^{22}\ \ldots\ \phi^{66}\ m^1\ m^2\ \ldots\ m^6)'$$

$$= (2\ 2\ \ldots\ 2\ \ 0.30\ 0.37\ 0.45\ 0.61\ 0.74\ 0.82\ -6.2\ 26.8\ -34.1\ 24.3\ -17.5\ 6.8)'$$

$$\hat{X}_{0/-1} = (1.5\ 1.5\ \ldots\ 1.5\ 1.0\ 1.0\ \ldots\ 1.0\ -10.0\ 20.0\ -40.0\ 30.5\ -20.0\ 10.0\ )'$$

$$C_{0/-1} = (\ X_0 - \hat{X}_{0/-1}\ )(\ X_0 - \hat{X}_{0/-1}\ )'$$

Cov w_k = Π : 6×6 unit matrix, Var v_k = 1.0

u_k = 5sin(kT) + 8 T = 1 sec

3.2. The Global Searching of the Multimodal Unknown Gain Function

To seek an optimal peak point in the multimodal system, the global searching of unknown gain function is performed by the second order filter. Suppose that this function is approximated by a linear combination of gauss functions with unknown parameters, i.e.

$$L(x_k,\underline{\alpha},\underline{r},\underline{P}) = \sum_{i=1}^{q} \alpha_i \exp-\frac{1}{2}\|x_k-r_i\|^2 {P_i}^{-1} \tag{46}$$

This approximation is only trial one but this function form seems to be adequate to represent the multimodal function. If the quadratic form is used, we cannot construct any multimodal form even if we try to represent it by any linear combination. In the Eq. (46), x_k is a n-dimensional system state and α_i, r_i, P_i are unknown parameters. $\underline{\alpha},\underline{r},\underline{P}$ are unknown parameter vectors composed of each component of α_i, r_i, P_i and the modal points are $x_k = r_i$. The system is governed by the dynamic equation and observed by the nonlinear scheme;

$$x_{k+1} = g(x_k,u_k) + w_k \tag{47}$$

$$z_k^x = h(x_k) + \xi_k^x \tag{48}$$

where w_k and ξ_k^x are white gaussian noises. And the gain output corresponding to the actual state x_k is observed with additive gaussian noise ξ_k^L;

$$z_k^L = L(x_k,\underline{\alpha},\underline{r},\underline{P}) + \xi_k^L \tag{49}$$

Define the augmented state vector X_k by $(x_k'\ \underline{\alpha}'\ \underline{r}'\ \underline{P}')'$, then this searching problem can be reduced to the state estimation problem in the nonlinear augmented system which is described by the following dynamic and observation equations;

$$X_{k+1} = f(X_k,u_k,w_,), \qquad y_k = H(X_k) + v_k$$

$$f(X_k,u_k,w_k) = \begin{bmatrix} g(x_k,u_k) + w_k \\ X_k^i \end{bmatrix} \qquad i = n+1,n+2,\ldots,N$$

$$H(X_k) = \begin{bmatrix} h(x_k) \\ L(x_k,\underline{\alpha},\underline{r},\underline{P}) \end{bmatrix}, \qquad N = \text{order of } X_k.$$

The estimation of X_k can be performed sequentially by using the observed output y_k corresponding to the actual state which is controlled to the appropriate position by the proper control input.

As the result of the parameter estimation, the unknown gain function is globally learned and the optimal peak can be searched. As an example, a two-modal gain function in a 2nd order system is taken and the searching results are shown in Figures 3 and 4.†

<u>Specifications for Figures 3 and 4.</u>

$$g(x_k,u_k) = \begin{bmatrix} x_k^1 \\ x_k^2 \end{bmatrix} + \begin{bmatrix} u_k^1 \\ u_k^2 \end{bmatrix}$$

$$L(x_k,\underline{\alpha},\underline{r},\underline{P}) = \alpha_1 \quad e^{-\{(x_k^1-r_1^1)^2/2P_1^{11}+(x_k^1-r_1^1)(x_k^2-r_1^2)/P_1^{12}+(x_k^2-r_1^2)^2/2P_1^{22}\}}$$
$$+ \alpha_2 \quad e^{-\{(x_k^1-r_2^1)^2/2P_2^{11}+(x_k^1-r_2^1)(x_k^2-r_2^2)/P_2^{12}+(x_k^2-r_2^2)^2/P_2^{22}\}}$$

$$X_0 = (x_0^1 \; x_0^2 \; r_1^1 \; r_1^2 \; \alpha_1 \; P_1^{11} \; P_1^{12} \; P_1^{22} \; r_2^1 \; r_2^2 \; \alpha_2 \; P_2^{11} \; P_2^{12} \; P_2^{22})'$$

$$= (7 \; 7 \; 4 \; 10 \; 40 \; 20 \; 50 \; 60 \; 18 \; 18 \; 50 \; 20 \; 50 \; 60)'$$

$$\hat{X}_{0/-1} = (0 \; 0 \; 0 \; 0 \; 10 \; 10 \; 20 \; 30 \; 0 \; 0 \; 5 \; 2 \; 5 \; 6)'$$

$$C_{0/-1} = (X_0 - \hat{X}_{0/-1})(X_0 - \hat{X}_{0/-1})', \; H(X_k) = L(X_k)$$

$\mathrm{Cov}\; w_k = \Pi$: 2×2 unit matrix, $\mathrm{Var}\; v_k = 1.0$

$x_1, x_2, \ldots x_5$: actual states given by uniform random inputs which are realized in the region

$$|u_k^1| \leqq 10.0, \qquad |u_k^2| \leqq 10.0 .$$

$x_6, x_7, \ldots, x_{50}$: actual states given by a proper policy.

3.3. Learning of a Nonstationary Unknown Function

Generally, the stochastic approximation of unknown functions is so slow in convergence as to be applicable to on-line learning processes and it has fewer feasibility for nonstationary functions. In this section, to accelerate the convergence, an optimal learning method is used based on the second order filter. Consider the nonstationary unknown function with the following sampling characteristic;

---

†In this example, a control input sequence is synthesized as the proper policy so that the system might transfer to the wide region around the estimated peak points where the gradient of the gain function will be comparatively steep.

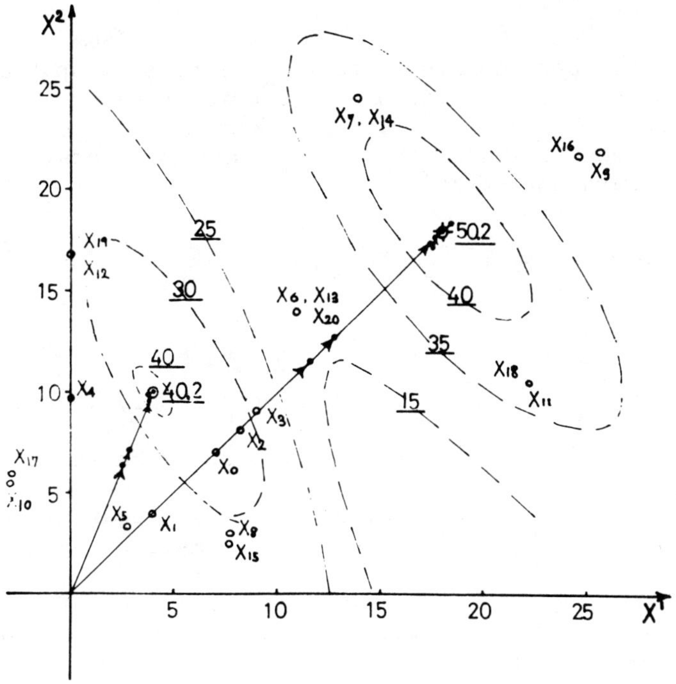

Figure 3. Searching Process of Modal Points.

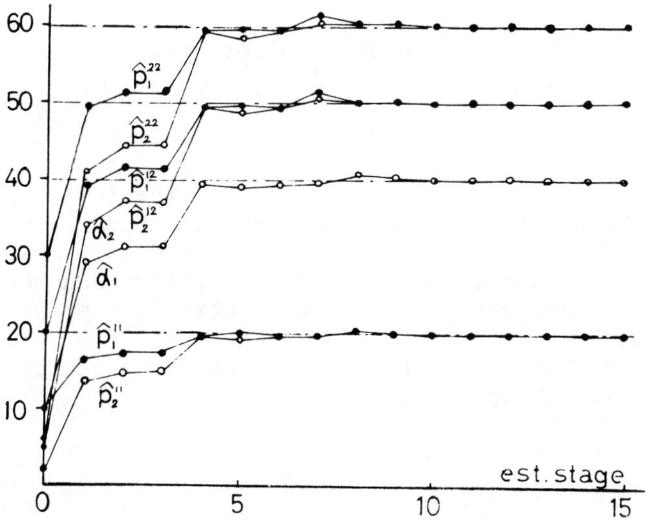

Figure 4. Estimation of Unknown Parameters.

$$y_k = F_k(\theta_k) + \xi_k \tag{50}$$

where θ_k is r×1 input sample vector, y_k is noisy output and ξ_k is white gaussian noise with zero mean and variance $\sigma^2_{\xi k}$. Now expand the function $F_k(\theta_k)$ by the given independent function series $\{\phi^i(\theta_k)\}$ with unknown coefficients which are assumed to transfer with unknown parametric characteristics;

$$F_k(\theta_k) = \sum_{i=1}^{q} z_k^i \, \phi^i(\theta_k) = M(\theta_k)Z_k \tag{51}$$

$$Z_{k+1} = H(Z_k, P) + w_k \tag{52}$$

$Z_k = (z_k^1 \; z_k^2 \; \ldots \; z_k^q)'$: q×1 nonstationary coefficient vector

$P = (p^1 \; p^2 \; \ldots \; p^s)'$: s×1 constant unknown parameter vector

$M(\theta_k) = (\phi^1(\theta_k)\phi^2(\theta_k) \; \ldots \; \phi^q(\theta_k))$: 1×q equivalent observation matrix,

and define the augmented state vector X_k composed of Z_k and P, then we obtain the following nonlinear augmented system dynamics and observation equation;

$$X_{k+1} = f(X_k, w_k), \qquad y_k = h(X_k) + v_k$$

$$f(X_k, w_k) = \begin{bmatrix} H(X_k) + w_k \\ X_k^i \end{bmatrix} \qquad i=q+1, q+2, \ldots, q+s \; ,$$

where w_k and v_k are white gaussian noises with zero means and variances σ^2_{wk} and $\sigma^2_{\xi k} + \delta^2$. δ^2 is considered as a relative increment of the variance in the measurement noise due to the approximation error of $F_k(\theta_k)$. Therefore the second order filter is available to the estimation of the augmented state, i.e. the learning of the nonstationary unknown function. Moreover in this case, the observation matrix $M(\theta_k)$ is a known function of the input sample θ_k. And so the optimal sample sequence can be determined through the minimization of the trace of parameter estimation error covariance matrix. As an example, a linear incremental type of nonstationary function is taken and the learning results using the optimal input sample are shown in Figure 5.

4. CONCLUSIONS

It has been shown that the nonlinear filter by the second order approximation is available to the linear system identification,

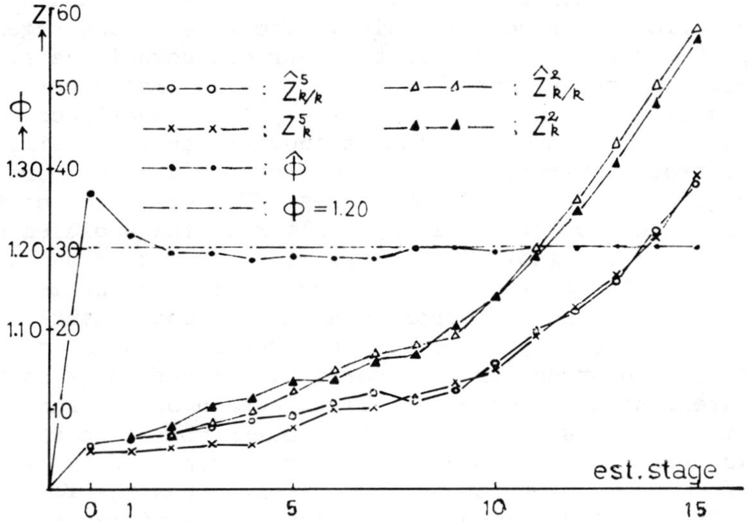

Figure 5. Estimation of Nonstationary Coefficients.

Specifications for Figure 5.

$F_k(\theta_k) = z_k^1/\sqrt{2\pi} + z_k^2\cos\theta_k + z_k^3\cos2\theta_k + z_k^4\sin\theta_k + z_k^5\sin2\theta_k$

$z_{k+1}^i = \phi z_k^i + w_k^i \qquad i=1,2,\ldots,5$

$X_k = (z_k^1\ z_k^2\ \ldots\ z_k^5\ \phi)'$

$X_0 = (20\ 5\ 5\ 5\ 5\ 1.2)'$

$M(\theta_k) = (1/\sqrt{2\pi}\ \cos\theta_k\ \cos2\theta_k\ \sin\theta_k\ \sin2\theta_k)$

$\hat{X}_{0/-1} = (0\ 0\ \ldots\ 0)', \quad C_{0/-1} = (X_0 - \hat{X}_{0/-1})(X_0 - \hat{X}_{0/-1})'$

$\Omega(\theta) = [-3.0,\ 3.0]$: quantized sample region with 0.2 unit

$\sigma_{\xi k}^2 = 1.0 \qquad \sigma_{wk^i}^2 = 1.0 \qquad \delta^2 = 0$

the search of an optimal peak in multimodal systems and the learning of a nonstationary unknown function. The rate of convergence of learning, however, depends upon the a-priori covariance strigently. The matrix composed by the product of true initial error vector elements seems to be a rather proper a-priori covariance matrix but the optimal selection of the initial covariance value will be very important problem hereafter, beacuse they are unknown more or less in the beginning of learning processes. The criterion of the estimation algorithm for the nonlinear filter is the unbiased estimate and the minimum estimation error covariance. And so the correction terms and the filter gain are selected as to make the zero mean of error and to minimize the error covariance. Consequently this filter is linear with respect to the observation. On the derivation of the filter construction, the normal distribution is assumed approximately to the probability distribution of the random variables. The consideration is taken up to the fourth moment so that the third moment is zero and the fourth moment is represented by the quadratic form of the variance but the higher moments than the fourth are neglected. The more sophisticated problem of identification or learning may include so higher order nonlinearity that the probability deviation from normal distribution or more general type of nonlinear filter with the nonlinear function of the innovation process should be considered.

The work reported on in this paper includes the results of efforts and calculations pursued by many graduate students in our control laboratory of Electrical Engineering Department of Kyushu University, especially K. Kubota, T. Tomimatsu, H. Nishikubo and S. Morinaga.

REFERENCES

1. R. E. Kalman, "A New Approach to Linear Filtering and Prediction Problems", ASME Trans., Journal of Basic Eng. 35(1960).
2. H. Cox, "On the Estimation of State Variables and Parameters for Noisy Dynamic Systems", IEEE Trans. on Automatic Control, Vol. AC-9, No. 5 (1964).
3. K. W. Jenkins and R. J. Roy, "A Design Procedure of Adaptive Control Systems", Preprints JACC, p. 624 (1966).
4. M. Athans, R. P. Wishner and A. Bertolini, "Suboptimal State Estimation for Continuous-Time Nonlinear Systems from Discrete Noisy Measurements", Preprints JACC, p. 364 (1968).
5. H. J. Kushner, "Approximation to Optimal Nonlinear Filters", IEEE Trans. on Automatic Control, Vol. AC-12, p. 546 (1967).
6. R. P. Wishner, J. A. Tabsczynski and M. Athans, "A Comparison of Three Nonlinear Filters", Automatica, Vol. 5, p. 487 (1969).

7. C. B. Chand and H. R. Martens, "A Second Order Filter for State Estimation in Nonlinear Discrete Systems", Seventh Annal. Allerton. Conference on Circuit and System Theory, (1969).

8. Laning and Battin, Random Processes in Automatic Control, McGraw-Hill (1956).

9. S. Tsuji and K. Kumamaru, "System Identifications and Function Learnings by a Nonlinear Filter", Memoirs of Faculty of Eng., Kyushu University, Fukuoka, Japan, Vol. 30, No. 4, (1971).

A LINEAR FILTER FOR DISCRETE SYSTEMS WITH

CORRELATED MEASUREMENT NOISE

Tzyh Jong Tarn* and John Zaborszky†

Washington University

Saint Louis, Missouri, U.S.A.

ABSTRACT

This paper introduces an optimal linear filter for discrete systems with correlated measurement noise by generalized least square method which is novel in its structure, its derivation and its simplicity. The equations reduce to the standard Kalman filter equations when the measurement noise is independent. The new filter avoids the increased order and other complexities of previously proposed methods particularly those based on augmented state and differencing approaches.

I. INTRODUCTION

The mathematical solution to the problem of linear autonomous discrete time filters for correlated noise has been known since 1960, see Kalman [1]. His method, known as the state augmentation method, involves the construction of a shaping filter with independent noise input. The dynamics of the shaping filter are included as a part of the system dynamics, and its states are estimated

This research was supported in part by National Science Foundation Grant Nos. GK-4394, GK-5570 and by the Air Force Office of Scientific Research under Themis Grant #F44620-69-C-0116.

*Assistant Professor, School of Engineering and Applied Science.

†Professor and Chairman of Control Systems Science and Engineering. Fellow IEEE.

along with the system states. In practice, state augmentation requires an increase in the state vector dimensions, such dimensional increase may be prohibitive because the computer storage space is limited. Also it has long been recognized that the numerical solutions of the Riccati equations of the augmented states filter are difficult to obtain because of the fact that this Riccati equation for the correlated noise case is not well conditioned and the solution can be a singular matrix.

Much effort has been expanded by many investigators to alleviate the ill-conditioned computations of the augmented state approach by reducing the number of state variables estimated and reduce the order of Riccati equation [2-6]. As pointed out by Bucy et al. [8], these papers have all had one idea in common, using differences (time derivatives in continuous time) of consecutive measurements. This results in a modified "measurement" equation which contains independent noise and hence allows the use of Kalman filtering but there remains a complex supplemental estimation problem. Bucy [11, Chapter 9] studied this problem; his approach is essentially based on Kalman's [1] orthogonal projections. In his derivation it is required to estimate both the states and the correlated measurement noises, then eliminating the estimate of the noises from the measurement equation. More recently, Johnson [7] obtained a filter for correlated measurement noise by solving the matrix Wiener-Hopf equations, which is not strictly optimal, yet the resulting equations are very complicated. Bucy, Rappaport and Silverman [8,9] analyzed the correlated noise filtering problem for time invariant systems by a mathematical analysis of invariant directions of the Riccati equation of the augmented states.

Here in this paper a filter which accounts for correlated measurement noise is derived in a way that requires neither the augmentation by a shaping filter of the filter dynmmic model nor the concepts of differencing of successive measurements. The filter equations are obtained by generalized least square method to minimize the quadratic form of the measurement errors weighted by the covariance matrix of the measurement noises.* This is an extension of the weighted regression analysis used by the authors in [10]. The derivation is straightforward and simple. It is not an approximation. The filter equations degenerate to the basic Kalman filter equations when it is assumed that the measurement noise is independent. Judging from the wide applications of the Kalman filter, the filter equations derived here should be very useful.

---

*At least one unsuccessful effort to solve the problem by this type of approach has been reported in the literature [11, Appendix A].

II. THE PROBLEM

Given: A discrete time plant or signal model

$$\underline{x}_{n+1} = \underline{\Phi}(t_{n+1},t_n)\underline{x}_n+\underline{u}_n \;, \qquad \underline{x}_n = \underline{x}(t_n) \;, \tag{1}$$

the measurement

$$\underline{y}_n = \underline{H}_n\underline{x}_n + \underline{w}_n \;, \tag{2}$$

and the measurement noise model

$$\underline{w}_{n+1} = \underline{S}(t_{n+1},t_n)\underline{w}_n + \underline{r}_{n+1} \tag{3}$$

where $\underline{x} \in R_i$ is the state vector, $\underline{y} \in R_j$ the measurement vector, $\underline{w} \in R_j$ the measurement noise vector, $\underline{\Phi}(t_{n+1},t_n)$ is the state transition matrix (ixi), $\underline{H}_n$ the measurement matrix (jxi), and $\underline{S}(t_{n+1},t_n)$ is the measurement noise process transition matrix (jxj). The $\underline{\Phi}$, $\underline{H}$ system is observable. $\underline{u}_n$ and $\underline{r}_n$ are independent zero mean random sequences with finite variances

$$\begin{aligned} &E\underline{u}_n = 0 \;, \quad E\underline{r}_n = 0 \quad \text{for all } n \;, \\ &E[\underline{u}_n\underline{u}_n'] = \underline{Q}_n \;, \quad E[\underline{r}_n\underline{r}_n'] = \underline{R}_n \;, \\ &E\underline{u}_n\underline{u}_m' = 0, \quad E\underline{r}_n\underline{r}_m' = 0 \quad \text{if } n\neq m \text{ and } E\underline{u}_n\underline{r}_m'=0 \text{ for any } n \text{ and } m \end{aligned} \tag{4}$$

where E is the expectation operator; the prime denotes the transpose.

Remark. The correlated noise of the measurements has been simulated (3) with an auxiliary linear dynamic system with independent noise inputs (sometimes called a "shaping filter"). But this dynamic system will not be adjoined to the signal model (1). This presents a point of departure from the augmented state approach.

Find:

$$\hat{\underline{x}}_n = E[\underline{x}_n|\underline{y}_k,\ 1 \leq k \leq n;\ \hat{\underline{x}}_1,\ \underline{P}(1|1);\ \hat{\underline{w}}_1,\ E[\underline{w}_1\underline{w}_1']] \text{ and} \tag{5}$$

$$\underline{P}(n|n) = E(\underline{x}_n-\hat{\underline{x}}_n)(\underline{x}_n-\hat{\underline{x}}_n)' \tag{6}$$

that is to find the expected value of the state $\underline{x}_n$ conditioned on the set of all past measurements, $\underline{y}_k$, and on the knowledge of

$$\begin{aligned} &E\underline{x}_1 = \hat{\underline{x}}_1, \quad E(\underline{x}_1-\hat{\underline{x}}_1)(\underline{x}_1-\hat{\underline{x}}_1)' = \underline{P}(1|1) \\ &E\underline{w}_1 = 0 \;, \quad E\underline{w}_1\underline{w}_1' = \underline{R}_1 \end{aligned} \tag{7}$$

$\underline{x}_1$ and $\underline{w}_1$ are assumed to be finite variance random processes, independent of $\underline{u}_n$ and $\underline{r}_n$ and $\underline{x}_1$ is independent of $\underline{w}_1$.

III. THEOREM

For the problem and assumptions stated in equations (1-7) the optimal estimate $\hat{\underline{x}}_{n+1}$ of $\underline{x}_{n+1}$ and the covariance matrix $\underline{P}(n+1|n+1)$ of $\underline{x}_{n+1}$ given the measurements $\underline{y}_0, \underline{y}_1, \ldots, \underline{y}_{n+1}$ is defined by the following recursive set of equations:

$$\hat{\underline{x}}_{n+1} = \underline{\Phi}(t_{n+1},t_n)\hat{\underline{x}}_n + \underline{K}_{n+1} [\underline{y}_{n+1} - \underline{H}_{n+1}\underline{\Phi}(t_{n+1},t_n)\hat{\underline{x}}_n - \underline{S}(t_{n+1},t_n)(\underline{y}_n - \underline{H}_n\hat{\underline{x}}_n)] \quad (8)$$

$$\underline{P}(n+1|n+1) = \underline{P}(n+1|n) - \underline{K}_{n+1} [\underline{H}_{n+1}\underline{P}(n+1|n) - \underline{S}(t_{n+1},t_n)\underline{H}_n\underline{P}(n|n)\underline{\Phi}'(t_{n+1},t_n)] \quad (9)$$

$$\begin{aligned}\underline{K}_{n+1} = {} & [\underline{P}(n+1|n)\underline{H}'_{n+1} - \underline{\Phi}(t_{n+1},t_n)\underline{P}(n|n)\underline{H}'_n\underline{S}'(t_{n+1},t_n)] \\ & [\underline{R}_{n+1} + \underline{H}_{n+1}\underline{P}(n+1|n)\underline{H}'_{n+1} - \underline{H}_{n+1}\underline{\Phi}(t_{n+1},t_n)\underline{P}(n|n)\underline{H}'_n\underline{S}'(t_{n+1},t_n) \\ & - \underline{S}(t_{n+1},t_n)\underline{H}_n\underline{P}(n|n)\underline{\Phi}(t_{n+1},t_n)\underline{H}'_{n+1} \\ & + \underline{S}(t_{n+1},t_n)\underline{H}_n\underline{P}(n|n)\underline{H}'_n\underline{S}'(t_{n+1},t_n)]^{-1}\end{aligned} \quad (10)$$

$$\underline{P}(n+1|n) = \underline{\Phi}(t_{n+1},t_n)\underline{P}(n|n)\underline{\Phi}'(t_{n+1},t_n) + \underline{Q}_n \quad (11)$$

Remark: To start the filter we may take $\underline{y}_1 = \underline{y}_2$ if the sampling is fast enough compared to the system's dynamics. Otherwise we could let $\underline{y}_1 = \underline{0}$.

IV. PROOF

For simplicity the proof will be first given for the simplified case.

$$\underline{x}_{n+1} = \underline{\Phi}(t_{n+1},t_n)\underline{x}_n \,, \quad (12)$$

$$\underline{y}_n = \underline{H}_n\underline{x}_n + \underline{w}_n \,, \quad (13)$$

$$\underline{w}_{n+1} = \underline{S}(t_{n+1},t_n)\underline{w}_n + \underline{r}_{n+1} \quad (14)$$

The results will be subsequently generalized to include $\underline{u}$ as in (1), (2) and (3).

When equation (12) is substituted into equation (13) and it is written out for the first n measurement points, the resulting set of equations can be condensed into

$$\underline{z}_n = \underline{B}_n\underline{x}_n + \underline{v}_n \tag{15}$$

where

$$\underline{z}_n = \begin{bmatrix} \underline{y}_1 \\ \underline{y}_2 \\ \cdot \\ \cdot \\ \cdot \\ \underline{y}_n \end{bmatrix}, \quad \underline{v}_n = \begin{bmatrix} \underline{w}_1 \\ \underline{w}_2 \\ \cdot \\ \cdot \\ \cdot \\ \underline{w}_n \end{bmatrix}, \quad \underline{B}_n = \begin{bmatrix} \underline{H}_1\underline{\Phi}(t_1,t_n) \\ \underline{H}_2\underline{\Phi}(t_2,t_n) \\ \cdot \\ \cdot \\ \cdot \\ \underline{H}_n \end{bmatrix} \tag{16}$$

and it can be shown (Appendix 1) from equation (14)

$$\underline{\Sigma}_n^{-1} = (E\underline{v}_n\underline{v}_n')^{-1} =$$

$$\begin{bmatrix} \underline{R}_1^{-1}+\underline{S}'(t_2,t_1)\underline{R}_2^{-1}\underline{S}(t_2,t_1), & -\underline{S}'(t_2,t_1)\underline{R}_2^{-1} & \cdots & 0 \\ -\underline{R}_2^{-1}\underline{S}(t_2,t_1) & & & \\ 0 & \ddots & & \vdots \\ \vdots & & & -\underline{S}'(t_n,t_{n-1})\underline{R}_n^{-1} \\ 0 & \cdots & -\underline{R}_n^{-1}\underline{S}(t_n,t_{n-1}), & \underline{R}_n^{-1} \end{bmatrix} \tag{17}$$

So that the generalized least square estimate $\hat{\underline{x}}_n$ of $\underline{x}_n$ is obtained by minimizing with respect to $\underline{x}_n$ the following

$$I_n = (\underline{B}_n\underline{x}_n - \underline{z}_n)'\underline{\Sigma}_n^{-1}(\underline{B}_n\underline{x}_n - \underline{z}_n) \tag{18}$$

Using calculus of extrema,

$$\hat{\underline{x}}_n = (\underline{B}_n'\underline{\Sigma}_n^{-1}\underline{B}_n)^{-1}\underline{B}_n'\underline{\Sigma}_n^{-1}\underline{z}_n \ . \tag{19}$$

After a new measurement $\underline{y}_{n+1}$ is added, the corresponding results can be obtained

$$\underline{z}_{n+1} = \underline{B}_{n+1}\underline{x}_{n+1} + \underline{v}_{n+1} \tag{20}$$

and

$$\hat{\underline{x}}_{n+1} = (\underline{B}_{n+1}'\underline{\Sigma}_{n+1}^{-1}\underline{B}_{n+1})^{-1}\underline{B}_{n+1}'\underline{\Sigma}_{n+1}^{-1}\underline{z}_{n+1} \tag{21}$$

where

$$\underline{z}_{n+1} = \begin{bmatrix} \underline{z}_n \\ \underline{y}_{n+1} \end{bmatrix}, \quad \underline{v}_{n+1} = \begin{bmatrix} \underline{v}_n \\ \underline{w}_{n+1} \end{bmatrix}, \quad \underline{B}_{n+1} = \begin{bmatrix} \underline{B}_n \underline{\Phi}(t_n, t_{n+1}) \\ \underline{H}_{n+1} \end{bmatrix} \tag{22}$$

and from Appendix 1

$$\underline{\Sigma}_{n+1}^{-1} = (E\underline{v}_{n+1}\underline{v}'_{n+1})^{-1} = \left[\begin{array}{c|c} \underline{\Sigma}_n^{-1} & 0 \\ \hline 0 & \underline{R}_{n+1}^{-1} \end{array}\right]$$

$$+ \left[\begin{array}{cc|c} 0 & 0 & 0 \\ 0 & \underline{S}'(t_{n+1}, t_n)\underline{R}_{n+1}^{-1}\underline{S}(t_{n+1}, t_n) & -\underline{S}'(t_{n+1}, t_n)\underline{R}_{n+1}^{-1} \\ \hline 0 & -\underline{R}_{n+1}^{-1}\underline{S}(t_{n+1}, t_n) & 0 \end{array}\right] \tag{23}$$

Now writing (21) in the form

$$\underline{B}'_{n+1}\underline{\Sigma}_{n+1}^{-1}\underline{B}_{n+1}\hat{\underline{x}}_{n+1} = \underline{B}'_{n+1}\underline{\Sigma}_{n+1}^{-1}\underline{z}_{n+1} \tag{24}$$

and expanding, the following is obtained:

$$[\underline{\Phi}'(t_n, t_{n+1})\underline{B}'_n\underline{\Sigma}_n^{-1}\underline{B}_n\underline{\Phi}(t_n, t_{n+1}) + \underline{M}'_{n+1}\underline{R}_{n+1}^{-1}\underline{M}_{n+1}]\ \hat{\underline{x}}_{n+1}$$

$$= \underline{\Phi}'(t_n, t_{n+1})\underline{B}'_n\underline{\Sigma}_n^{-1}\underline{z}_n + \underline{H}'_{n+1}\underline{R}_{n+1}^{-1}\underline{y}_{n+1} \tag{25}$$

where

$$\underline{M}_{n+1} = \underline{H}_{n+1} - \underline{S}(t_{n+1}, t_n)\underline{H}_n\underline{\Phi}^{-1}(t_{n+1}, t_n) \ . \tag{26}$$

It can be shown (Appendix 2) that

$$\underline{P}(n|n) = E[\underline{x}_n - \hat{\underline{x}}_n)(\underline{x}_n - \hat{\underline{x}}_n)'|\underline{y}_k, \quad 1 \leqq k \leqq n]$$

$$= (\underline{B}'_n\underline{\Sigma}_n^{-1}\underline{B}_n)^{-1} \tag{27}$$

$$\underline{P}(n+1|n) = E[(\underline{x}_{n+1} - \hat{\underline{x}}_{n+1})(\underline{x}_{n+1} - \hat{\underline{x}}_{n+1})'|\underline{y}_k, \ 1 \leqq k \leqq n]$$

$$= \underline{\Phi}(t_{n+1}, t_n)\underline{P}(n|n)\underline{\Phi}'(t_{n+1}, t_n) \tag{28}$$

$$\underline{P}(n+1|n+1) = E[\underline{x}_{n+1} - \hat{\underline{x}}_{n+1})(\underline{x}_{n+1} - \hat{\underline{x}}_{n+1})'|\underline{y}_k, \ 1 \leqq k \leqq n+1]$$

$$= [\underline{\Phi}'(t_n,t_{n+1})\underline{B}_n'\underline{\Sigma}_n^{-1}\underline{B}_n\underline{\Phi}(t_n,t_{n+1})+\underline{M}_{n+1}'\underline{R}_{n+1}^{-1}\underline{M}_{n+1}]^{-1}$$

$$= [\underline{P}^{-1}(n+1|n)+\underline{M}_{n+1}'\underline{R}_{n+1}^{-1}\underline{M}_{n+1}]^{-1} \ . \tag{29}$$

From (26), (28), and (29) also

$$\underline{P}(n+1|n+1) = \underline{P}(n+1|n)-\underline{P}(n+1|n)\underline{M}_{n+1}'$$

$$\cdot \ (\underline{R}_{n+1}+\underline{M}_{n+1}\underline{P}(n+1|n)\underline{M}_{n+1}')^{-1}\underline{M}_{n+1}\underline{P}(n+1|n) \tag{30}$$

$$= \underline{P}(n+1|n)-\underline{K}_{n+1}[\underline{H}_{n+1}\underline{P}(n+1|n)-\underline{S}(t_{n+1},t_n)\underline{H}_n\underline{P}(n|n)\underline{\Phi}'(t_{n+1},t_n)]$$

where $\underline{K}_{n+1}$ is as given in equation (10). This is easily proven by direct substitution of (29) and (30) into

$$\underline{P}(n+1|n+1)\underline{P}^{-1}(n+1|n+1) = I \ . \tag{31}$$

Utilizing these notations in (25), the resulting equation is

$$\hat{\underline{x}}_{n+1} = \underline{\Phi}(t_{n+1},t_n)\hat{\underline{x}}_n \tag{32}$$

$$+ \underline{K}_{n+1}[\underline{y}_{n+1}-\underline{H}_{n+1}\underline{\Phi}(t_{n+1},t_n)\hat{\underline{x}}_n-\underline{S}(t_{n+1},t_n)(\underline{y}_n-\underline{H}_n\hat{\underline{x}}_n)] \ .$$

Remark 1. From Aitken's generalized Gauss-Markov theorem [12] which states that "In the generalized linear regression model the Best Linear Unbiased Estimator is the generalized least-square estimator; it immediately follows that this filter is the optimal filter."

Remark 2. From (32) it is required for the existence of this filter that the inverse of

$$\underline{R}_{n+1}+\underline{H}_{n+1}\underline{P}(n+1|n)\underline{H}_{n+1}'-\underline{H}_{n+1}\underline{\Phi}(t_{n+1},t_n)\underline{P}(n|n)\underline{H}_n'\underline{S}'(t_{n+1},t_n)$$

$$-\underline{S}(t_{n+1},t_n)\underline{H}_n\underline{P}(n|n)\underline{\Phi}'(t_{n+1},t_n)\underline{H}_{n+1}'$$

$$+\underline{S}(t_{n+1},t_n)\underline{H}_n\underline{P}(n|n)\underline{H}_n'\underline{S}'(t_{n+1},t_n) \tag{33}$$

exists.

It is known that the covariance matrix of a random vector is positive definite if and only if the elements of the random vector are not linearly dependent random variables [12, p. 87]. It can be shown that (33) is the extrapolated covariance matrix of $\underline{H}_{n+1}\underline{x}_{n+1}+\underline{S}(t_{n+1},t_n)\underline{w}_n+\underline{r}_{n+1}$. Now suppose that $\underline{R}_{n+1}$ is invertable. This implies that the elements of $\underline{r}_{n+1}$ are not linearly dependent random variables. However $\underline{r}_{n+1}$ is independent of $\underline{x}_{n+1}$ and $\underline{w}_n$ even though $\underline{H}_{n+1}\underline{x}_{n+1}+\underline{S}(t_{n+1},t_n)\underline{w}_n$ is a linear function of $\underline{x}_{n+1}$ and $\underline{w}_n$. It follows that the elements of $\underline{H}_{n+1}\underline{x}_{n+1}+\underline{S}(t_{n+1},t_n)\underline{w}_n+\underline{r}_{n+1}$ are not linearly dependent random variables. Hence (33) is invertable. This shows that the existence of R_{n+1}^{-1} guarantees the existence of the filter.

Remark 3. The presence of an independent noise term $\underline{u}_n$ on the right hand side of (12) as in (1) is easily shown to result simply in the appearance of an additive term $\underline{Q}_n$ on the right hand side of (28) and in all $\underline{P}(n+1|n)$ terms in (30) and (32). This is obvious from the fact that the expectation in (28) is extrapolated from data available at t_n so the addition of $\underline{u}_n$ which is independent of the observations up to t_n will simply increase the covariance of the extrapolated $\underline{x}$ by $\underline{Q}_n$.

Remark 4. If the independent noise processes $\underline{u}_n$ and $\underline{r}_n$ are normally distributed, then the estimate $\hat{\underline{x}}$ of $\underline{x}$ is also the maximum likelihood and minimax estimate.

V. CONCLUSION

A filter is developed for correlated measurement noise by generalized least square method to eliminate the potential ill-conditioned computation and increased dimensionality of the augmented state approach. It excells over previous approaches not only by its simplicity but also by being a complete optimal solution. The filter equations reduce to the basic Kalman filter equations when the measurement noise is independent. Judging from the wide application of the Kalman filter, this filter could prove very useful.

APPENDIX 1

The measurement noise model is

$$\underline{w}_{n+1} = \underline{S}(t_{n+1},t_n)\underline{w}_n + \underline{r}_{n+1} \tag{A1}$$

where $\underline{w}_1$ is independent of $\underline{r}_n$ for all n and

$$E\underline{w}_1 = 0, \quad E\underline{w}_1\underline{w}_1' = \underline{R}_1 . \tag{A2}$$

$$E\underline{r}_n = 0, \quad E[\underline{r}_n\underline{r}_n'] = \underline{R}_n; \quad \text{for all } n . \tag{A3}$$

$$E\underline{r}_n\underline{r}_m' = 0 \text{ if } n \neq m . \tag{A4}$$

For the first n measurement points, the set of equations (A1) can be condensed into

$$\underline{J}_n\underline{v}_n = \underline{c}_n \tag{A5}$$

where

$$\underline{J}_n = \begin{bmatrix} \underline{I} & 0 & & 0 \\ -\underline{S}(t_2,t_1) & & & 0 \\ 0 & \ddots & & 0 \\ \vdots & & & \vdots \\ 0 & 0 \ldots & -\underline{S}(t_n,t_{n-1}) & \underline{I} \end{bmatrix}, \quad \underline{v}_n = \begin{bmatrix} \underline{w}_1 \\ \underline{w}_2 \\ \\ \vdots \\ \underline{w}_n \end{bmatrix}, \quad \underline{c}_n = \begin{bmatrix} \underline{w}_1 \\ \underline{r}_2 \\ \\ \vdots \\ \underline{r}_n \end{bmatrix} \tag{A6}$$

The assumptions about $\underline{w}_1$ and $\underline{r}_n$ ensure that $\underline{v}_n$ has expected value zero

$$E\underline{v}_n = 0 \ . \tag{A7}$$

Now, from (A5)

$$E[\underline{J}_n\underline{v}_n\underline{v}_n'\underline{J}_n'] = \underline{J}_n\underline{\Sigma}_n\underline{J}_n' = \begin{bmatrix} \underline{R}_1 & & 0 \\ & \underline{R}_2 & \\ 0 & & \ddots & \underline{R}_n \end{bmatrix} . \tag{A8}$$

From (A8)

$$[\underline{J}_n\underline{\Sigma}_n\underline{J}_n']^{-1} = \begin{bmatrix} \underline{R}_1^{-1} & & 0 \\ & \underline{R}_2^{-1} & \\ 0 & & \ddots & \underline{R}_n^{-1} \end{bmatrix} . \tag{A9}$$

Hence

$$\underline{\Sigma}_n^{-1} = \underline{J}_n' \begin{bmatrix} \underline{R}_1^{-1} & & 0 \\ & \underline{R}_2^{-1} & \\ 0 & & \ddots & \underline{R}_n^{-1} \end{bmatrix} \underline{J}_n \ . \tag{A10}$$

Expanding (A10), equation (17) is obtained. Similarly,

$$\underline{\Sigma}_{n+1}^{-1} = (E\underline{v}_{n+1}\underline{v}_{n+1}')^{-1} \tag{A11}$$

$$= \begin{bmatrix} \underline{R}_1^{-1}+\underline{S}'(t_2,t_1)\underline{R}_2^{-1}\underline{S}(t_2,t_1), & -\underline{S}'(t_2,t_1)\underline{R}_2^{-1}, & \cdots & & 0 \\ -\underline{R}_2^{-1}\underline{S}(t_2,t_1) & , & & & \vdots \\ 0 & & \ddots & & \\ \vdots & & & \underline{R}_n^{-1}+\underline{S}'(t_{n+1},t_n)\underline{R}_{n+1}^{-1}S(t_{n+1},t_n), & -\underline{S}'(t_{n+1},t_n)\underline{R}_{n+1}^{-1} \\ 0 & \cdots & & -\underline{R}_{n+1}^{-1}\underline{S}(t_{n+1},t_n), & \underline{R}_{n+1}^{-1} \end{bmatrix}$$

From (17) and (A11) it is easy to see that

$$\underline{\Sigma}_{n+1}^{-1} = \left[\begin{array}{c|c} \underline{\Sigma}_n^{-1} & 0 \\ \hline 0 & \underline{R}_{n+1}^{-1} \end{array}\right] \tag{A12}$$

$$+ \left[\begin{array}{cc|c} 0 & 0 & 0 \\ 0 & \underline{S}'(t_{n+1},t_n)\underline{R}_{n+1}^{-1}\underline{S}(t_{n+1},t_n) & -\underline{S}'(t_{n+1},t_n)\underline{R}_{n+1}^{-1} \\ \hline 0 & -\underline{R}_{n+1}^{-1}\underline{S}(t_{n+1},t_n) & 0 \end{array}\right];$$

this is equation (23).

APPENDIX 2

The signal model and measurement equations are

$$\underline{x}_{n+1} = \underline{\Phi}(t_{n+1},t_n)\underline{x}_n \tag{B1}$$

$$\underline{y}_n = \underline{H}_n\underline{x}_n + \underline{w}_n \tag{B2}$$

$$\underline{w}_{n+1} = \underline{S}(t_{n+1},t_n)\ \underline{w}_n + \underline{r}_{n+1}\ , \tag{B3}$$

and from (15) and (19),

$$\underline{B}_n\underline{x}_n = \underline{z}_n-\underline{v}_n \tag{B4}$$

$$\hat{\underline{x}}_n = (\underline{B}_n'\underline{\Sigma}_n^{-1}\underline{B}_n)^{-1}\underline{B}_n'\underline{\Sigma}_n^{-1}\underline{z}_n\ . \tag{B5}$$

Multiplying (B4) by $(\underline{B}_n'\underline{\Sigma}_n^{-1}\underline{B}_n)^{-1}\underline{B}_n'\underline{\Sigma}_n^{-1}$ and using the fact of (B5), the following is obtained:

$$\underline{x}_n = \hat{\underline{x}}_n-(\underline{B}_n'\underline{\Sigma}_n^{-1}\underline{B}_n)^{-1}\underline{B}_n'\underline{\Sigma}_n^{-1}\underline{v}_n \tag{B6}$$

Hence,

$$E\underline{\tilde{x}}_n = E[\underline{\hat{x}}_n-\underline{x}_n] = E(\underline{B}_n'\underline{\Sigma}_n^{-1}\underline{B}_n)^{-1}\ \underline{B}_n'\underline{\Sigma}_n^{-1}\underline{v}_n = 0\ . \tag{B7}$$

Now the covariance matrix of $\underline{x}_n$ given all the information up to time t_n will be

$$\begin{aligned}E[\underline{\tilde{x}}_n\underline{\tilde{x}}_n'|\underline{y}_k,\ 1\leq k\leq n] &= E[(\underline{B}_n'\underline{\Sigma}_n^{-1}\underline{B}_n)^{-1}\underline{B}_n'\underline{\Sigma}_n^{-1}\underline{v}_n\underline{v}_n'\underline{\Sigma}_n^{-1}\underline{B}_n(\underline{B}_n'\underline{\Sigma}_n^{-1}\underline{B}_n)^{-1'}]\\ &= (\underline{B}_n'\underline{\Sigma}_n^{-1}\underline{B}_n)^{-1}(\underline{B}_n'\underline{\Sigma}_n^{-1}\underline{B}_n)'(\underline{B}_n'\underline{\Sigma}_n^{-1}\underline{B}_n)^{-1'}\\ &= (\underline{B}_n'\underline{\Sigma}_n^{-1}\underline{B}_n)^{-1} = \underline{P}(n|n)\ ,\end{aligned} \tag{B8}$$

which is equation (27). Similarly it can be shown that

$$\begin{aligned}&E[\underline{\tilde{x}}_{n+1}\underline{\tilde{x}}_{n+1}'|\underline{y}_k,\ 1 \leq k \leq n+1] = (\underline{B}_{n+1}'\underline{\Sigma}_{n+1}^{-1}\underline{B}_{n+1})^{-1}\\ &= [\underline{\Phi}'(t_n,t_{n+1})\underline{B}_n'\underline{\Sigma}_n^{-1}\underline{B}_n\underline{\Phi}(t_n,t_{n+1})+\underline{M}_{n+1}'\underline{R}_{n+1}^{-1}\underline{M}_{n+1}]^{-1}\\ &= \underline{P}(n+1|n+1)\end{aligned} \tag{B9}$$

which is equation (29) and where $\underline{M}_{n+1}=\underline{H}_{n+1}-\underline{S}(t_{n+1},t_n)\underline{H}_n\underline{\Phi}^{-1}(t_{n+1},t_n)$. From (B1) it is clear that

$$E[\underline{\tilde{x}}_{n+1}\underline{\tilde{x}}_{n+1}'|\underline{y}_k,\ 1 \leq k \leq n] = \underline{\Phi}(t_{n+1},t_n)\underline{P}(n|n)\underline{\Phi}'(t_{n+1},t_n) \tag{B10}$$

which is equation (28).

REFERENCES

1. Kalman, R. E., "A New Approach to Linear Filtering and Prediction Problems," Trans. ASME, J. Basic Eng., Vol. 82, pp. 35-44, March 1960.
2. Bryson Jr., A. E. and Johansen, D. E., "Linear Filtering for Time-Varying Systems Using Measurements Containing Colored Noise," IEEE Trans. on Automatic Control, pp. 4-10, January 1965.
3. Bucy, R. S., "Optimal Filtering for Correlated Noise," Journal of Math. Anal. and Appl. 20, pp. 1-8, 1967.
4. Bryson Jr., A. E. and Henrikson, L. J., "Estimation Using Sampled Data Containing Sequentially Correlated Noise," J. Spacecraft and Rocket, Vol. 5, No. 6, pp. 662-665, June 1968.
5. Deyst, J. J., "A Derivation of the Optimum Continuous Linear Estimator for Systems with Correlated Measurement Noise," AIAA Journal, Vol. 7, No. 11, pp. 2116-2119, November 1969.
6. Kailath, T. and Geesey, R., "Covariance Factorization," Proceedings of 2nd Asilomar Systems Conference, 1968.

7. Johnson, D. J., "A General Linear Sequential Filter," Paper Presented at 1970 JACC, Atlanta, Georgia, June 24-26, 1970.

8. Bucy, R. S., Rappaport, D. and Silverman, L. M., "Correlated Noise Filtering and Invariant Directions for the Riccati Equation," To be published in IEEE Trans. on Automatic Control.

9. Bucy, R. S., Rappaport, D. and Silverman, L. M., "A New Approach to Singular Discrete Time Filtering and Control," Proceedings of the 3rd Hawaii Conf. on Systems Science, Honolulu, Hawaii, February 1970.

10. Tarn, T. J. and Zaborszky, J., "A Practical, Nondiverging Filter," AIAA Journal, Vol. 8, No. 6, pp. 1127-1133, June 1970.

11. Bucy, R. S. and Joseph, P. D., Filtering for Stochastic Processes with Applications to Guidance, Interscience Publishers, New York, 1968.

12. Goldberger, A. S., Econometric Theory, John Wiley and Sons, Inc., New York, 1964, pp. 232-233.

PART II
LEARNING PROCESS AND LEARNING CONTROL

STOCHASTIC LEARNING BY MEANS OF CONTROLLED STOCHASTIC PROCESSES

Seigo Kano

Kyushu University

Fukuoka, Japan

1. INTRODUCTION

There are two categories of the Learning Processes. One is machine Learning and the other is organism Learning. Pattern recognition by a machine is an example of the first category. The Learning Process of a system which is composed of man and machine is an example of the second category. In both categories, the responses of organisms or machines are fundamental. We have many ways to evaluate the responses in each Learning Process. In this study we shall consider a system of quality controls as a Learning Process for organism. Quality controls of manufactured products, which are considered as an effective system of quality maintenance and quality improvement, are essentially the same as the Learning Processes performed by management systems or organisms. The Controlled Stochastic Process, which has been defined as a mathematical model of quality control by T. Kitagawa in 1959, is generalized to cover Stochastic Learning Process in psychology. The convergence theorem corresponding to M. F. Norman's learning model in psychology is proven in the linear case where Martingale transforms play a fundamental role. Since our Controlled Stochastic Process is defined with finite memories, the corresponding Stochastic Learning Process is defined with finite memories also. It is explained that certain dependencies between the state of the subject and the aim of learning is effective for the Learning Process on the analogy of the experimenter-subject controlled events in psychology.

2. LINEAR CONTROLLED STOCHASTIC PROCESSES (L.C.S.P.)

Let $U = (U_1, U_2,...)$ be any discrete parameter Stochastic Process and is supposed to be unknown. Let $A=(\xi_1, \xi_2,...)$ be any real vector which is called target vector and is given in advance. We consider a set of transformations by which U is successively transformed so as to make $E\{U\}$ come near A. If $U_1=u_1$ is observed then U_i is transformed to

$$U_i^{(2)} = U_i + c_{11}(\xi_1-u_1), \quad i=2,3,\ldots, \tag{2.1}$$

where c_{11} is any real number. In general, if $U_m^{(m)}=u_m^{(m)}$ is observed then $U_i^{(m)}$ is transformed to

$$U_i^{(m+1)} = U_i^{(m)}+\sum_{k=1}^{m} c_{mk}(\xi_k-u_k^{(k)}), \quad i=m+1,\ m+2,\ldots, \tag{2.2}$$

where c_{mk}, $k=1,2,\ldots, m$, are any real numbers. Consequently, we obtain the stochastic process $V=(U_1,\ U_2^{(2)},\ U_3^{(3)},\ldots)$ which is called L.C.S.P. The real infinite matrix $C=(c_{ij})$, where $c_{ij}=0$, $i<j$, is chosen to make $E\{V\}$ come near A in some sense. The following theorem which was given by T. Kitagawa [7] is fundamental.

Theorem 2.1. If L.C.S.P. V is obtained from U by the matrix C then

$$U_{m+1}^{(m+1)} = U_{m+1}+ \sum_{k=1}^{m} e_{mk}(\xi_k-U_k) \tag{2.3}$$

where

$$e_{mk} = \sum_{i=k}^{m} c_{ik}-e_{mm}\sum_{i=k}^{m-1} c_{ik}-\cdots -e_{mk+1}c_{kk}, \quad k=1,2,\ldots,m$$

We illustrate the Theorem by some examples.

Example 1. Let U be independent stochastic process with $E\{U_i\}=\mu$ and $\mathrm{Var}\ \{U_i\}=\sigma^2$, $i=1,2,\ldots$, and C is given by

$$\begin{pmatrix} 1 & 0 & 0 & 0 & 0 & . & . & . \\ 1/2 & 1/2 & 0 & 0 & 0 & . & . & . \\ 0 & 1/2 & 1/2 & 0 & 0 & . & . & . \\ 0 & 0 & 1/2 & 1/2 & 0 & . & . & . \\ . & . & . & . & . & . & . & . \end{pmatrix}.$$

Let A be $(\xi,\ \xi,\ .\ .\ .\ )$. Then

$$e_{m,m-k}=\sqrt{8/7}\ (1/\sqrt{2})^k \sin\ (k\theta+\alpha), \quad k=0,1,\ldots,m-1,$$

where $\tan\ \alpha=\sqrt{7/5}$ and $\tan\ \theta=\sqrt{7}$. Therefore we obtain

$$\lim_{m\to\infty} E\ \{U_{m+1}^{(m+1)}\}=\xi, \qquad \text{and} \qquad \lim_{m\to\infty} \mathrm{Var}\ \{U_{m+1}^{(m+1)}\}=2\sigma^2 .$$

Example 2. U and A are the same as given in Example 1. C is given by

$$\begin{pmatrix} 1 & 0 & 0 & 0 & 0 & . & . & . \\ 1/2 & 1/2 & 0 & 0 & 0 & . & . & . \\ 1/2 & 1/3 & 1/3 & 0 & 0 & . & . & . \\ 0 & 1/3 & 1/3 & 1/3 & 0 & . & . & . \\ . & . & . & . & . & . & . & . \end{pmatrix}.$$

Then $e_{m,m-k}$ are given by

$$\alpha x^k + \beta y^k + \gamma z^k, \quad k=0,1,\ldots,m-1,$$

where

$$x = A + B + 2/9, \quad y = A + B + 2/9, \quad z=\bar{y}$$

$$\begin{matrix} A \\ B \end{matrix} = (1/16^2)(-31\pm\sqrt{28341}/9)^{1/3}, \ \omega^3=1,$$

$$\alpha = \frac{(16/27)-(5/9)(y+z)+yz/3}{(x-z)(x-y)},$$

$$\beta = \frac{(16/27)-(5/9(x+z)+xz/3}{(y-x)(y-z)},$$

$$\gamma = \bar{\beta},$$

Therefore, we obtain $\lim_{m\to\infty} E \ U_m^{(m)} = \quad ,$

and $\lim_{m\to\infty} \mathrm{Var} \ U_m^{(m)} = (2.45299\ldots)\sigma^2.$

Example 3. Let C be given by

$$\begin{matrix} 1 \\ 0 \\ 0 \\ \cdot \end{matrix} \begin{pmatrix} 0 & 0 & \cdot & \cdot & \cdot \\ 1/2 & 0 & \cdot & \cdot & \cdot \\ 0 & 1/3 & \cdot & \cdot & \cdot \\ \cdot & \cdot & \cdot & \cdot & \cdot \end{pmatrix}.$$

U and A are the same as given in Example 1. Then we obtain $e_{mk}=1/m$, $k=0,1,\ldots,m-1$, and

$$\lim_{m\to\infty} U_m^{(m)} = U_1+\xi-\mu$$

with probability one. So that $E\{U_m^{(m)}\}=\xi$ and $\lim_{m\to\infty} \mathrm{Var} \ \{U_m^{(m)}\}=\sigma^2$.

3. STOCHASTIC LEARNING PROCESSES (S.L.P.) IN VIEW OF L.C.S.P.

We shall give a formulation of S.L.P. in which a subject and reinforcements are contained. Suppose that a subject is repeatedly exposed to an experimental situation in which various responses are possible and suppose that each such response receives a penalty or reinforsement. Learning of the subject is measured by the states of the subject. We assume that when the subject with state S_n is exposed to an experiment in the n-th step, he has response X_n and then reinforcement Y_n transforms his state S_n to S_{n+1}. This expressed as

$$S_{n+1} = O_n(S_1, X_1, Y_1, \ldots, S_n, X_n, Y_n)\cdot S_n,$$

where the transformation O_n is supposed to be a function of all states, responses and reinforecments in the past. In many Learning Processes, S_n are taken as the correct response probabilities conditioned on the past responses $X_1,\ldots,X_{n-1}$ and reinforcements $Y_1,\ldots,Y_{n-1}$. Now we consider two cases in which transformations O_n are given by $O_n(Y_n)$ and $O_n(X_n, Y_n)$.

Case 1. We consider the case $O_n(S_1, X_1, Y_1, \ldots, S_n, X_n, Y_n) = O_n(Y_n)$. Let random variables X_n be two valued, say correct responses x_{n1} and incorrect responses x_{n2} and we put $P_r\{X_n=x_{n1}\}=P_n$, $n=1,2,\ldots$. Let Y_n, random variables of reinforcements, be two valued, say reinforcements of incorrect responses y_{n2} and we put $\Pi_i=P_r\{Y_n=y_{ni}\}$, $i=1,2$, $n=1,2,\ldots$.

Now we assume that Y_n are independent of $X_1,\ldots,X_n$, $n=1,2,\ldots$. We note that X_n depend upon Y_{n-1}, $n=1,2\ldots$. Hence we have

$$S_{m+1} = P_r\{X_{n+1}=x_{n+1,1} \mid Y_1,\ldots,Y_n\}\ , \quad n=1,2,\ldots, \text{ and } S_1=P_1.$$

When the reinforcement Y_1 is carried out, S_2 is supposed to be

$$S_2 = \alpha p_1+(1-\alpha)y_{1i}\ , \quad i=1,2, \tag{3.1}$$

where $0 \leqq \alpha < 1$, $y_{11}=1$ and $y_{12}=0$. In general, if the first n trials $X_1, Y_1,\ldots,X_n, Y_n$, are carried out then S_{n+1} is supposed to be

$$S_{n+1} = \alpha S_n+(1-\alpha)y_{ni}\ , \quad i=1,2, \tag{3.2}$$

where we put $y_{n1}=1$ and $y_{n2}=0$. Expressions (3.1) and (3.2) are represented by

$$S_{n+1} = \alpha S_n+(1-\alpha)Y_n\ , \quad n=1,2,\ldots . \tag{3.3}$$

Does the S_{n+1} converge as n tends to infinity? We evaluate $E\{S_n\}$.

$$\begin{aligned} E\{S_n\} &= \sum_{Y_1,\ldots,Y_{n-1}} P_r\{X_n=x_{n1} \mid Y_1,\ldots, Y_{n-1}\}\cdot P_r\{Y_1,\ldots, Y_{n-1}\} \\ &= \sum_{Y_1,\ldots,Y_{n-1}} P_r\{X_n=x_{n1}, Y_1,\ldots, Y_{n-1}\} \\ &= P_n\ . \end{aligned}$$

Therefore, taking the expectation of (3.3), we have

$$p_{n+1} = \alpha p_n+(1-\alpha)\Pi_1\ ,$$

which converges to Π_1 as n tends to infinity. This means that learning is effective.

Case 2. Here we have the case $O_n(S_1, X_1, Y_1,\ldots, S_n, X_n, Y_n) = O_n(X_n, Y_n)$. In this case, reinforcements Y_n depend on X_n, that is; if the n-th response of the subject X_n is x_{ni} then the reinforcement Y_n he receives is $y_{n,ij}$, $j=1,2$, $i=1,2$, where $y_{n,ij}$ is reinforcement of x_{nj}. Let Π_{ij} be $P_r\{Y_n=y_{n,ij} \mid x_{ni}\}$, $i,j=1,2$, therefore $\Pi_{i1}+\Pi_{i2}=1$, $i=1,2$. If the n-th response and reinforcement are performed, say $(X_n, Y_n)=(x_{ni}, y_{n,ij})$ then S_{n+1} is given by the following four operators,

$$S_{n+1} = (1-\alpha_{ij})S_n+\alpha_{ij}\delta_{ji}, \quad i,j=1,2, \tag{3.4}$$

where $0 \leqq \alpha_{ij} < 1$ and δ_{ji} is Kronecker δ. In this model we have six parameters α_{ij}, $i,j=1,2$, Π_{11} and Π_{22}.

M. F. Norman [10] obtained the following theorem in his Stochastic Learning models with distance diminishing operators.

Theorem 3.1. For any non-periodic distance deminishing four operators model, S_n converges to a state S_∞ with probability one. For some real numbers $c < \infty$, $\alpha < 1$, and any positive integer n, we have

$$||E\{S_n^{\upsilon}\}-E\{S_\infty^{\upsilon}\}|| \leqq c(\upsilon+1)\alpha^n, \quad \upsilon \geqq 1,$$

where

$$||f(p)|| = \sup_{0\leqq p\leqq 1} |f(p)|+\sup_{p'\neq p} \frac{|f(p)-f(p')|}{|p-p'|}$$

Now we can rewrite (3.4) as follows,

$$S_{n+1} = S_n-(1-\alpha)(S_n-Y_n). \tag{3.5}$$

This is the same form as (2.2) in which U, $U^{(n)}$, C and A are given by

$$U = (S_1, S_1, \ldots), \quad U^{(n)} = (S_n, S_n, \ldots),$$

$$C = \begin{pmatrix} 1-\alpha & 0 & 0 & \cdot & \cdot & \cdot \\ 0 & 1-\alpha & 0 & \cdot & \cdot & \cdot \\ 0 & 0 & 1-\alpha & \cdot & \cdot & \cdot \\ \cdot & \cdot & \cdot & \cdot & \cdot & \cdot \end{pmatrix},$$

and $A = (Y_1, Y_2, \ldots)$. Therefore we have to make generalized form of L.C.S.P. with target stochastic process $\{Y_n\}$.

The expression (3.4) is also written as

$$S_{n+1} = (1-\alpha)S_n+\alpha\delta \quad , \tag{3.6}$$

where α and δ are random variables assuming the values α_{ij}, $i,j=1,2$, and δ_{j1}, $j=0,1$, with some probabilities respectively. Equation (3.6) is reduced to $S_{n+1} = S_n-\alpha(S_n-\delta)$. Again we have generalized form of L.C.S.P. in which C and A matrix and vector of random variables. Now, we shall give the convergence of generalized L.C.S.P. in which we consider the stochastic processes U and V instead of U-A and V-A respectively and we rewrite the expression (2.3) as

$$V_{m+1} = \sum_{k=1}^{m+1} d_{m+1,k}(U_k-U_{k-1}) , \tag{3.7}$$

where we put $d_{m+1,k} = 1-e_{mm}-e_{m,m-1}-\dots-e_{m,k}$, $k=1,2,\dots,m+1$, and $U^{(m+1)}_{m+1}=V_{m+1}$. The transformations from U to V are considered as the generalized form of the Martingale transforms given by D. L. Burkholder [2].

Theorem 3.2. We assume the following five conditions:

(1) $U=(U_1, U_2,\dots)$ is L_1 bounded Martingale where U_i is F_i -measurable, $i=1,2,\dots$.

(2) d_{ij} is bounded and F_i -measurable, $j=1,2,\dots,i$, $i=1,2,\dots$.

(3) $d_{ii}=1$, $i=1,2,\dots$.

(4) $E\{d_{mk} \mid d_{ij}, j=1,2,\dots, ,i=1,2,\dots, m-1\}=d_{m-1,k}$', $k=1,2,\dots,m-1$.

(5) $E\{(d_{mk}-d_{m-1,k})^2 \mid d_{ij}, j \leqq i, i \leqq m-1\}=O(m^{-(4+\delta)})$,

$\delta > 0$, $k=1,2,\dots, m-1$.

Then V converges with probability one.

Proof. Any Martingale process is expressed as the difference of two non-negative Martingale processes, by a result due to Krickeberg [9]. Then we may assume $U \geqq 0$ without any loss of generality.

For any positive number c, we define $\hat{U}_n$ by $\hat{U}_n \equiv -\min(U_n, c)$, $n=1,2,\dots$. Obviously, $\hat{U}_n$ are uniformly bounded.

$$|\hat{U}_n| = \min(U_n, c) \leqq c.$$

Furthermore, $\hat{U}_n$ are semi-Martingales, since

$$E\{\hat{U}_n \mid \hat{U}_1,\dots,\hat{U}_{n-1}\} = -c\, P\{U_n>c \mid U_1,\dots,U_{n-1}\}- \tag{3.8}$$

$$\Big(\sum_{U_n<c} U_n \frac{P\{U_n|U_1,\dots,U_{n-1}\}}{P\{U_n<c \mid U_1,\dots,U_{n-1}\}}\Big)P\{U_n<c \mid U_1,\dots,U_{n-1}\}$$

$$= -c\, P\{U_n\geqq c \mid U_1,\dots,U_{n-1}\}- \sum_{U_n<c} U_n P\{U_n \mid U_1,\dots,U_{n-1}\}$$

$$\geqq -c\, P\{U_n\geqq c \mid U_1,\dots,U_{n-1}\}-c\, P\{U_n<c \mid U_1,\dots,U_{n-1}\} = -c,$$

and the third expression of (3.8) is reduced to

$$E\{\hat{U}_n \mid \hat{U}_1,\dots,\hat{U}_{n-1}\}= -c\, P\{U_n>c \mid U_1,\dots,U_{n-1}\}-$$

$$\Big(\sum_{U_n} - \sum_{U_n\geqq c}\Big)U_n\, P\{U_n \mid U_1,\dots,U_{n-1}\}\geqq -c\, P\{U_n\geqq c \mid U_1,\dots,U_{n-1}\}-$$

$$\sum_{U_n} U_n P\{U_n \mid U_1,\dots,U_{n-1}\}+ \sum_{U_n\geqq c} c\, P\{U_n\geqq c \mid U_1,\dots,U_{n-1}\}$$

$$= - \sum_{U_n} U_n P\{U_n \mid U_1,\dots,U_{n-1}\}=-U_{n-1}$$

Let $\hat{V}$ be the transform of $\hat{U}$, that is,

$$\hat{V}_n = \sum_{k=1}^{n} d_{nk}(\hat{U}_k-\hat{U}_{k-1}) \quad .$$

If $\sup_n |U_n|<c$ then $V=\hat{V}$. From the fact that U is L_1 bounded, the sequence $\{U_n\}$ converges almost everywhere and for some positive constant k,

$$E\{| \lim_{n\to\infty} U_n|\} \leqq k \ .$$

Therefore we have $P\{\sup_n |U_n|<\infty\}=1$, and this implies $V=\hat{V}$ almost everywhere.

Consequently we have to probe the convergence of $\{\hat{V}_n\}$ with probability one. Since $\hat{U}$ is a uniformly bounded semi-Martingale, we may assume $\hat{U}>0$ without any loss of generality. If we put $\hat{d}_n=\hat{U}_n-\hat{U}_{n-1}$, $n=1,2,\ldots$, and $\hat{U}_0=0$, then we see $E\{\hat{U}_{n-1}\hat{d}_n\}\geqq 0$ because $\hat{U}$ is semi-Martingale. Accordingly

$$E\{\hat{U}_n^2\} = E\{(\sum_{k=1}^{n}\hat{d}_k)^2\} \geqq E\{ \sum_{k=1}^{n} \hat{d}_k^2\} \ , \quad n \geqq 1.$$

If $\hat{\hat{U}}_n$, $n=1,2,\ldots$, are defined by

$$\hat{\hat{U}}_n = \sum_{k=1}^{n} [\hat{d}_k-E\{\hat{d}_k \mid \hat{U}_1,\ldots,\hat{U}_{k-1}\}] \quad n=1,2,\ldots,$$

then

$$E\{\hat{\hat{U}}_n \mid \hat{\hat{U}}_1,\ldots,\hat{\hat{U}}_{n-1}\} = E[\hat{d}_n-E\{\hat{d}_n \mid \hat{U}_1,\ldots,\hat{U}_{n-1}\} \mid\hat{\hat{U}}_1,\ldots,\hat{\hat{U}}_{n-1}]+$$

$$\sum_{k=1}^{n-1} E[\hat{d}_k-E\{\hat{d}_k \mid \hat{U}_1,\ldots,\hat{U}_{k-1}\} \mid\hat{\hat{U}}_1 k\ldots,\hat{\hat{U}}_{n-1}] \tag{3.9}$$

$$= \sum_{k=1}^{n-1} [\hat{d}_k-E\{d_k \mid \hat{U}_1,\ldots,\hat{U}_{k-1}\}] = \hat{\hat{U}}_{n-1} \quad .$$

Furthermore, if we put $\hat{\hat{d}}_n=\hat{\hat{U}}_n-\hat{\hat{U}}_{n-1}$, then

$$\begin{aligned} E\{\hat{\hat{d}}_n^2\} &= E\{[\hat{d}_n-E\{\hat{d}_n \mid \hat{U}_1,\ldots,\hat{U}_{n-1}\}]^2\} \\ &= E\{\hat{d}_n^2\}-E\{[E\{\hat{d}_n \mid \hat{U}_1,\ldots,\hat{U}_{n-1}\}]^2\} \leqq E\{\hat{d}_n^2\} \ . \end{aligned} \tag{3.10}$$

The relations (3.9) and (3.10) make $\hat{\hat{U}}$ L_2 bounded Martingale, since

$$E\{\hat{\hat{U}}_n^2\} = E\{(\sum_{k=1}^{n} \hat{\hat{d}}_k^2)\} = \sum_{k=1}^{n} E\{\hat{\hat{d}}_k^2\}\leqq \sum_{k=1}^{n} E\{\hat{d}_k^2\}\leqq E\{\hat{U}_n^2\} \quad .$$

Let the transform of $\hat{\hat{U}}$ by D be $\hat{\hat{V}}$, that is $\hat{\hat{V}}_n = \sum_{k=1}^{n} d_{nk}\hat{\hat{d}}_k$. Then we can prove that $\hat{\hat{V}}$ is L_2 bounded Martingale. In fact

$$E\{\hat{\hat{V}}_{m+1} \mid \hat{\hat{V}}_1,\dots,\hat{\hat{V}}_m\} = E\{\hat{U}_{m+1}-\hat{U}_m \mid \hat{U}_1,\dots,\hat{U}_m\}+$$

$$E\{\sum_{k=1}^{m} d_{m+1,k}(\hat{U}_k-\hat{U}_{k-1}) \mid \hat{U}_1,\dots,\hat{U}_m\} = \sum_{k=1}^{m} d_{mk}(\hat{U}_k-\hat{U}_{k-1}) = \hat{\hat{V}}_m ,$$

and

$$E\{\hat{\hat{V}}_n^2\} = E\{[\sum_{k=1}^{n}(\hat{\hat{V}}_k-\hat{\hat{V}}_{k-1})]^2\} = \sum_{k=1}^{n} E\{(\hat{\hat{V}}_k-\hat{\hat{V}}_{k-1})^2\} \tag{3.11}$$

$$= \sum_{k=1}^{n} E\{[\hat{\hat{d}}_k + \sum_{i=1}^{k-1}(d_{ki}-d_{k-1,i})\hat{\hat{d}}_i]^2\}$$

$$= E\{(\sum_{k=1}^{n}\hat{\hat{d}}_k)^2\} + \sum_{k=2}^{n} E\{2\hat{\hat{d}}_k \sum_{i=1}^{k-1}(d_{ki}-d_{k-1,i})\hat{\hat{d}}_i\} +$$

$$\sum_{k=2}^{n} E\{[\sum_{i=1}^{k-1}(d_{ki}-d_{k-1,i})d_i]^2\} ,$$

where we used the property, $E\{(\hat{\hat{V}}_i-\hat{\hat{V}}_{i-1})(\hat{\hat{V}}_j-\hat{\hat{V}}_{j-1})\} = 0$, $i > j$.
In the last expression of (3.11) we put

$$E\{\hat{\hat{V}}_n\} = I_1 + I_2 + I_3 .$$

Then we have $|I_1|=E\{\hat{\hat{U}}_n^2\}<M$ since $\hat{\hat{U}}$ is L_2 bounded. I_2 is also bounded since

$$|I_2| \leqq 2\sum_{k=2}^{n}\sum_{i=1}^{k-1} | E\{\hat{\hat{d}}_k \hat{\hat{d}}_i(d_{ki}-d_{k-1,i})\}|$$

$$\leqq 2\sum_{k=2}^{n}\sum_{i=1}^{k-1} [E\{\hat{\hat{d}}_k^2\}E\{\hat{\hat{d}}_i^2(d_{ki}-d_{k-1,i})^2\}]^{1/2}$$

$$\leqq 2\sqrt{2M}\sum_{k=2}^{n}\sum_{i=1}^{k-1} [E\{E\{(d_{ki}-d_{k-1,i})^2|\hat{\hat{d}}_1,\dots,\hat{\hat{d}}_{k-1}\}\hat{\hat{d}}_i^2\}]^{1/2}$$

$$< 4M\sum_{k=2}^{n}\sum_{i=1}^{k-1} O(k^{-(2+\delta/2)}) < M_0.$$

Similarly, I_3 is evaluated as follows,

$$|I_3| \leqq \sum_{k=2}^{n}\sum_{i,j=1}^{k-1} | E\{(d_{ki}-d_{k-1,i})(d_{kj}-d_{k-1,j})\hat{\hat{d}}_i\hat{\hat{d}}_j\}|$$

$$\leqq \sum_{k=2}^{n}\sum_{i,j=1}^{k-1} | E\{[E\{(d_{ki}-d_{k-1,i})^2\hat{\hat{d}}_i|\hat{\hat{d}}_1,\dots,\hat{\hat{d}}_{k-1}\}$$

$$E\{(d_{kj}-d_{k-1,j})^2\hat{\hat{d}}_j^2|\hat{\hat{d}}_1,\dots,\hat{\hat{d}}_{k-1}\}]^{1/2}\}|$$

$$\leq \sum_{k=2}^{n} \sum_{i,j=1}^{k-1} O(k^{-(4+\delta)}) E\{|\hat{\hat{d}}_i \hat{\hat{d}}_j|\}$$

$$\leq 2M \sum_{k=2}^{n} O(k^{-(2+\delta)}) < M_1 .$$

Therefore, $\hat{\hat{V}}_n$ are L_2 bounded, that is, $\sup_n E\{\hat{\hat{V}}_n^2\} \leq M+M_0+M_1$, and $\hat{\hat{V}}$ converges almost everywhere.

Now, we consider the equations

$$\hat{\hat{U}}_n = \sum_{k=1}^{n} \hat{\hat{d}}_k = \sum_{k=1}^{n} [\hat{d}_k - E\{\hat{d}_k \mid \hat{U}_1, \ldots, \hat{U}_{k-1}]$$

$$= \hat{U}_n - \sum_{k=2}^{n} E\{\hat{d}_k \mid \hat{U}_1, \ldots, \hat{U}_{k-1}\} .$$

In these equations, $\hat{\hat{U}}_n$ and $\hat{U}_n$ converge with probability one, therefore, the last term in the right hand side converges with probability one.

On the other hand, $\hat{\hat{V}}_n$ is expressed as

$$\hat{\hat{V}}_n = \sum_{k=1}^{n} d_{nk} [\hat{d}_k - E\{\hat{d}_k \mid \hat{U}_1, \ldots, \hat{U}_{k-1}\}]$$

$$= \hat{V}_n - \sum_{k=2}^{n} d_{nk} E\{\hat{d}_k \mid \hat{U}_1, \ldots, \hat{U}_{k-1}\} .$$

$\hat{\hat{V}}_n$ in the left hand side and last term in the right hand side converge hence $\hat{V}_n$ converges with probability one. This completes the proof of theorem 3.1.

Some final notes on the Learning Process are: (1) In this study the aims of Learning are given by the target stochastic processes which depend upon the states of the subject. It is obvious that the aims of Learning are affected by the states of the subject. An organism does not study an aim of learning which is too difficult for the organism. An easy aim is immediately mastered and it's changed to an aim of a higher level step by step. Therefore, aims of learning have to be given in accordance with the states of the organisms. This may be considered that the organism responds to the given aim of learning and changes to fit the situation. (2) In the Learning Processes of organisms, data of the past responses, reinforcements and states of the subject are naturally taken into account by the organisms. And they are, therefore, used to obtain stability and efficiency of the Learning Process. S.L.P. by means of L.C.S.P. which utilizes all the past data in each stage is therefore effective for learning. (3) Some motivations on the organisms which are selected so as to expedite the learning, play an important role in psychology. These are essentially different from the reinforcements and targets imposed on the organisms. Nevertheless, some motivations can be expressed by multi-dimensional stochastic processes which contain targets and reinforcements. This, also, applies to the multi-dimensional L.C.S.P. stated in [5] of the references.

REFERENCES

1. R. C. Atkinson, G. H. Bower, and E. J. Crothers, An Introduction to Mathematical Learning Theory, Wiley,(1965).
2. D. L. Burkholder,"Martingale Transforms", A.M.S. Vol. 37, No. 6, pp. 1494-1504,(1966).
3. T. Indo, Mathematical Psychology, Tokyo Univ. Press. (1969).
4. S. Kano, "Linear Controlled Stochastic Processes," Sci. Rep. Kagoshima Univ., No. 8, (1959).
5. S. Kano, "Multi-dimensional Linear Controlled Stochastic Processes with Discrete Parameter", Sci. Rep. Kagoshima Univ., No. 10, (1961).
6. S. Kano, "Martingale Transforms and Linear Controlled Stochastic Processes," Res. Rep. of Res. Inst. Fund. Inf. Sci., No. 5, (1969).
7. T. Kitagawa, "Successive Processes of Statistical Controls", (2). Mem. Fac. Sci., Kyushu Univ. Ser. A, Vol. 13, No. 1, pp. 1-16, (1959).
8. T. Kitagawa, "Successive Processes of Statistical Controls (3)", Mem. Fac. Sci., Kyushu Univ. Ser. A, Vol. 14, No. 1, (1960).
9. K. Krickeberg, "Convergence of Martingales with a Directed Index Set," Trans. Amer. Math. Soc. 83, pp. 313-337, (1956).
10. M. F. Norman, "Some Convergence Theorems for Stochastic Learning Models with Distance Diminishing Operators," J.M.P. Vol. 5, No. 1, (1968).

LEARNING PROCESSES IN A RANDOM MACHINE

Sadamu Ohteru, Tomokazu Kato, Yoshiyuki Nishihara and Yasuo Kinouchi

Waseda University

Tokyo, Japan

1. INTRODUCTION

A random machine, [1 , 2] in which the random pulse frequency is used as the machine variable, may be considered to be somewhere between an analog and a digital computer in speed and accuracy of operation and to perform functions similar to those of the brain. The authors have developed an experimental random machine which can be effectively applied to the analysis of a system governed by algebraic or differential equations.

In this paper, some of the learning processes involved in the application of this random machine are reported--for example, its use as an Adaline-type learning machine and as automatic image processor.

2. PRINCIPLE OF THE RANDOM MACHINE

In the machine, an analog quantity E is expressed by probability P and has the following relation

$$P(E) = \left(\frac{E}{V_E} + 1\right)/2 \qquad (1)$$

in the range $-V_E \leq E \leq V_E$.

The transformation given by eq. 1 is carried out by the digital-to-stochastic converter in which the frequency of random pulses on a clocked sequence is compared with digital quantities by the Monte Carlo Method as shown in Figure 1. That is, when a digital quantity is larger than random number at that time, it assumes the level ON,

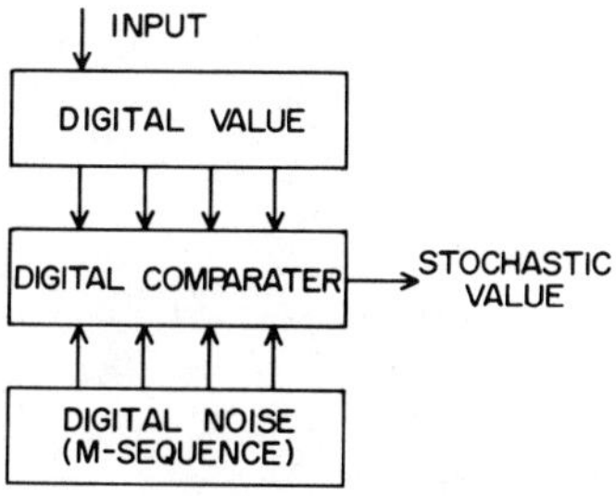

Figure 1. Digital to Stochastic Converter

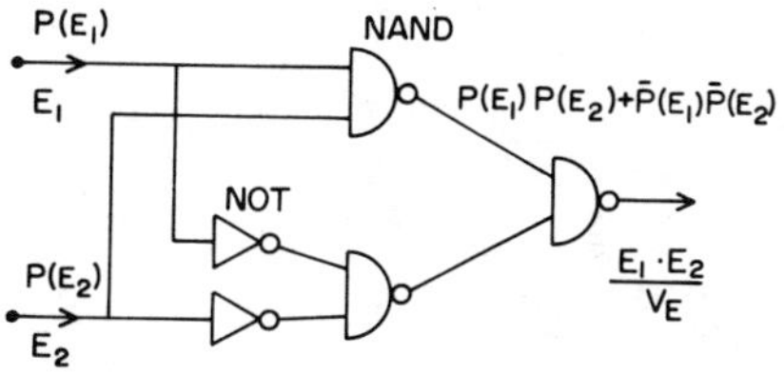

Figure 2. Stochastic Multiplier.

otherwise it is OFF. In our case, the analog-quantity is first converted into a digital-quantity, since the digital sequence of the maximum-length linear shift register (M-sequence) is used instead of analog random noise source. The stochastic-to-digital converter is composed simply of the ordinary digital reversible counter (up-down counter), and the stochastic-to-analog conversion is easily performed by a simple low pass filter.

Since the probability that two independent events will occur simultaneously is given by the product of the probability of each event, the multiplier can be easily constructed using logic coincidence as shown in Figure 2. The adder works by randomly sampling each input pulse sequence by the digital noise as shown in Figure 3(a). When the pulse frequency of each input i is given by the probability $\overset{o}{P}_i$. The weighted sum

$$\sum_i Pa_i \overset{o}{P}_i$$

is easily obtained by randomly sampling each input i according to probabilities Pa_i ($\sum Pa_i=1$) as shown in Figure 3(b). The inverter in Figure 4 works as an ordinary logic NOT to reverse a pulse. The stochastic integrator is consisted of the reversible counter together with the digital to stochastic converter as shown in Figure 5.

Some features of the random pulse machine are: (1) The random machine is scarcely affected by the noise and does not make any gross errors. (2) The machine configuration is very simple and the multiplier and the analog memory are extremely accurate, since they are essential digital. (3) The link between the stochastic and

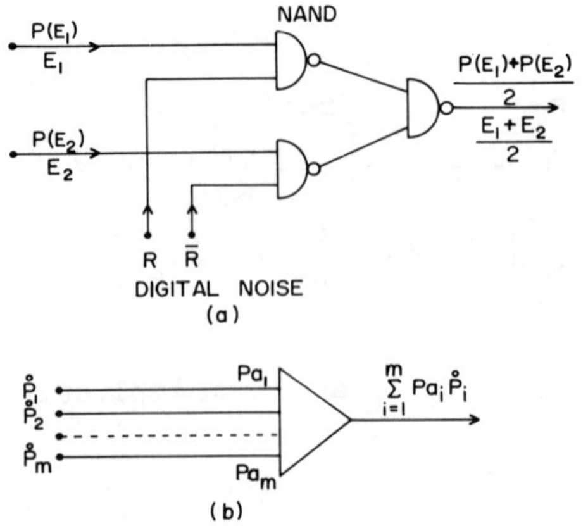

Figure 3. Stochastic Adder
(a) Simple Adder
(b) Weighted Sum.

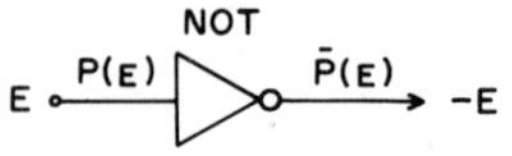

Figure 4. Stochastic Inverter.

digital computers is easily realized. (4) The operating time does not depend on the number of operations. (5) The weighted sum in the space domain is easily obtained in the time domain by adding the probability of exclusive input events. (6) The accuracy of the stochastic computer increases with its operating time, that is, each single pulse contains some fractions of information. When used as a stochastic learning machine, some portions of the learning process are completed in a period as short as 10^{-8} seconds.

3. STOCHASTIC ADALINE

We consider the learning machine without a quantizer, [4] when it is driven by stochastic pulses. A block diagram of a stochastic Adaline is shown in Figure 6. The vectors $X_j = (x_{ij})$ indicate the analog input patterns, and D_j the desired output. Their elements x_{ij} are converted by eq. 1 to the probability $P(x_{ij})$ and $P(D_j)$. For the weight the digital reversible counter is used. The delay d_k is inserted in order to assure the independency of the two input sequences, that is the $P(x_{kj})$ sequence and the $\bar{P}_2$ sequence as shown in Figure 6. The multiplier G_k requires that the two input sequences be stochastically independent of each other.

At each clock time of an input M-sequence, the same learning procedure is performed as in the ordinary Adaline. The value of

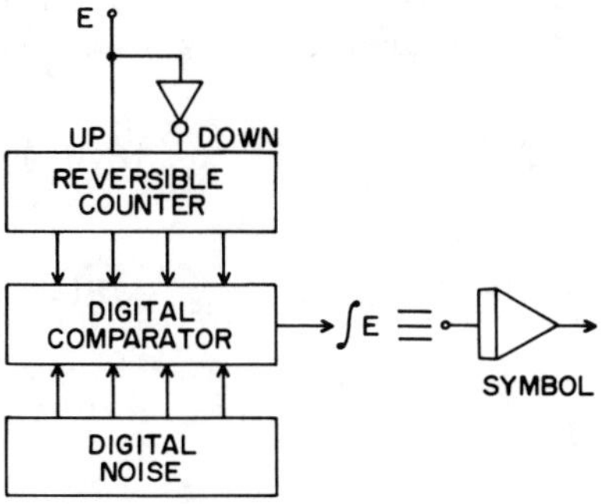

Figure 5. Stochastic Integrator.

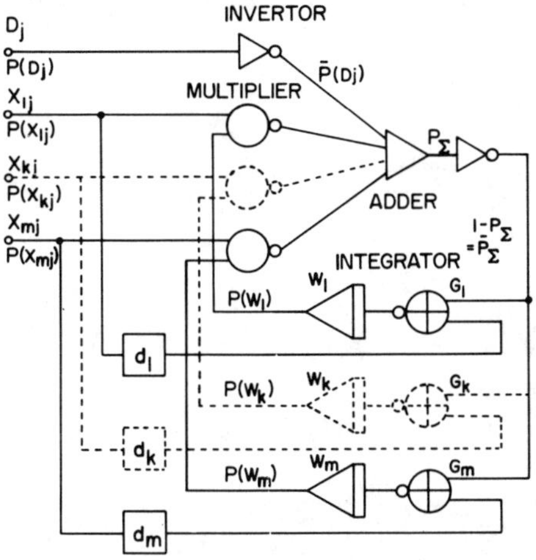

Figure 6. Stochastic Adaline.

reversible counter increases or decreases according to the input pattern sequence. Therefore, the changes in weight vector $\underline{W}$ during the learning process are formulated as an m dimensional random walk.

$$P_{t+1}(\underline{W}) = \sum_{\underline{\alpha}} P_t(\underline{W}-\underline{\alpha})P(\underline{W}/\underline{W}-\underline{\alpha}) \tag{2}$$

where $P_{t+1}(\underline{W})$ is the probability of $\underline{W}$ at t+1 and $\underline{\alpha}=(\alpha_1,..,\alpha_m)$, $\alpha_i=\pm\frac{V_W}{n}$.

$P(\underline{W}/\underline{W}-\underline{\alpha})$ is the probability of the transition from $\underline{W}-\underline{\alpha}$ to W and is determined by input pattern probability $P(x_{ij})$. (Appendix I)

$$P(\underline{W}+\underline{\alpha}/\underline{W}) = \frac{1}{m+1}\sum_j P_j\{[P(D_j)-\sum_i^m(P(x_{ij})P(W_i)+\bar{P}(x_{ij})\bar{P}(W_i)+m]$$
$$\prod_i^m f(P(x_{ij}),\ \alpha_i)+[-P(D_j)+\sum_i^m(P(x_{ij})P(W_i)+\bar{P}(x_{ij})\bar{P}(W_i)+1]$$
$$\prod_i^m f(P(x_{ij}),\ -\alpha_i)\} \tag{3}$$

where each pattern X_j occurs with the probability P_j, $\bar{P}$ means 1-P, and

$$f(P,\alpha) = P \quad \text{for } \alpha>0, \qquad f(P,\alpha) = 1-P \text{ for } \alpha<0 \tag{4}$$

Let us consider the expected value $\langle W_k\rangle$ over the space $\underline{W}$

$$\langle W_k\rangle_{t+1} = \sum_{\underline{W}} W_k\ P_{t+1}(\underline{W}) \tag{5}$$

$$= \sum_{\underline{W}} W_k \sum_{\underline{\alpha}} P_t(\underline{W}-\underline{\alpha})P(\underline{W}/\underline{W}-\underline{\alpha}) \tag{6}$$

set,

$$\underline{W} - \underline{\alpha} = \underline{w} \tag{7}$$

$$\langle W_k\rangle_{t+1} = \sum_{\underline{w}}\sum_{\underline{\alpha}} (w_k + \alpha_k)\ P_t(\underline{w})P(\underline{w}+\underline{\alpha}/\underline{w}) \tag{8}$$

$$= \sum_{\underline{w}} w_k\ P_t(\underline{w}) \sum_{\underline{\alpha}} P(\underline{w}+\alpha/\underline{w})$$

$$+ \sum_{\underline{w}} P_t(\underline{w})\cdot\frac{V_W}{n}\cdot\Big(\sum_{\underline{\alpha},\alpha_k=\frac{V_W}{n}} P(\underline{w}+\alpha/\underline{w}) - \sum_{\underline{\alpha},\underline{\alpha}_k=\frac{V_W}{n}} P(\underline{w}+\alpha/\underline{w}) \tag{9}$$

Considering the relation

$$\sum_{\underline{\alpha}} P(\underline{w}+\underline{\alpha}/\underline{w}) = 1 \tag{10}$$

$$\sum_{\underline{\alpha},\alpha_k=\frac{V_W}{n}} P(\underline{w}+\underline{\alpha}/\underline{w}) - \sum_{\underline{\alpha},\alpha_k=-\frac{V_W}{n}} P(\underline{w}+\underline{\alpha}/\underline{w}) = C\sum_j P_j(D_j-\sum_i X_{ij}w_i)X_{kj} \tag{11}$$

where

$$C = 1/\{V_x^2\ V_W(m+1)\}\ , \qquad V_xV_W = V_D$$

$$\langle W_k \rangle_{t+1} = \sum_{\underline{w}} P_t(\underline{w}) w_k + \frac{V_W C}{n} \sum_{\underline{w}} P_t(\underline{w}) \sum_j P_j (D_j - \sum_i X_{ij} w_i) X_{kj} \qquad (12)$$

$$= \langle W_k \rangle_t + \frac{V_W \cdot C}{n} \sum_j P_j (D_j - \sum_i X_{ij} \langle W_i \rangle_t) X_{kj} \qquad (13)$$

On the other hand, the mean-square error of the random machine is

$$L(\underline{W}) = \sum_j P_j (D_j - \sum \langle W_i \rangle X_{ij})^2$$

and the partial derivative for $\langle W_k \rangle$ is

$$\frac{\partial L}{\partial \langle W_k \rangle} = Z \sum_j P_j (D_j - \sum_i \langle W_i \rangle X_{ij}) X_{kj} \qquad (14)$$

Therefore using eq. 13, eq. 14 is rewritten as

$$\langle W_k \rangle_{t+1} = \langle W_k \rangle_t + \frac{V_W \cdot C}{2n} \cdot \frac{\partial L}{\partial \langle W_k \rangle} \qquad (15)$$

Eq. 15 shows that, as far as the expectation is concerned, the learning process in this machine is carried out by the method of steepest descent for the mean-square error.

When the learning process is finished for a group of patterns, the values of the weight $\underline{W}$ fluctuate around the mean values $\langle \underline{W} \rangle$ with the definite variances. The variances are determined by the correlation matrix of patterns and, in the contrast to the ordinary Adaline, take the same values whether $L(\underline{W})$ attains zero or not. (Appendix II)

4. THE AUTOMATIC IMAGE PROCESSOR

Now, let us consider an example of a stochastic learning machine. The problem of image restoration for information preprocessing is reduced to solving automatically and rapidly the integral equation

$$F(t) = \int_a^b K(t,\tau) f(\tau) d\tau \qquad (16)$$

We assume that the kernel k (t,τ), expresses input patterns, the known function $F(t)$ represents their desired outputs and the unknown function $f(\tau)$, the weights. We can obtain the solution of the above equation using the learning machine technique [5]. But here, we explain the other method by using our new type function generator. The application of random machine techniques to this problem gives not only a simple configuration and high accuracy, but also an interesting learning process.

The $f(\tau)$ and $K(\tau,t)$ are expanded as

$$f(\tau) = \sum_i a_i \, \phi_i \, (\tau) \tag{17}$$

$$K(t,\tau) = \sum_i b_i(t)\cdot\phi_i(\tau) \tag{18}$$

using the orthogonal function $\phi_i(\tau)$. Then we can obtain from eq. 16

$$F(t) = \sum_i a_i b_i(t) \tag{19}$$

Solving this equation for a_i is equivalent to solving eq. 16. In general, however, $b_i(t)$ is not the orthogonal function. It is not easy to calculate analytically, but we can determine a_i by the stochastic learning machine technique. But in this case, a_i corresponds to the weight.

Next, let us consider how to obtain $b_i(t)$. The coefficient $b_i(t)$ can be expressed as

$$b_i(t) = \int_a^b K(t,\tau)\phi_i(\tau)d\tau \tag{20}$$

Therefore, $b_i(t)$ can be generated by a random sampling of $\phi_i(\tau)$ with probability $P_k(t/\tau)$, which is proportional to $K(t,\tau)$.

But the summation must be performed in eq. 20 over τ and in the original eq. 19 over i. Therefore in order to apply this method to an automatic image processor, we must perform the summations over two variables in real time. This is the reason why a conventional analog computer is not fitted for solving an integral equation.

At first, for an orthogonal function, it is convenient for us to introduce the Walsh orthogonal system which takes on only two values, +1 or -1. Furthermore it is sufficient, in our case, to use the Rademacher orthogonal function $R_i(t) = \text{sign}(\sin 2^i\pi t)$ which is composed of simple cascade connections of flip-flop circuits as shown in Figure 7. The Rademacher system is not complete, but it is easily possible to transform the Rademacher function into a Walsh function, by controlling the probabilities of the occurrences of the Rademacher function and its combination by the gate controller. [6].

Next, for a kernel function, we propose a simple and interesting circuit, "probability density oscillator" which spontaneously generates within itself, for example the normal distribution with the desired expected value and variance. This is carried out using

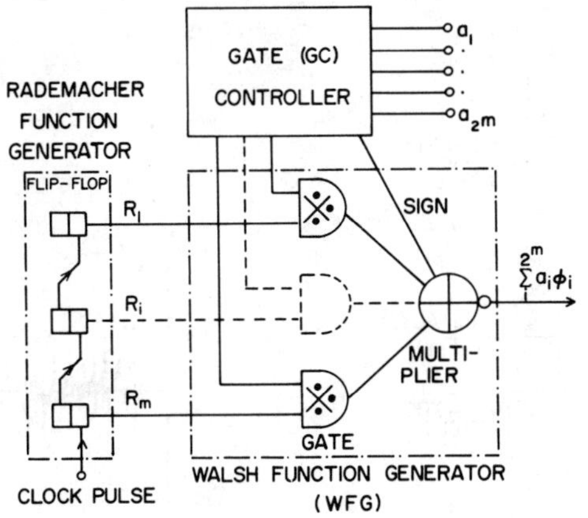

Figure 7. Stochastic Function Generator.

an extremely simple circuit composed of a reversible counter with negative feedback, and driven by random pulse as shown in Figure 8(a). The changes in the integrator are regarded as a random walk

$$
\begin{aligned}
q_{t+1}(Z) = {} & q_t(Z+1)P(Z/Z+1) \\
& + q_t(Z)P(Z/Z) \\
& + q_t(Z-1)P(Z/Z-1)
\end{aligned}
\tag{21}
$$

where, q(Z) is the probability of occurrence of state A in an integrator whose capacity is n. The distribution in the steady state of the integrator is the binominal distribution truncated by boundaries as

$$
q(Z) = C(p)\,\frac{(Zn)!}{(n-np+Z)!(n+np-Z)!}
\tag{22}
$$

in which the expected value is np and the variance n/2 and C(P) is constant value for Z. Figure 9 shows an experimental result. It is possible to control easily the variance and expected value at the desired values. Furthermore, it is also possible to get the other type distributions. Figure 8(b) shows the circuit for a Poisson distribution.

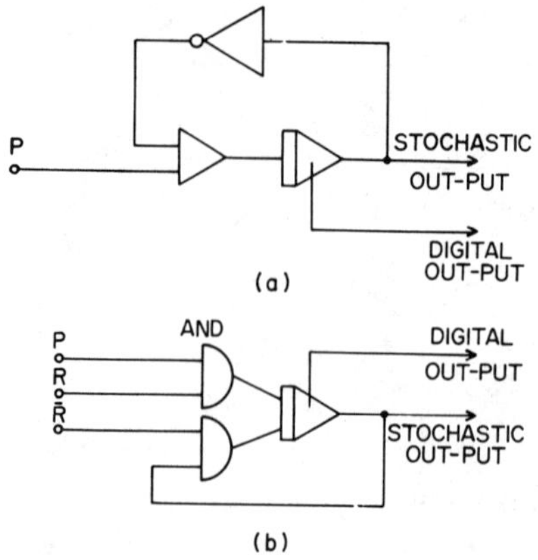

Figure 8. Probability Density Oscillator (a) Normal Distribution (b) Poisson Distribution.

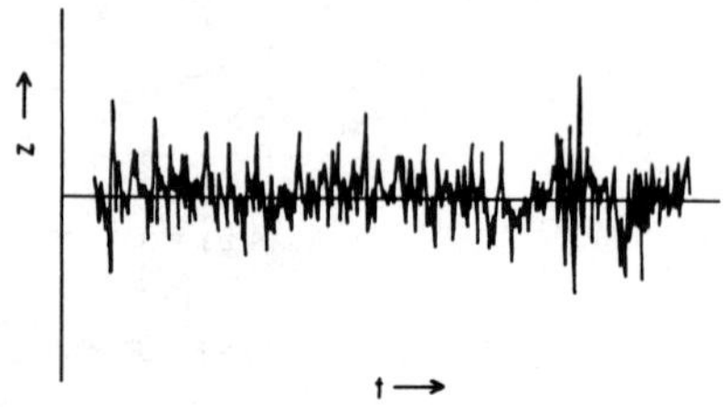

Figure 9. Experimental Result

Our whole automatic image processor is shown in Figure 10. WFG1 is the Walsh function generator used to generate $\sum_i a_i \phi_i$ by controlling the probability of occurrence a_i over i by GC1. At the same time, the probability of occurrence $\Sigma a_i \ \phi_i(\tau)$ over τ is controlled by the probability density oscillator $K(t,\tau)$. Therefore the output of the WFG1 is expressed as

$$\int_a^b K(t,\tau) \ \Sigma \ a_i \phi_i(\tau) d\tau$$

and equals

$$\sum_i a_i b_i(t)$$

because of the orthogonality of $\phi_i(\tau)$. In order to generate $b_j(t)$, the separate function generator WFG2 is used by which the stochastic independence between the two lines is maintained.

Figure 11 shows an example of the image processing obtained by computer simulation. The dotted line shows the real image. The difference between calculated value and real value depends on the small capacity of the integrator used in simulation. We can expect the accuracy of at least more than 0.1% by the use of our experimental machine.

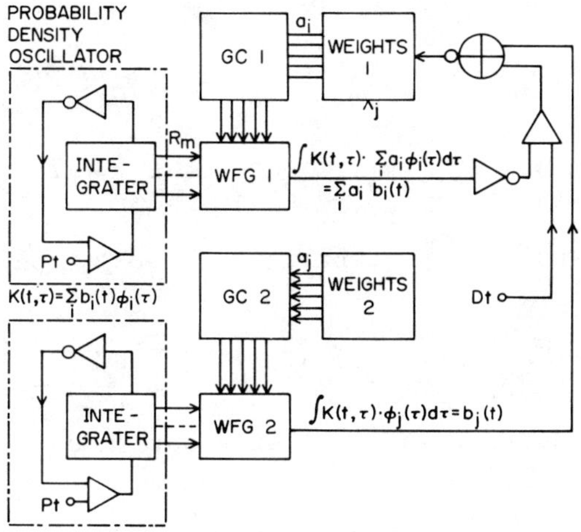

Figure 10. Stochastic Image Processor.

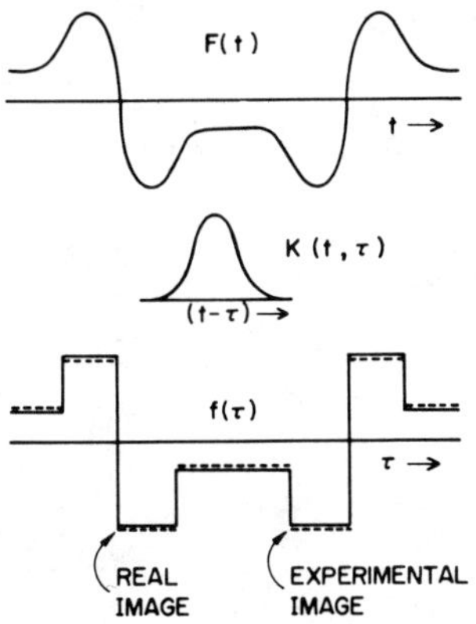

Figure 11. Result of Image Processor.

5. CONCLUSION

Learning process in the stochastic Adaline has been discussed as a random walk problem. As an example of its application, an integral equation analyzer is proposed with a new type of function generator and probability density oscillator.

APPENDIX I. DERIVATION OF TRANSITION PROBABILITY

The outputs P_Σ and P_{EK} of adder and multiplier in Figure 6 can be written as

$$P_\Sigma = \frac{1}{m+1}\{\bar{P}(D_j) + \sum_i (P(X_{ij})P(W_i) + \bar{P}(X_{ij})\,\bar{P}(W_i))\} \tag{A-1}$$

$$P_{EK}(W_k+\alpha_k/W_k) = \bar{P}_\Sigma \cdot f(P(X_{kj}),\ \alpha_k) + P_\Sigma \cdot F(P(X_{kj}), -\alpha_k) \tag{A-2}$$

Therefore the transition probability P_E for jth pattern is expressed as

$$P_E(\underline{W}+\underline{\alpha}/\underline{W}) = \bar{P}_\Sigma \prod_i^m f(P(X_{ij}),\alpha_i) + P_\Sigma \prod_i^m f(P(X_{ij}),-\alpha_i) \tag{A-3}$$

From the above equation, we can easily obtain eq. 3.

APPENDIX II. VARIANCES OF STOCHASTIC ADALINE

Let the covariance $\sigma^t_{k\ell}$ of the weights be

$$\sigma^t_{k\ell} = \sum_{\underline{W}} (W_k - \langle W_k\rangle_t)(W_\ell - \ W_{\ell\ t})P_t(\underline{W}) \tag{A-4}$$

Using eqs. (3), (4) and (5), the equations for $\sigma^{t+1}_{k\ell}$ are obtained as follows.

$$\sigma^{t+1}_{k\ell} = \sigma^t_{k\ell} - \frac{V_W \cdot C}{n}\sum_i \sigma^t_{i\ell}\xi_{ik} - \frac{V_W \cdot C}{n}\sum_i \sigma^t_{ik}\ \xi_{i\ell}$$

$$-\left(\frac{V_W \cdot C}{n}\right)^2 \left(\sum_i \langle W'_i\rangle_t\ \xi_{ik}\right)\left(\sum_i \langle W'_i\rangle_t\ \xi_{i\ell}\right) + \frac{V_W^{\ 2}}{n^2} H_{k\ell} \quad k,\ell=1,\sim m. \tag{A-5}$$

where

$$\xi_{k\ell} = \sum_j P_j\ X_{kj}\ X_{\ell j} \tag{A-6}$$

$$H_k = \begin{cases} \xi_{k\ell}/V_X^2 & \text{for } k \neq \ell \\ 1 & \text{for } k=\ell, \end{cases} \tag{A-7}$$

$$\underline{W}' = \underline{W} - \underline{\tilde{W}} \tag{A-8}$$

and $\underline{\tilde{W}}$ is the weight vector which minimizes the mean-square error. The fluctuation of weight in learning process is calculated from eq. (A-5).

The covariance $\sigma_{k\ell}$ in the limit of $t\to\infty$ is given

$$\sum_i \sigma_{ik}\,\xi_{i\ell} + \sum_i \sigma_{i\ell}\,\xi_{ik} = \frac{H_{k\ell}V_W}{nC} \qquad \text{(A-9)}$$

Quantities corresponding to $\sigma_{k\ell}$ in the ordinary Adaline are determined by the minimum value of mean-square error. On the other hand, in the stochastic Adaline the magnitude of the fluctuation is not determined by the mean-square error value but is determined by the matrix $\xi_{k\ell}$ and coefficient

$$\frac{V_W}{nC}$$

as eq. (A-9).

REFERENCES

1. J. R. Gaines, "Stochastic Computing," AFIPS Spring Joint Computer Conf. 30. p. 149, 1967.

2. S. T. Ribeiro, "Random Pulse Machines," IEEE Vol. EC-16 No. 3, p. 261, 1967.

3. S. Ohteru, T. Kato, Y. Nishihara, Y. Kinouchi, "Stochastic Differential Analyzer," Proceeding ACTES, (Munich), 1970.

4. B. Widrow, M. E. Hoff, "Adaptive Switching Circuits," WESCON Convention Record, Part IV, 1960

5. S. Ohteru, T. Kato, T. Nagano, Y. Fujii, "Image Restoration Utilizing Learning Machine Techniques," IMEKO V (Paris), 1970.

6. S. Ohteru, T. Kato, Y. Nishihara, S. Hashimoto, "Stochastic Function Generator," Joint Convention Record of Four IEE, Japan, 1970.

LEARNING PROCESS IN A MODEL OF ASSOCIATIVE MEMORY

Kaoru Nakano

University of Tokyo

Tokyo, Japan

SUMMARY

The excellent information processing in a human brain is considered to depend upon its association mechanisms. To simulate this function, we propose in this paper a model of the neural network named "Associatron" which operates like a human brain in some points. Associatron stores many entities at the same place of its structure, and recalls the whole of any entity from a part of it. From that mechanism some properties are derived, which are expected to be utilized for human-like information processing. After the properties of the model have been analyzed, an Associatron with 180 neurons is simulated by a computer and is applied to simple examples of concept formation and game playing. Hardware realization of an Associatron with 25 neurons and thinking process by the sequence of associations are mentioned, too.

INTRODUCTION

The purpose of this paper is to outline the work of simulating the association mechanism of a human brain and of applying it to human-like information processing.

The first aspect of this work is to design an associative memory device that is considered to be reasonable as a model for association mechanisms in the human brain. Association was studied mainly in the field of psychology, and some semantic models for association were presented in the last few years. On the other hand, biological studies have gradually revealed the structure of the nervous system. Although knowledge about it is still not enough to construct the structure artificially, work having been presented by Post [1] is concerning rather biological associative memory. His model is a distributed memory device which memorizes triplets of entities and recalls one of these entities from two other entities in the specific triplet. The model will be helpful for realizing life-like information processing.

Associatron [2,3] proposed here is one of this kind of models, whose structure is more similar to the nervous system and more general so that larger part of the function of the brain will possibly be explained by the use of the model. In this model a lot of entities are stored in the same region of its structure, and any stored entity can be recalled from parts of it without any sequential search. The more parts that are fed into the memory device, the more accurately the entity will be recalled. The principle is based on the application of the auto-correlation functions. If the auto-correlation function of an entity is held in the memory, the entity can be easily reproduced from only a small part of it. Since storing all these auto-correlation functions is too redundant to be practical, they are linearly added in memorizing process. Therefore, if the stored entities are numerous, entities cannot always be recalled completely, but they are expected to be reproduced only probabilistically.

This kind of associative memory has the following properties, differing from the conventional memory.

(1) Reliability is improved in the sense that the failure of some components of the memory device does not cause the loss of the whole of any entity, but cause only inaccuracy in recalling.

(2) Although some uncertainties appear in recalling, they are rather useful for realizing artificial intelligence with flexibility and applicability. This is because strictly logical comparison, for example, is apt to discard useful data which resembles a certain datum. The uncertainty confusing similar data to the same data will make the device extract automatically, from a lot of entities, the essence useful for a certain job.

(3) Since information is accessed by using the relation of meaning without any sequential search, the speed of access or information retrieval is very high.

By these properties, information processing using the associative memory is different from that of the conventional computers. To demonstrate the learning ability of this model, a simple game is used. In a lot of work concerning game playing, the learning program for the game of checkers [4,5] is well known. The model in this paper is also related to such heuristics in game playing, general problem solving and thinking process in the nervous system.

PRINCIPLE OF DISTRIBUTED ASSOCIATIVE MEMORY

A model of associative memory realized in the following way is named "Associatron." Presume that an entity is represented by a row vector

$$x = (x_1, x_2, \ldots, x_n) , \tag{1}$$

where $x_i = -1, 0, 1$. An entity is composed of several patterns and of neutral areas. Each pattern is composed of -1's, 0's and 1's, while the neutral area is composed only of 0's. Each pattern has a simple or complex meaning, and the entity represents the association of these patterns.

Let x^t denote the transposed vector x. Inner state of the memory, after k entities $x^{(1)}, x^{(2)}, \ldots, x^{(k)}$ have been stored, is defined as an n×n matrix

$$M = x^{(1)t} x^{(1)} + x^{(2)t} x^{(2)} + \ldots + x^{(k)t} x^{(k)} . \tag{2}$$

To recall entities, first we define the quantizing function,

$$\phi(t) = \begin{cases} -1 & \text{if } t < 0 , \\ 0 & \text{if } t = 0 , \\ 1 & \text{if } t > 0 . \end{cases} \tag{3}$$

Suppose that it can also be applied to a vector $u = (u_j)$ and a matrix $A = (a_{ij})$, as

$$\begin{aligned} \phi(u) &= (\phi(u_j)) , \\ \phi(A) &= (\phi(a_{ij})) . \end{aligned} \tag{4}$$

From an input $y = (y_1, y_2, \ldots, y_n)$, the same kind of vector x, the memory device recalls a row vector z, where

$$z = \phi(y\ \phi(M)) .$$

If y is composed of a neutral area and a few patterns which are also the components of a stored entity x, then it is expected that y is nearly equal to x, even when a lot of entities are memorized. This means the realization of recalling the whole of a stored entity from a part of it. Consequently, a few patterns previously mentioned can be associated with the rest of the patterns of entity x.

Figure 1 depicts the behavior of the Associatron as a neural network. Any pair of neurons x_i and x_j is connected through a memory unit m_{ij}, which corresponds to the synapse. This model differs from general models of neural networks in the following points: (1) all neurons are mutually connected; (2) each neuron has three possible states; (3) for any pair of neurons, the values of synaptic conductance are assumed to be equal for both directions. The value of the memory unit m_{ij} is multi-valued and is modified during a memorizing process by adding the product $x_i x_j$ to the

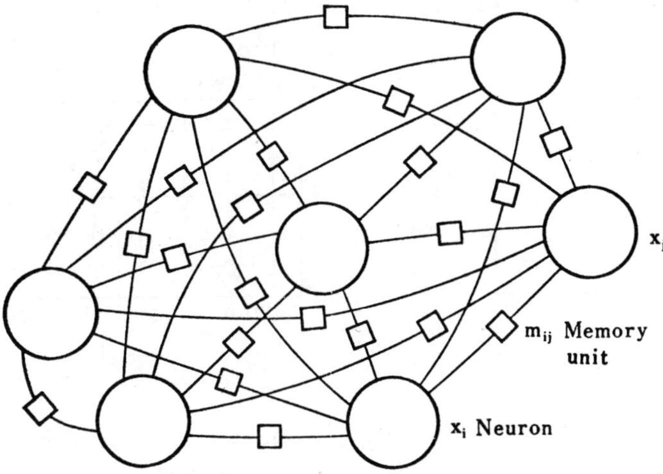

Figure 1. Associative Memory as a Neural Network.

previous value. However, in propagation of a stimulus, the quantized value is used; (4) refractory time is not considered.

From these items, it can be said that the Associatron is a considerably simplified model of the neural network. This model is similar to Perceptron [6] and Adaline [7] in a sense, but it differs from them in the point that it covers functions not only of learning but of life-like memory.

PROPERTIES OF THE ASSOCIATIVE MEMORY

In this section, properties of memorizing and recalling entities are discussed. It is previously mentioned that an entity is composed of patterns, each of which has a meaning. Here we consider it a little more in detail. Table 1 illustrates the relation of an entity and its component patterns; that is, Table 1 (a) shows the array of components of vector x and (b) shows an example of entities. The entity is composed of patterns 1, 2, 3, and a neutral area, where, pattern 1 has the meaning of "red", pattern 2 the meaning "spherical", etc. Now it is natural to consider that a pattern or a few patterns construct a concept. In the present example, pattern 1 has the concept of "red", pattern 2 the concept of "spherical", the set of patterns 1 and 2 has the complex concept of "red and spherical", the set of patterns 1, 2, and 3 has the concept of "apple", and so on. In the Associatron, when entities or associations of patterns are stored, not only can patterns be recalled, but also various concepts are gradually formed.

Table 1. Configuration of an Entity.

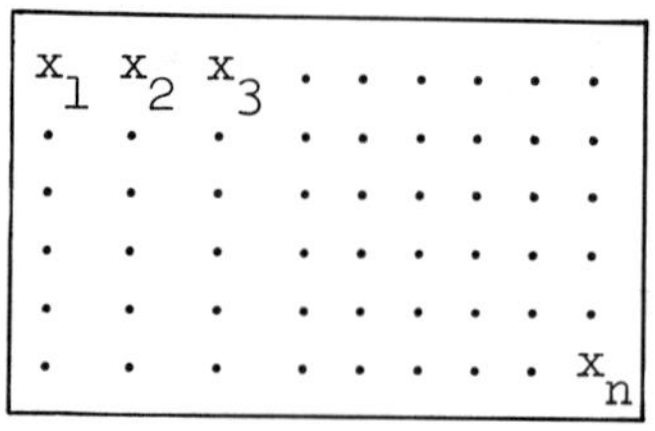

(a)

Red Pattern 1 — Spherical Pattern 2

1	1	1	-1	-1	0	0	-1	-1
-1	1	-1	0	1	0	0	1	-1
-1	1	-1	1	-1	0	0	1	-1
-1	1	1	-1	0	0	0	1	1
1	1	-1	-1	0	0	0	0	0
1	0	-1	1	0	0	0	0	0

Pattern 3 — Neutral Area

(b)

These properties of memorizing and recalling process are discussed as follows. Presume that Q denotes the set of all n-dimensional vectors

$$x = (x_1, x_2, \ldots, x_n) \ , \tag{6}$$

where x_i=-1,0,1. Now, we consider the recalling function α: $Q \to Q$. If $x \in Q$, and α is defined as

$$z = \alpha(x) = \phi(x\ \phi(M)) \ , \tag{7}$$

then z is also a member of Q. Now let us introduce the index vector

$$v = (v_1, v_2, \ldots, v_n) \ , \tag{8}$$

where v_i is equal to 0 or 1. This vector represents a certain area of the neural network. Defining elementwise multiplication * as

$$v * x = (v_1 x_1, v_2 x_2, \ldots, v_n x_n) \ , \tag{9}$$

we call v * x the concept of x at the area v. Presume that both x and y are members of Q. If

$$v * x = v * y \ , \tag{10}$$

then x and y are said to have the same concept at area v. The measure of area v is defined by

$$l(v) = \sum_i^n v_i \ . \tag{11}$$

Now it is presumed that the entity x or the association of patterns A and B is stored in the memory device, as in Figure 2(a), whose representation is the same as Table 1. If the order of the elements of x is changed to put together the elements constructing the same pattern, the entity x is represented by a row vector as

$$x = (A, B, 0) \ , \tag{12}$$

where A and B are row vectors, and 0 denotes zero vector. Then the matrix M which stores only x will be

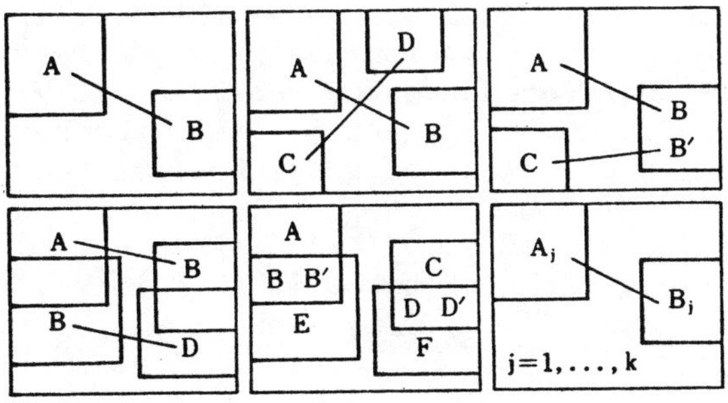

Figure 2. Association Mode

$$M = (A,B,0)^t\ (A,B,0) = \begin{bmatrix} A^tA & A^tB & 0 \\ B^tA & B^tB & 0 \\ 0 & 0 & 0 \end{bmatrix} . \tag{13}$$

If the index vectors of A and B are v_A and v_B, respectively, the concept at area v_B of the recalled pattern from $v_A * x$ is

$$v_B * \phi((v_A * x)\phi(M)) = v_B * \phi(A\phi(A^tA), A\phi(A^tB), 0)$$
$$= (0, A\phi(AB), 0) = (0,B,0) . \tag{14}$$

This means that patterns or concept B is completely recalled from A.

As the associations A-B and C-D are stored independently, as shown in Figure 2(b), in the same way it can be clarified that a pattern of each pair is completely recalled from another.

In the case of Figure 2(c), two associations, A-B and C-B', are stored in such a way that pattern B and B' are overlapping. Stored vectors are

$$x = (A,B,0,0), \qquad y = (0,B',C,0) . \tag{15}$$

The matrix is

$$M = \begin{bmatrix} A^tA & A^tB & 0 & 0 \\ B^tA & B^tB+B'^tB' & B'^tC & 0 \\ 0 & C^tB' & C^tC & 0 \\ 0 & 0 & 0 & 0 \end{bmatrix} \tag{16}$$

The recalling process is as follows,

$$\begin{aligned} v_B * \alpha(A,0,0,0) &= \phi(0, A\phi(A^tB), 0, 0) = (0,B,0,0) , \\ v_A * \alpha(0,B,0,0) &= \phi(B\phi(B^tA), 0,0,0) = (A,0,0,0) . \end{aligned} \tag{17}$$

In the case of Figure 2(d), where two associations, A-B and C-D, are stored, A and C, B and D partially overlap, respectively.

This is rewritten in Figure 2(e), where the overlapping areas are dealt with as new patterns. From the previous result, it is clear that E and F are indifferent to the recalling AB → CD, where AB means the complex concept of A and B, etc., so they are eliminated. Therefore, stored vectors can be considered as

$$x = (A,B,C,D,0), \qquad y = (0,B',0,D',0) \,. \tag{18}$$

The recalling process from AB to CD is as follows:

$$\begin{aligned} v_{CD} * \alpha(v_{AB} * x) &= (0,0,\phi(A\phi(A^tC)+B\phi(B^tC)),\phi(A\phi(A^tD)+B\phi(B^tD+B'^tD')),0) \\ &= (0,0,C,D,0) \,. \end{aligned} \tag{19}$$

Thus, CD can be completely recalled from AB, because only two patterns are overlapping at v_B. When this number is arbitrary, the following argument will hold.

Let k pairs of A_j-B_j be stored as shown in Figure 2(f). For simplicity of calculation, k is presumed to be an odd number. Besides, it is presumed that $l(v_A) = s$ is odd and that the entities are random patterns at areas v_A and v_B. These assumptions do not necessarily hold in actual use of the associative memory. But it matters little, because this only causes a small error in estimation of accuracy of recalling. In any recalling process, the probability that a memory unit votes for the right state of a neuron for the specific entity stored in the memory device is

$$(k + 1)/2k \,. \tag{20}$$

The probability that the state of a neuron decided by majority is right, is represented by the sum of the first $(s + 1)/2$ terms of the binomial expansion of

$$\left(\frac{k+1}{2k} + \frac{k-1}{2k} \right)^s \tag{21}$$

That is,

$$P = \left(\frac{1}{2k}\right)^s \sum_{i=0}^{(s-1)/2} {}_sC_i (k+1)^{s-i}(k-1)^i \,, \tag{22}$$

where ${}_sC_i$ is the number of combination. As this summation cannot be written as a simple form, to determine the behavior of this equation, the graphical expression is taken by calculating the values of P for various values of k and s. This is shown in Figure 3. From the graph, it is found that completely accurate recalling can be done when the number of stored entities is very small or when the number of elements constructing the input pattern is very large. But it is important that the ability of an associative memory is not determined only by the accuracy of memory.

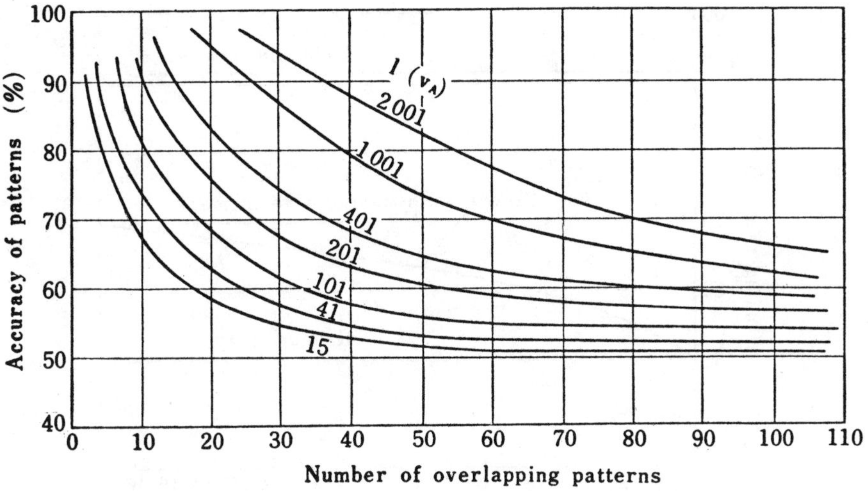

Figure 3. Accuracy of Recalling.

SOME EXPERIMENTAL RESULTS

The Associatron composed of 180 neurons is simulated by the digital computer HITAC 5020E. This simulated model is used for performing a few experiments. The neurons are arranged in 12 rows and 15 columns. Although input patterns for recalling should also be memorized, for convenience of evaluation of the behavior of the memory, they are not memorized in these experiments.

Example 1. Three areas A, B, and C, each of which consists of 40 neurons are taken on the neural plane. Each of the names of 6 things, 3 colors and 3 shapes is coded into a 40 bit pattern of 1's and -1's. A thing, its color and its shape are stored at areas A, B, and C, respectively. After 6 triplets are stored, the memory device is made to recall the color and the shape from a thing, and vice versa. The results are shown in Figure 4. The device recalls the color and the shape completely from the thing, but the accuracy of reverse recalling is about 92 percent.

Example 2. An experiment concerning simple concept formation was accomplished. In this case, for example, red things and the word "red" such as apple-red spherical-word "red," brick-red box-like-work "red" are shown one by one to the memory device. At first, the memory device might take the word "red" for "spherical," but as it is trained, it forms the concept "red" as shown in Table 3, where symbols are used instead of things, shapes, colors, etc. Thus, the pattern of "red" is associated with pattern of word "red," and the memory device recalls one from another. Besides, if entities apple-red and apple-spherical are stored at other times, the device

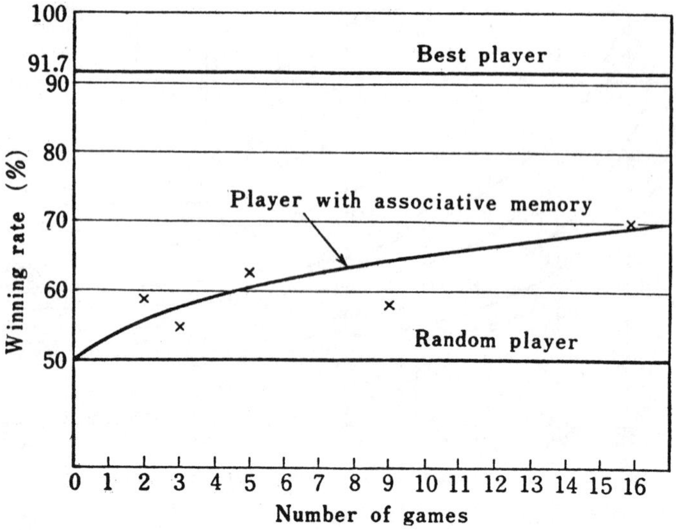

Figure 4. Graphical Expression of Learning.

Table 2. Concept Formation 1.

Memorizing	Recalling	Accuracy
1. Apple-Spherical-Red	Apple → Red, Spherical	(100%)
2. Watermelon-Spherical-Green	Watermelon → Spherical, Green	(100%)
3. Banana-Stick-like-Yellow	Banana → Stick-like, Yellow	(100%)
4. Wooden box-Box-like-Yellow	Wooden box → Box-like, Yellow	(100%)
5. Cucumber-Stick-like-Green	Cucumber → Stick-like, Green	(100%)
6. Brick-Box-like-Red	Brick → Box-like, Red	(100%)
	Red, Spherical → Apple	(90%)
	Spherical, Green → Watermelon	(88%)
	Stick-like, Yellow → Banana	(90%)
	Box-like, Yellow → Wooden box	(95%)
	Stick-like, Green → Cucumber	(90%)
	Box-like, Red → Brick	(95%)

recalls "red" and "spherical" from "apple" and vice versa. That is, although the triplet apple-red-spherical is not a stored entity, it is formed in the memory device.

Example 3. When the associative memory device is applied to games, a set of patterns effective for winning the game is expected to be extracted automatically. The following game is used for demonstration of learning process in the associative memory.

Table 3. Concept Formation 2.

Memorizing

	Thing-Shape-Color-Word
1.	$A_1 - B_1 - C_1 - C_1'$
2.	$A_2 - B_2 - C_1 - C_1'$
3.	$A_3 - B_3 - C_1 - C_1'$
4.	$A_4 - B_2 - C_2 - C_2'$
5.	$A_5 - B_1 - C_2 - C_2'$
6.	$A_6 - B_3 - C_2 - C_2'$
7.	$A_1 - B_1 - C_1 - B_1'$
8.	$A_2 - B_2 - C_1 - B_2'$
9.	$A_4 - B_2 - C_2 - B_2'$
10.	$A_5 - B_1 - C_2 - B_1'$

Recalling

Word → Concept	Concept → Word
$C_1' \to C_1$	$C_1 \to C_1'$
$C_2' \to C_2$	$C_2 \to C_2'$
$B_1' \to B_1$	$B_1 \to B_1'$
$B_2' \to B_2$	$B_2 \to B_2'$

Thing → Shape, Its word, Color, Its word

$A_1 \to B_1, B_1', C_1, C_1'$

$A_2 \to B_2, B_2', C_1, C_1'$

Every recalling is completely accurate.

There are n chips on the board initially. Two players take any number of one to three chips alternately from these chips. The player who is forced to take the last chip loses the game. The restriction is that each player must not take the number which the opponent took at the previous move. For simplicity, presume that there are only six chips on the initial board. To play the game using the Associatron, first the components of vector x are assigned as shown in Table 4 to four patterns of the board position, the move, the mutual effect of the board position and the move, and the image of the result. Learning is performed in such a way that all sets of patterns of board position, strategy, mutual effect and the result, which have appeared in the game, are memorized one by one. The player with the associative memory (Player A), at every move, recalls the image of the result from the board position, a possible strategy

Table 4. Allocation of the Entity in Game Playing.

Board position 1, 2, 3, 4, 5, 6
Strategy 1, 2, 3
Board position × Strategy
Image of the result   Win, Lose

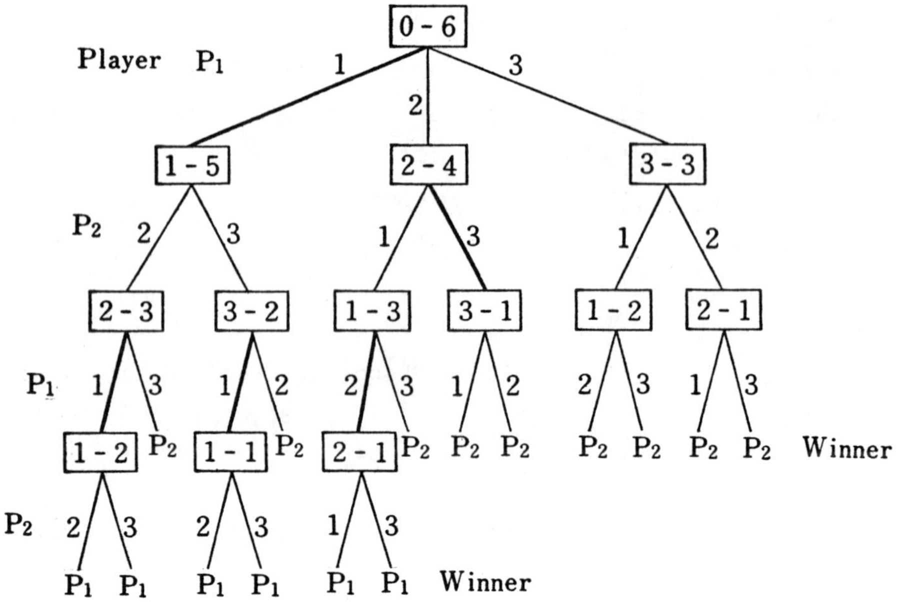

Figure 5. Game Tree.

and the mutual effect. The strategy, from which the image of "win" is recalled, is taken. If the images are the same through possible strategies, one strategy is chosen at random.

In this simple way, Player A is expected to make progress in developing his skill in the game, utilizing the property of the associative memory. In this experiment, it is assumed that the opponent (Player R) chooses one of the possible strategies at random. In training after each game, for the opponent's move, the image of the result is reversed from "win" to "lose" and vice versa, and is used. Therefore, Player A makes progress using his opponent's strategies, too.

Eighteen games have been played. As the result, the winners are R, A, A, A, A, A, A, A, A, R, A, A, A, R, R, A, R, A. Examination of the game tree yields the best player's strategy at every board position (Figure 5). Strategies of Player A are examined after a certain number of games, as shown below. At every possible board position, a check is made as to which image of "win" or "lose" Player A recalls from the set of board position, possible strategy at the board and their mutual effect. The result of the examination is shown in Table 5. The winning rate of Player A to Player R is also shown. Figure 4 shows it graphically. The graph shows that Player A has made progress in his skill through the learning process.

Table 5. Learning Process in Game Playing.

Board Strategy	After the nth game 2	3	5	9	16	Best Player
0 - 6 - 1	X	0	0	0	0	0
2	X	X	X	X	X	X
3	X	X	X	X	X	X
1 - 5 - 2	X	X	-	-	-	X
3	X	X	X	X	X	X
2 - 4 - 1	X	0	0	X	X	X
3	X	X	X	X	X	0
3 - 3 - 1	0	0	0	0	0	0
2	0	0	X	0	=	0
2 - 3 - 1	0	0	0	0	0	0
3	=	0	0	0	0	X
1 - 3 - 2	0	0	X	0	=	0
3	X	0	=	0	0	X
3 - 2 - 1	X	X	0	X	=	0
3	X	X	X	X	X	X
1 - 2 - 2	X	X	X	X	X	X
3	X	X	X	X	X	X
Winning rate %	58.3	54.2	62.5	58.3	70.9	91.7

0: Win, X: Lose, =: Neither, -: Reverse image of "Win"

HARDWARE REALIZATION

The hardware can be constructed as an iterative circuit shown in Figure 6(a). Since the matrix in eq. (2) is symmetrical, only half of the n×n memory units are required. The memory unit is represented by a simple automaton operating synchronously. Presume that the notation is assigned as in Figure 6(b). The automaton is described as

$$\begin{aligned} S(t+1) &= S(t) + X_1(t)X_2(t), \\ Y_1(t+1) &= \phi(S(t))\,X_2(t) + X_3(t), \\ Y_2(t+1) &= \phi(S(t))\,X_1(t) + X_4(t), \\ Y_3(t+1) &= X_1(t), \\ Y_4(t+1) &= X_2(t), \end{aligned} \tag{23}$$

where S corresponds to the value of memory unit and t is time. For the value of the unit, few levels are enough to operate the device specifically.

In this way, a memory device with 25 neurons, that is, 325 memory units, has been constructed for trial. Each memory unit is composed of 20 integrated circuit elements. For convenience, input and output

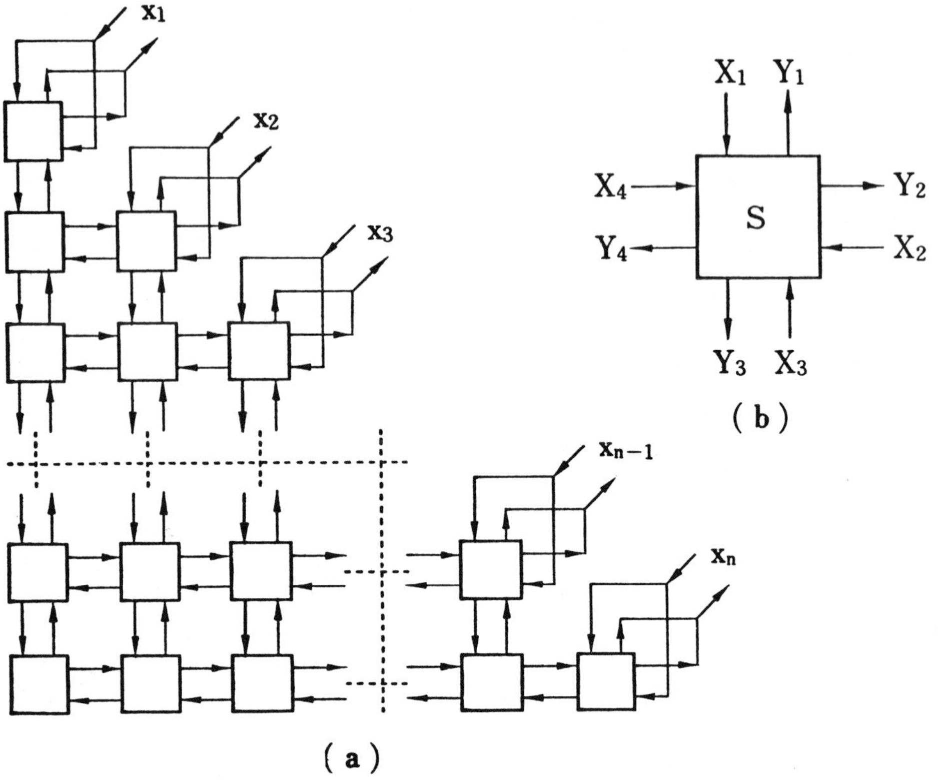

Figure 6. Hardware Realization.

parts are clearly separated in this device, and any part of the output pattern can be transferred to the input register manually or automatically. Thus, the recalled pattern can be immediately modified and be used as the next input. Although this device is very small in number of neurons, it is effective for some experiments concerning sequential recalling. The importance of this kind of experiments is shown in the next section.

THINKING BY THE SEQUENCE OF ASSOCIATIONS

If the recalling process is repeated in the method where the recalled pattern is used as the next input, a chain of associations may be traced. The process is considered as one of the thinking processes, rather than recalling the stored entities, because the structure of the associations was not directly memorized but has been formed in the memory through the experience from a piece of associations.

Though the real thinking process may be far more complicated, here we consider the simplest loop A → B → C → A by setting

$$x = (A,B,0), \quad y = (0,B,C), \quad z = (A,0,C) , \tag{24}$$

where $l(v_A) = l(v_B) = l(v_C)$ 60. After a number of such loops have been stored, an initial input of u = (A',0,0) is used to recall patterns sequentially, where A' is an arbitrary pattern. In the case of the stored loop number less than six, the loops can be separated completely. That coincides with the result of theoretical evaluation. When the number of loops is larger, various kinds of confusion take place, which is rather interesting from the point of view of machine intelligence. The behavior is being studied, but any remarkable conclusions are not yet obtained.

CONCLUSIONS

Information processing based on associative memory was studied. The model, named Associatron, memorizes entities distributively and recalls them associatively. Consequently, it has different properties than conventional memory devices. Those properties were analyzed and found useful for realizing machine intelligence. The ability of Associatron increases as the number of neurons increases. A few examples using the model simulated by a digital computer show that concept formation and game playing with learning are possible. In the example of game playing, it is found that the application of the method is not restricted to a special game, because it fully utilizes the properties of associative memory and the method itself is very primitive and simple. Hardware of the Associatron with 25 neurons, made for trial purposes, used integrated circuit elements. Although consideration is limited to "static" properties, or the properties of single recalling, it is suggested that "dynamic" behavior, or the sequence of associations, is more important.

ACKNOWLEDGMENT

This work was done under the supervision of Dr. Jin-ichi Nagumo, Professor of the University of Tokyo, and was supported by a grant from the Science and Technology Agency of Japan.

REFERENCES

1. P. B. Post, "A Lifelike Model for Association Relevance," Proc. Int. Joint Computer Conf. on Artificial Intelligence, pp. 271-280, May 1969.

2. K. Nakano, "'Associatron' and Its Applications," Conf. on Information Theory of Inst. of Electronics and Communication Engineers of Japan, Sept. 1969.

3. K. Nakano and J. Nagumo, "Studies on Associative Memory Using a Model of Neural Network," Conf. on Medical Electronics of Inst. of Electronics and Communication Engineers of Japan, June 1970.

4. A. Samuel, "Some Studies in Machine Learning Using the Game of Checkers," IBM J. Res. Develop., Vol. 3, No. 3, pp. 210-229, July 1950.

5. A. Samuel, "Some Studies in Machine Learning Using the Game of Checkers II - Recent Progress, IBM J. Res. Develop., Vol. 11, No. 6, p. 601, Nov. 1967.

6. F. Rosenblatt, "The Perceptron - A Probabilistic Model for Information Storage and Organization in the Brain," Psychol. Rev., Vol. 65, pp. 386-407, 1958.

7. B. Widrow, "Generalization and Information Storage in Networks of Adaline 'Neurons'," Self-organizing Systems, pp. 435-461, Spartan Books, Washington, D. C., 1962.

ADAPTIVE OPTIMIZATION IN LEARNING CONTROL

George J. McMurtry

The Pennsylvania State University

University Park, Pennsylvania, U.S.A.

INTRODUCTION

Adaptive and learning control systems normally require methods which search automatically for the set of parameters which optimize a performance index. Since environmental or other conditions may change in a learning situation, the optimization must occur continuously and adapt to the existing conditions. If the performance index is unimodal and noisy, but non-stationary, a combination of gradient and random methods yields a search method which exhibits the advantages of both the random and stochastic approximation methods. The overall rate of convergence is improved and under certain conditions can be optimized. In addition, the random component of such a search provides a capability for multimodal search applications.

Adaptive and learning control systems require methods which search automatically for the set of control parameters which optimize a performance index (IP). This IP is normally an unknown but measurable function of these parameters. Since environmental conditions may change or a priori information may be inaccurate and inadequate in a learning situation, the optimization must occur continuously and adapt to the existing conditions. Thus search procedures in a learning system may be labeled as adaptive optimization techniques. (McMurtry, 1970)

Search techniques may be classified according to whether the function (surface) to be optimized is unimodal or multimodal, single or nulti-dimensional, deterministic or noisy, and discrete or continuous. Since learning systems are frequently found to have an IP which is so complex that a mathematical solution is not feasible, it seems reasonable to assume that multimodal functions will be encountered. Wilde (1964) suggests that consideration of multi-dimensional problems also weakens belief in the concept of unimodality. If the IP is multimodal, the search procedure should be capable of finding the global extremum and not terminating on a local optimal point. In addition, each measurement of the IP may be subject to random errors, thus necessitating optimization of the expected value of the IP. The rate of convergence is an important consideration in applying any search procedure and the required searching time obviously increases with dimensionality, modality, and noise.

Most search techniques developed to date are sequential unimodal methods. A class of such methods called stochastic approximation has been specifically designed to converge to a unimodal optimum when noise is present. Random search techniques (Karnopp, 1963) have been proposed and used primarily as exploratory tools on unknown surfaces. Their simplicity and ease of implementation are appealing, but their lack of convergence properties have reduced their effectiveness. In recent years, there have been developed several sequential methods which have a random component in the search algorithm, but which are not entirely random in nature. Such methods will be called "controlled random" searches and it is the purpose of this paper to indicate their generality and to suggest that they possess the potential of being effective multimodal techniques.

Assume the IP is denoted as $I(\underline{y})$, and it is desired to find the unique value of $\underline{y}$, $\underline{y}_o$, which minimizes this index of performance, i.e.,

$$I_o = I(\underline{y}_o) < I(\underline{y}) \quad \text{for all } \underline{y} \varepsilon Y$$

where $\underline{y} = (y_1, y_2, \ldots, y_n)'$ is the vector of parameter states by which the system or process may be governed. The region Y is the subset of the y-space in which the function I(y) is defined.

Gradient techniques are perhaps the best known unimodal methods. The method of steepest descent measures the gradient of a unimodal $I(\underline{y})$ at each trial and makes the next step in the negative gradient direction. The gradient is defined as the vector

$$\underline{\nabla_I}(\underline{y}_k) = \left(\frac{\partial I}{\partial y_1}, \frac{\partial I}{\partial y_2}, \ldots, \frac{\partial I}{\partial y_n}\right)_k \qquad (1)$$

where $\frac{\partial I}{\partial y_i}$ is the partial derivative of $I(\underline{y})$ with respect to y_i evaluated at the kth trial, $\underline{y}_k$, of the search. (Since it is assumed that the expression for $I(\underline{y})$ is not known, the gradient components, $\frac{\partial I}{\partial y_i}$, may be measured at each trial by making a small test step in each direction, ∇y_i, and observing the resultant change in I(y), Δ_I. Then $\frac{\partial I}{\partial y_i}$ is estimated by $\frac{\Delta_I}{\Delta y_i}$.) The next trial is conducted at

$$\underline{y}_{k+1} = \underline{y}_k - a_k \underline{\nabla I}(\underline{y}_k) \tag{2}$$

where a_k is an arbitrary value which may or may not change with k. Thus the step size decreases with a reduction in the gradient magnitude, and in the limit when $\underline{\nabla I}(\underline{y}) = 0$ at the minimum, I_o, no change is made in $\underline{y}$.

Kiefer and Wolfowitz (1952) proposed a stochastic approximation method to search for the minimum of a bounded unimodal (but not necessarily convex) one-dimensional noisy surface I(y). The algorithm is basically that of the steepest descent method using noisy gradient measurements. Let

$$\tilde{I}(y) = I(y) + v \tag{3}$$

where v is a random error having zero mean and finite variance. Using two measurements made at $(y_k + \Delta_k)$ and $(y_k - \Delta_k)$, where Δ_k is a small test step, the algorithm is

$$y_{k+1} = y_k - a_k \tilde{\nabla} I(y_k) \tag{4}$$

where

$$\tilde{\nabla} I(y_k) = \frac{[\tilde{I}(y_k+\Delta_k) - \tilde{I}(y_k-\Delta_k)]}{2\Delta_k} \tag{5}$$

This algorithm converges to y_o with probability one and in mean square if the Dvoretsky conditions on a_k and Δ_k are satisfied. Blum (1954) generalized the Kiefer-Wolfowitz method to the n-dimensional case by using n+1 measurements to obtain

$$\tilde{\Delta} I_j = \tilde{I}(\underline{y}_k+\Delta_k \underline{e}_j) - \tilde{I}(\underline{y}_k) \qquad j=1,\ldots,n \tag{6}$$

where $\underline{e}_j$ is the unit vector in the y_j direction. Then

$$\underline{\tilde{\nabla} I}(\underline{y}_k) = \frac{1}{\Delta_k} (\tilde{\Delta} I_1, \tilde{\Delta} I_2, \ldots, \tilde{\Delta} I_n) \tag{7}$$

and

$$\underline{y}_{k+1} = \underline{y}_k - a_k \underline{\tilde{\nabla} I}(\underline{y}_k) \tag{8}$$

Gradient techniques have the advantages of using information obtained at each trial of the search and converging to the optimum under rather general unimodal conditions. They are easily applied and conceptually appealing. They are, of course, not convergent on multimodal surfaces. In addition, they have the disadvantage of requiring the measurement of the gradient at each trial, and each such measurement normally requires a minimum of n test steps. Each test step presumably requires as much time as each gradient search step. Rastrigin (1963) and Gurin and Rastrigin (1965) have shown that random search methods are faster than gradient techniques as the dimension (n) of the surface increases. Saridis from Purdue has indicated at this meeting that he has experimental results which favor the random search techniques for high dimensional spaces (p. 204).

Pure random search may be defined as a method in which the trial points, $\underline{y}$, are chosen by means of a random process defined by a probability density function, $p(\underline{y})$, representing the probability of choosing $\underline{y}$ in Y. A set of trials is then conducted with trial points, $\underline{y}_k$, selected according to the scan distribution, $p(\underline{y})$. The point yielding the smallest measured value of I is assumed to be $\underline{y}_o$. Random search techniques have the distinct advantages of simple implementation, insensitivity to discontinuities in I, and high efficiency when little is known about I. In addition, much general information is gained about I (and the corresponding system) during a random search. The random method makes no assumptions about the form of the response surface; it is applicable when the surface is multimodal, and, if n is not too large, the random search may be employed until a region of unimodality has been found and then one of the unimodal searches could be initiated. The principal disadvantage of the pure random search procedure is that it is conceptually a simultaneous search and therefore it does not take advantage of any previous results.

Brooks (1958) suggested two sequential variations of the random method. In the first procedure a pure random search is employed in a sequential manner. As each new point is randomly selected it is rejected as a test point if it is within a distance, d, of any previously selected point. Thus a set of non-overlapping hyperspheres of diameter d are tested. The choice of d is a design problem which must consider the resulting number of hyperspheres (or trial points) and the need for a near-uniform coverage of the parameter space. A similar procedure called clustering is frequently utilized in pattern recognition, while Waltz and Fu (1965) have used the clustering concept to partition a control system measurement space. In the second procedure, called the creeping random method, knowledge gained from previous trials is utilized. An initial trial is conducted at random and is followed by k trials selected from a normal scan distribution about the initial point.

Of these (k+1) trials, the best one is selected as the center of the distribution for the next (k+1) trials and if the variance of the normal scan distribution is reduced every (k+1) trials, then the method should converge to a local minimum.

Two controlled random search methods (Matyas and gradient biased random search) are now discussed and a relationship to the unimodal methods of steepest descent and stochastic approximation is shown. It is then noted that the controlled random search procedures have the additional capability of being used for multimodal search without altering their basic form.

Matyas (1965) suggests the simple random optimization method in which a random vector, $\underline{\zeta}_{k+1} = (\zeta_1, \zeta_2, \ldots, \zeta_n)_{k+1}$, is selected from a multivariable normal distribution with zero mean and unit correlation matrix. Then the next parameter vector is given by

$$\underline{y}_{k+1} = \underline{y}_k - \rho_k \underline{\zeta}_k + \underline{\zeta}_{k+1} \tag{9}$$

$$\rho_k = \begin{cases} 0 \text{ if } I(\underline{y}_k) < I^* (\underline{y}_{k-1}) & \text{(success)} \\ 1 \text{ if } I(\underline{y}_k) \geqq I^* (\underline{y}_{k-1}) & \text{(failure)} \end{cases} \tag{10}$$

where $I^*(\underline{y}_{k-1})$ is the smallest value of I found in the first (k-1) trials. The procedure is repeated until the optimum state, $\underline{y}_o$, is reached, or until a suitable stopping rule is executed. Matyas next suggests the adaptive random method which attempts to accelerate the search by using the information gained in previous trials to bias each new step. The algorithm now becomes

$$\underline{y}_{k+1} = \underline{y}_k + \underline{d}_{k+1} + T_{k+1} \underline{\zeta}_{k+1} \tag{11}$$

where T_{k+1} is a variable correlation matrix, $\underline{\zeta}_{k+1}$ is the same as before, and the vector $\underline{d}_{k+1}$ serves as the bias. Then $\underline{d}_{k+1}$ and T_{k+1} are made dependent upon the directions and lengths of previous steps and upon the outcome of those steps.

Pensa (1969) proposed the gradient biased random search (GBRS) in which the estimate of the negative gradient of $I(\underline{y})$ is used as $\underline{d}_{k+1}$. In the absence of a priori information concerning the function $I(\underline{y})$, the search algorithm may also be simplified, with no sacrifice of effectiveness, letting the covariance matrix become $T_{k+1} = b_{k+1}J$, where J is the identity matrix and b_{k+1} is the common standard deviation.

The search algorithm may then be written as:

$$\underline{y}_{k+1} = \underline{y}_k - a_k \underline{\tilde{\nabla} I}(\underline{y}_k) + b_{k+1}\underline{\zeta}_{k+1} \tag{12}$$

and it is noted that if $b_{k+1} = 0$, equation (12) yields equation (8). Thus GBRS can also be viewed as a stochastic approximation method with a random step added.

GBRS is specifically designed to converge to a unimodal optimum when measurement noise is present. In addition it allows for the specification of the matrix, T_{k+1}. Furthermore, the algorithm may, under certain conditions, be controlled to yield an optimum convergence rate (Pensa and McMurtry, 1970).

It is assumed that the gradient measurement errors are unbiased, i.e. $\{E\ \underline{\tilde{\nabla I}}(\underline{y}_k)\} = \underline{\nabla I}(\underline{y}_k)$, where the operator, E, denotes the mathematical expectation in the usual sense. Then the noisy measurement of the gradient may be written as:

$$\underline{\tilde{\nabla} I}(\underline{y}_k) = \underline{\nabla I}(\underline{y}_k) + \underline{\eta}_{k+1} \tag{13}$$

where $\underline{\eta}_{k+1}$ is an n-dimensional random vector representing the accumulative noise associated with the measurements of the index of performance used to estimate the gradient. It is assumed that each of the components of $\underline{\eta}_{k+1}$ is mutually stochastically independent and normally distributed with zero mean and finite variance. Thus the algorithm for gradient biased random search may be written as:

$$\underline{y}_{k+1} = \underline{y}_k - a_{k+1}\underline{\nabla I}(\underline{y}_k) - a_{k+1}\ \underline{\eta}_{k+1} + b_{k+1}\ \underline{\zeta}_{k+1} \tag{14}$$

Thus a direct application of the Dvoretzky conditions for the convergence of a stochastic approximation search yields a proof of convergence to $\underline{y}_o$ for equation (14) subject to the following conditions:

$$\begin{aligned} &a_{k+1} \sim k^{-(\varepsilon_A+1/2)} && 0<\varepsilon_A<1/2 \\ &b_{k+1} \sim k^{-(\varepsilon_B+1/2)} && 0<\varepsilon_B<1/2 \\ &E\{\underline{\eta}_{k+1}\} = 0 \text{ for all } k \\ &E\{\underline{\eta}_{k+1}\ \underline{\eta}_{k+1}^T\}\ \ \sigma^2 < \infty \quad \text{for all } k \end{aligned} \tag{15}$$

The rate of convergence of GBRS is thus a function of ε_A and ε_B. Pensa (1969) investigated the optimization of the rate of convergence on a noisy, unimodal, one-dimensional IP, using the same approach as Kabalevskii (1965). The discrete difference equation of the search algorithm is approximated by a continuous stochastic differential equation, the solutions of which are found to be trajectories of a continuous Markovian diffusion process. Using the time required to reach y_o as a search cost function to be minimized, it was found that the optimum value of ε_A is 0.5, while the optimum value of ε_B is a function of the surface being searched. Pensa suggests the use of an adaptive control policy on the choice of ε_B, although a fixed value of 0.5 will generally be quite satisfactory.

When viewed as a random search with a gradient bias, GBRS retains the rapid initial improvement of the random method while providing a faster rate of convergence in the later stages of a search. When viewed as a stochastic approximation method with an added random component, the GBRS improves the inherently slow initial convergence rate of stochastic approximation and, in addition, permits more mobility on large flat areas (plateaus) where gradient methods tend to become ineffective. The random component associated with controlled random search methods also provides the means by which they can be used for multimodal search without modifying the basic search algorithm or destroying the unimodal convergence properties.

Future research on controlled random search methods should include the extension of the convergence rate optimization of Kabalevskii to the multidimensional case. Determination of a suitable set of boundary conditions is necessary. Such an extension might include an investigation of multidimensional Markovian optimization in the discrete as well as the continuous time domain. Investigation of the multimodal convergence properties of controlled random search might include a Markov analysis such as that used by Vaysbord and Yudin (1968). Sugiyama of Osaka University has reported at this meeting that he has analyzed an algorithm identical to GBRS for multimodal performance and will soon be publishing his results.

In many applications, it is assumed that only one extremum exists. It seems entirely reasonable to assume that since few effective multimodal techniques exist, IP are often purposely designed so that they will have only one minimum and can be easily searched. It is therefore suggested that if more effective multimodal search techniques were available, more meaningful IP might be proposed and utilized without concern for either their analytical niceties or their desirable search properties. Consequently, investigation of the multimodal convergence properties of controlled random search methods is considered to be an important area for continued serious research.

REFERENCES

1. J. Blum, "Multidimensional Stochastic Approximation Method", Anals of Math. Stat., 25, pp. 400-407 (1954).

2. S. H. Brooks, "A Discussion of Random Methods for Seeking Maxima", Operations Research 6, No.2, pp. 244-251 (1958).

3. A. Dvoretsky, "On Stochastic Approximation", Proc. 3rd Berkeley Symp. on Math. Stat. and Prob., 1956.

4. L. S. Gurin and L. A. Rastrigin, "Convergence of the Random Search Method in the Presence of Noise", Automation and Remote Control 26, No.9, pp. 1505-1511 (1965).

5. A. N. Kabalevskii, "Analysis of Search Under Noisy Conditions by Means of Markov Processes", Automation and Remote Control 26, No.11, pp. 1938-1946 (1965).

6. D. C. Karnopp, "Random Search Techniques for Optimization Problems", Automatica 1, pp. 111-121 (1963).

7. J. Kiefer, and J. Wolfowitz, "Stochastic Estimation of the Maximum of a Regression Function", Annals of Math. Stat. 23, pp. 462-466 (1952).

8. J. Matyas, "Random Optimization", Automation and Remote Control 26, No. 2, pp. 244-251 (1965).

9. G. J. McMurtry, Adaptive, Learning, and Pattern Recognition Systems, Mendel and Fu, ed., Academic Press, 1970.

10. A. F. Pensa, "Gradient Biased Random Search", Ph.D. Thesis, Department of E.E., Penn State University, 1969.

11. A. F. Pensa and G. J. McMurtry, "Gradient Biased Random Search", Proc. IEEE 1970 Sys. Science and Cybernetics Conf., 1970.

12. L. A. Rastrigin, "The onvergence of the Random Search Method in the Extremal Control of a Many-Parameter System", Automation and Remote Control 24, No. 11, pp. 1337-1342 (1963).

13. E. M. Vaysbord and D. B. Yudin, "Multiextremal Stochastic Approximation", Engineering Cybernetics 6, No. 5, pp. 1-11 (1968).

14. M. D. Waltz and K. S. Fu, "A Heuristic Approach to Learning Control Systems", IEEE Trans. Auto. Control AC-10, No. 4, pp. 390-398 (1965).

15. D. J. Wilde, Optimum Seeking Methods, Prentice-Hall, Englewood Cliffs, New Jersey, 1964.

16. G. N. Saridis, "On a Class of Performance-Adaptive Self-Organizing Control Systems," this volume.

LEARNING CONTROL OF MULTIMODAL SYSTEMS BY FUZZY AUTOMATA

Kiyoji Asai
University of Osaka Prefecture
Osaka, Japan

Seizo Kitajima
Osaka City University
Osaka, Japan

INTRODUCTION

The random search method is well known as an optimization technique by which a global search of multimodal systems can be executed, but its convergence characteristics are not good. In order to improve the convergence characteristics of the random search method, an idea of the modification of the search probability may be used. There is the method of learning control using stochastic automata [1] as a method based on this idea. The proposed method of learning control using fuzzy automata in which the membership function [2] is used instead of the probability is more simple in the learning algorithm and able to realize more clear self-organizing operation as compared with that using stochastic automata.

W. G. Wee and K. S. Fu [3] have proposed a fuzzy automaton in which the transition from a state to another state may be executed on the basis of the membership function between these two states, and they have shown a learning system using this automaton. The authors have formulated a kind of fuzzy automata which have many outputs in a branch or a state, and have shown the learning behavior of new optimizing control systems using these automata. A part of these works have been presented in preliminary reports [4,5].

The ideas of the higher-order transition in the automata and the partition of the domain of objective function are introduced to the method of learning control, and the control systems can hold the true optimum at small hunting loss without staying at any local optimum point.

FUZZY AUTOMATA FOR LEARNING CONTROL SYSTEMS

A fuzzy automaton is a kind of automaton which will transfer from a state to another state or the same state via the branch whose membership function is the largest one among those of all branches diverging from the state, when an input is applied.

Two classes of fuzzy automata similar to the ordinary automata may be defined as follows: (i) Mealy type of fuzzy automata - the output will be sent out when a transition is executed between two states via a branch; and (ii) Moore type of fuzzy automata - the output will be sent out when a transition has arrived at the next state from a state. The transformable characteristics between these two automata have been shown by the authors [5].

These fuzzy automata may be applied to a learning control system with the controlled system whose characteristics are unknown. In this case, on the basis of the objective function, the membership functions which are entries of the transition matrix may be modified by a learning operation and the optimum may be searched. This modification is executed by the following equation:

$$f_{ij}(n+1) = \alpha f_{ij}(n) + (1-\alpha)\lambda \,, \qquad (1)$$

where $f_{ij}(n)$ is the membership function for the transition from state s_i to state s_j when the nth input is applied; λ is the limiting value (1 or 0) of $f_{ij}(n)$; α is the coefficient ($1 > \alpha > 0$); and n is the trial number.

In this equation, $f_{ij}(n)$ will converge to λ as $n \to \infty$, but the convergence speed depends on the value of α. The convergence behaviors are shown in Figure 1. From the figure it is clear that $f_{ij}(n+1)$ will rapidly converge to λ accordingly as the value of α becomes smaller. Therefore, $f_{ij}(n+1)$ may rapidly converge to λ with small hunting loss because the value of α will be small at the early stage of learning and be large with the progress of learning by modifying the value of α corresponding to the objective functions.

The value of α in eq. (1) is given as follows:

$$\alpha_n = 1 - |(I_n-\bar{I})/\bar{I}| \qquad (2)$$

where I_n is the nth value of objective function and $\bar{I}$ is the mean value of objective functions thus far obtained. Since $0 < \alpha_n < 1$ when $0 < I_n < \bar{I}$ or $\bar{I} < I_n < 2\bar{I}$, $f_{ij}(n+1)$ will converge to λ. Since $\alpha_n = 1$ when $I_n = \bar{I}$, $f_{ij}(n+1)$ will not be varied. Thus, the difference $(I_n-\bar{I})$ is large in the early stage, but the difference will be small with the progress of the learning. Consequently, $f_{ij}(n+1)$ is largely varied in the early stage, but $f_{ij}(n+1)$ is slightly varied in the final stage and rapidly converges to λ.

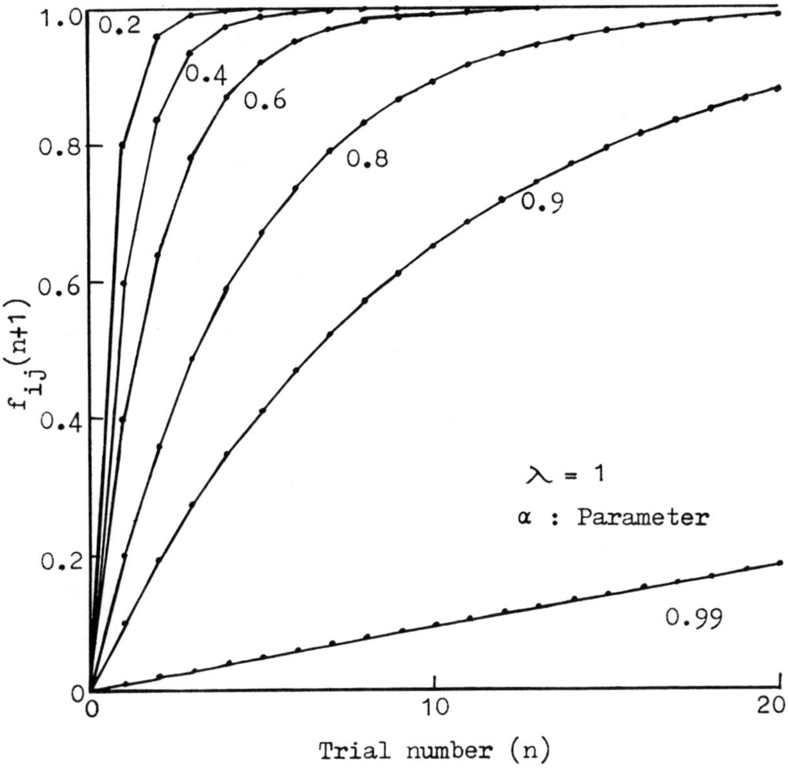

Figure 1. Convergence Behaviors of Membership Function $f_{ij}(n+1)$.

LEARNING CONTROL SYSTEMS

The block diagram of a learning control system using the fuzzy atuomaton is shown in Figure 2. The fuzzy automaton with ν states and $(\nu\times\nu)$ outputs is a class of the Moore type of fuzzy automaton with many outputs at a state. The transition is executed via the branch of the membership function given as

$$f_{ij} = \max_{h} \{\min(f_{ij}, g_{jh})\} \tag{3}$$

where g_{jh} is the membership function for the choice of an output U_h at the state s_j; and $h = 1,2,\ldots,\nu$. The outline of performance of the control system is as follows.

(i) When the nth input $x(n)$ arrives at the fuzzy automaton, a transition will be executed from state s_i to state s_j on the basis of the membership function $f_{ij}(n)$, and the nth output $u(n)$ which is generally called "control variable" will be sent out.

(ii) The nth output $y(n)$ will be sent out from the controlled system whose input-output characteristic is unknown, corresponding to the control variable $u(n)$.

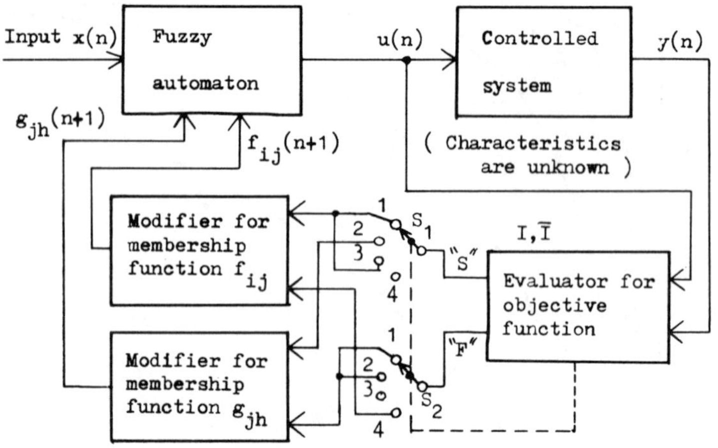

"S" : Success signal , "F" : Failure signal , S_1 and S_2 : Switches , 1 : Global search , 2 : Local search , 3 : Self loop , 4 : Selection .

Figure 2. Block Diagram of Learning Control System.

(iii) The nth objective function will be calculated from u(n) and y(n) in the evaluator for objective function. Also, the evaluator decides whether the nth objective function approaches the optimum value (success) or not (failure), and sends out the success signal or failure signal.

(iv) The n+1th membership function $f_{ij}(n+1)$ will be determined by modifying $f_{ij}(n)$ on the basis of the success or the failure signal. By increasing or decreasing the membership function of the branch, the grade of transition by the branch of success or failure will be raised or made low, respectively.

For simplicity of explanation, the system performance has been mentioned above on the case of learning control by the first order transition in the automaton, without considering the subdomains, but the authors have introduced the ideas of the higher-order transition and the partition of the domain of objective function in order to realize a global search and the control at small hunting loss, the details of which will be given below.

(i) The domain of the objective function for optimization is divided into some number of subdomains corresponding to the numbers of states in the fuzzy automaton, and every subdomain is also divided into some unit domains corresponding to the outputs from a state.

(ii) In the first, the membership function f_{ij} and g_{jh} are set in the middle value 0.5 between 0 and 1 so that transition from a state to an arbitrary state can be executed.

(iii) For the purpose of finding a path that loops from a state s_i to the same state s_i via k branches, a kth-order transition matrix is calculated, and the maximum value [$f^k_{ii,max}$] of membership function on the diagonal entries in the matrix and the number [M] of these maximum membership functions are found.

(iv) Comparing M with k, if $M < k$, Mth-order transition matrix is calculated in order to remove overlapping branches, and then the same procedure as (ii) is executed. This procedure is repeated until all overlapping branches are removed from the loop path. Let the maximum value of membership function and the number of the maximum membership functions thus obtained be $f^k_{ii,max}$ and m, respectively.

(v) Since the k branches on the single loop leading from s_i to s_i should have the membership functions that are equal to or larger than $f^k_{ii,max}$, the loop path may be found by comparing $f^k_{ii,max}$ with the membership functions for k branches leading from s_i to s_i. In this case, if the single loop leading from s_i to s_i cannot be found, the procedures (ii)-(iv) should be repeated for the (k-1)th-order transition matrix or the transition matrix of lower order than the (k-1)th until the single loop is obtained.

(vi) When the transition is executed on the path obtained in (iv), k control variables are sent from the automaton to the controlled system from which k outputs will be sent out, and then k objective functions $I_k(n)$ and the mean value $\bar{I}(n)$ of all objective functions thus far obtained are calculated.

(vii) Comparing k objective functions $I_k(n)$ with the mean value $\bar{I}(n)$, the (n+1)th membership functions $f_{ij,k}(n+1)$ and $g_{jh,k}(n+1)$ are respectively determined from the nth membership functions $f_{ij,k}(n)$ and $g_{jh,k}(n)$ by using the algorithm as shown in the following:

$$\left.\begin{aligned} f_{ij,k}(n+1) &= \alpha f_{ij,k}(n) + (1-\alpha) \quad \text{if } I > \bar{I} \quad \text{(success)} \\ g_{jh,k}(n+1) &= \alpha g_{jh,k}(n) \qquad\qquad\quad \text{if } I \leqq \bar{I} \quad \text{(failure)} \end{aligned}\right\} \tag{4}$$

where $\alpha = 1 - |(I-\bar{I})/\bar{I}|$, but we regard α as 0.99 if the calculated value of α is equal to 1. In this case, the global search may be executed and the switches S_1 and S_2 are connected to the terminal "1" in Figure 2.

(viii) Since the optimum output is probably included in the state s_j, if the membership functions f_{ij} modified by the algorithm are larger than some value (e.g., 0.8) in case of success, the outputs whose membership functions g_{jh} are larger than some value (e.g., 0.45) will be applied to the controlled system, and then the membership functions of the branches that have not been modified will be set to the maximum value of g_{jh} in order to facilitate moving to the state s_j. In this case, the local search may be executed and the switches S_1 and S_2 are connected to the terminal "2" in

Figure 2. The membership function f_{jj} will be magnified if f_{ij} grows very large (e.g., 0.9). In Figure 2, S_1 and S_2 are connected to the terminal "3".

(ix) Since the optimum output is probably not included in the state s_j if $g_{jh}(n+1)$ grows less than some value (e.g., 0.2), f_{ij} will be made small. In Figure 2, S_1 and S_2 are connected to the terminal "4".

From the explanation mentioned above, it will be seen that the operation of self-organization has been performed to optimize the system in the automaton as follows. The membership functions of the branches that contribute to success will approach to 1, while those of the branches that cause failure will approach to 0. Therefore, the proper ones for getting the optimum are selected from among all the branches in the automaton while the unsuitable ones are weeded out.

SIMULATION RESULTS

A simulation study has been carried out for the purpose of investigating the behaviors of the optimizing control using the system described above. In this computer simulation, the following equation [6] has been used as an objective function that includes the characteristics of controlled system.

$$I(u_1,u_2) = (1 + 8u_1 - 7u_1^2 + \frac{7}{3} u_i^3 - \frac{1}{4} u_1^4)\, u_2^2 \cdot e^{-u_2} . \qquad (5)$$

This objective function is the function with two variables u_1,u_2 as control variables and includes an optimum point, a local optimum point, and a saddle point. We assume that the measured value of $I(u_1,u_2)$ includes an observational error in proportion to random numbers with normal distribution (standard deviation σ) in the calculated value of (5).

An example of the results of simulation study are shown in Figure 3. From Figure 3 it may be seen that a learning control has been executed as follows. A global search was executed at an early stage of learning, but the operation of reducing the domain of search was executed with the progress of the learning. The states (subdomains) 3 and 4 were searched over all outputs (unit domains) since the optimum is probably included in some domain among these two domains, and then a loop path which connects the states 3 and 4 was formed. At the final stage of learning, the state 4 was selected and the optimum was maintained by the transition by the self loop at the state 4 .

In order to show the convergence characteristics of the learning control, the convergence behavior of the membership functions f_{ij}

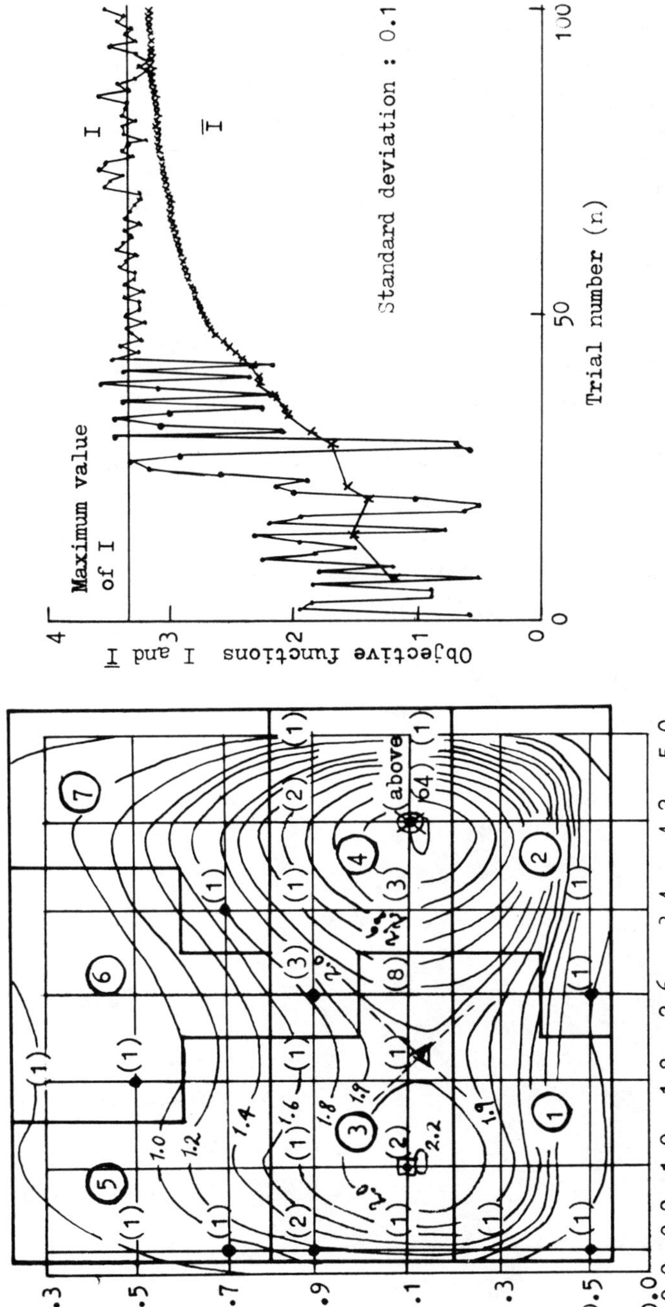

Figure 3. An Example of Results of Simulation Study.

and g_{jh} of the fuzzy automaton in the simulation study is shown in Figure 4. In this figure, f_{ij} is the membership function of the branch which arrived at the state including the optimum point and g_{jh} is the membership function of the choice of the output in the state. Also shown is the case of constant α in comparison with the case of variable α. As is evident from the figure, the smaller the value of α grows, the faster the membership functions converge to λ, when the value of α is constant. Also, the membership functions may converge rapidly to λ when the value of α is varied.

CONCLUSIONS

In the preceding paragraphs, the learning control systems using the fuzzy automata have been described. From the results of experimental investigation, it has been concluded that the learning control systems using fuzzy automata are able to avoid the continuation of trials in the vicinity of a local optimum by global search

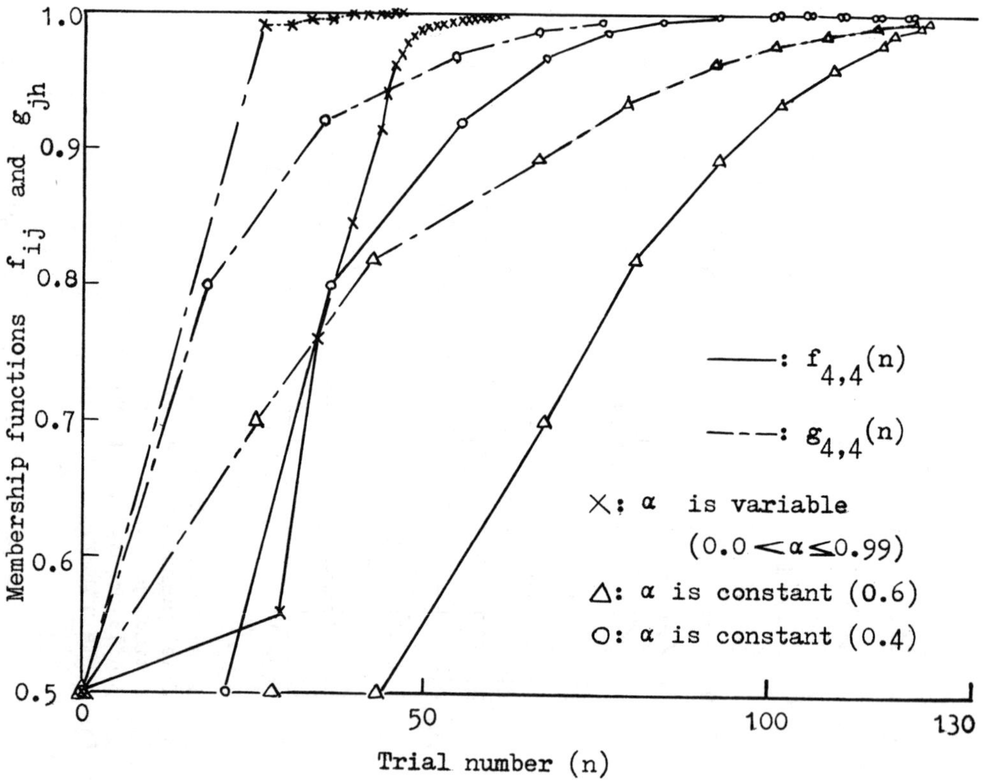

Figure 4. An Example of Convergent Characteristics of Learning Control.

and to get the performance of small hunting loss at the steady state. These merits are due to the ideas of the higher-order transition of the fuzzy automata and the partition of the domain of objective function.

It has also been clarified that the convergence characteristics of the learning control can be improved by using α varied with respect to the objective functions. In addition, it has been seen that the operation of self-organization is performed to optimize the system in the automata. This operation is more clearly performed than in the case of the learning control system using the stochastic automata because, in the stochastic automata, the transition characteristics are at random in the whole duration of learning, while in the fuzzy automata, the transition is randomly executed at the initial period, but the transition path is gradually determined with the progress of learning, and the transition path is perfectly decided at the final period of learning.

The authors would like to express their appreciation to Professor L. A. Zadeh and Professor K. S. Fu for their valuable discussions.

REFERENCES

1. G. J. McMurtry and K. S. Fu, "A Variable Structure Automaton Used as a Multimodal Searching Technique," IEEE Trans. on Automatic Control, Vol. AC-11, No. 3, July 1966, pp. 379-387.

2. L. A. Zadeh, "Fuzzy Sets," Information and Control, Vol. 8, No. 3, June 1965, pp. 338-353.

3. W. G. Wee and K. S. Fu, "A Formulation of Fuzzy Automata and Its Application as a Model of Learning Systems," IEEE Trans. on Systems Science and Cybernetics, Vol. SSC-5, No. 3, July 1969, pp. 215-223.

4. H. Hirai, K. Asai and S. Kitajima, "Fuzzy Automaton and Its Application to Learning Control Systems," Memoirs of the Faculty of Engrg., Osaka City Univ., Vol. 10, Dec. 1968, pp. 67-73.

5. K. Asai and S. Kitajima, "A Method for Optimizing Control of Multimodal Systems Using Fuzzy Automata," Information Sciences, Vol. 3, No. 1, 1971, pp. 1-11.

6. H. E. Zellnik, N. E. Sondak and R. S. Davis, "Gradient Search Optimization," Chem. Engrg. Prog., Vol. 58, No. 8, August 1962, pp. 35-41.

ON A CLASS OF PERFORMANCE-ADAPTIVE SELF-ORGANIZING CONTROL SYSTEMS

George N. Saridis

Purdue University

Lafayette, Indiana, U.S.A.

I. INTRODUCTION

The purpose of this article is twofold: first to define rigorously the discipline called Self-Organizing Control which deals mainly with the on-line control of systems with completely or partially unknown dynamics operating in an unknown stochastic environment; second to propose a solution to a class of performance-adaptive self-organizing controls which presents a mathematical interpretation of a "behavioral" approach to the problem. Self-organizing control is a discipline of the class of "adaptive and learning" control systems which deals on-line with systems with inherent uncertainties which are either impossible or too costly to be measured directly. The performance-adaptive self-organizing control attempts to control the system and at the same time reduce on-line the uncertainties pertaining to the improvement of the performance of the system by subsequent observations. The performance-adaptive self-organizing procedure proposed here is defined as follows. A "subgoal" criterion is defined over a sequence of expanding intervals of time, the improvement of which renders a "global" random search to converge in the limit to the min.value of a given "overall" performance criterion. The random search updates the coefficients of a specific controller formed by a linear combination of independent properly chosen functions of the measurable output. A special case of the method has been successfully applied by the author to the stochastic fuel regulator problem to yield an asymptotically optimal controller [1]. Definite advantages over other existing approximations to the optimal solution are discussed.

II. SELF-ORGANIZING CONTROL SYSTEMS

The scope of the discipline defined as self-organizing control is the on-line solution of a control problem of inherently higher complexity and in the presence of uncertainties regarding the system. Such uncertainties may appear because of poor modelling in the processes that are too costly to be tested off-line or because of variations in the system parameters due to unpredictable changes in the environment. Such are the situations that arise in many industrial, space, and bioengineering control problems.

In order to establish the discipline on a rigorous basis, the following definitions are appropriate.

Definition 1. A control problem will be called self-organizing (SOC) if it is possible to reduce the *a priori* uncertainties pertaining to the effective control of the process through subsequent observations of inputs and outputs as the process evolves.

Definition 2. A controller for a self-organizing control problem will be called self-organizing if it accomplishes on-line reduction of the *a priori* uncertainties pertaining to the effective control of the process as it evolves.

The above definitions introduce a new feature to the control problem, whether it is implemented by a classical or an optimal controller. This new feature is the reduction of uncertainties, and it may involve parameter or system identification to be used in the adaptation of a prestructured controller, or it may estimate the parameters pertaining directly to the improvement of the performance of the system in a deterministic or stochastic environment indiscriminately. Thus, self-organizing control can handle more effectively systems with completely or partially unknown dynamics operating in a stochastic environment with unknown statistics than Stochastic Optimal Control for which the systems uncertainties are irreducible as the process evolves or the Differential Game approach which takes the "worst case design" point of view.

SOC takes a "behavioral" approach to the problem very similar to Learning Control. It differs, however, from the latter discipline because it does not admit any "off-line" training period for the controller and it provides room for some "classical" adaptive control techniques as well as on-line structural adjustments in the controller. In order to distinguish between these two cases, the following four definitions are given.

Definition 3. A self-organizing control process will be called parameter-adaptive if it is possible to reduce the *a priori* uncertainties of a parameter vector characterizing the process through subsequent observations of the inputs and outputs as the control process evolves.

Definition 4. A self-organizing controller will be called parameter-adaptive if it accomplishes on-line reduction of the uncertainties of the parameter vector through subsequent observations of the inputs and outputs as the control process evolves.

Definition 5. A self-organizing control process will be called performance-adaptive if it is possible to reduce directly the uncertainties pertaining to the improvement of the performance of the process through subsequent observations of the inputs and the outputs as the control process evolves.

Definition 6. A self-organizing controller will be called performance-adaptive if it accomplishes on-line reduction of the undertainties related to the direct improvement of the performance of the process through subsequent observations of the inputs and the outputs as the control process evolves.

Examples of parameter-adaptive and performance-adaptive SOC approaches are found in references [2] and [1], respectively.

III. THE GENERALIZED EXPANDING-SUBGOAL PERFORMANCE-ADAPTIVE SOC

A specific performance-adaptive SOC approach is proposed here for systems with completely or partially unknown dynamics operating in an unknown stochastic environment where asymptotic optimization of an infinite-time process is required. Its major features may be summarized as follows.

(1) The controller does not depend structurally or parametrically on the explicit form of the plant's dynamics or the noise statistics. Therefore, minimal information about the structure of the plant is required.

(2) The combined estimator and control depends only on the input-output measurements of the plant.

(3) There is a definite relation between the per-interval cost or subgoal improved during the evolution of the process and the overall performance index of the system. Therefore, the proposed algorithm may be used for the direct asymptotic minimization of the overall performance index.

(4) The search technique possesses global properties and is very easy to implement.

A mathematical formulation of the problem is given here, and a performance-adaptive, self-organizing controller designed to improve a "subgoal" measured on a sequence of expanding intervals is proposed. This "subgoal" is properly related to the overall performance index such that an overall minimization in the limit is obtained.

Consider a process governed by the following relations:

$$\dot{x}(t) = f(x(t),u(t),\xi(t),t); x(t_o) = x_o, \; u(t) \in \Omega_u \tag{1}$$

$$z(t) = g(x(t),\eta(t),t) \tag{2}$$

where $x(t)$ is an ℓ-dimensional state vector in the state space Ω_x; $u(t)$ is an m-dimensional control vector in a closed and bounded set Ω_u of admissible controls; x_o, $\xi(t)$ and $\eta(t)$ are three vector-valued random variables of dimensions ℓ, q, and r, respectively, belonging to independent stationary processes with unknown statistics; $z(t)$ is an s-dimensional output vector, $s \leq p$; $f(\cdot)$ and $g(\cdot)$ are unknown nonlinear continuous and bounded functions of their arguments of appropriate dimensions. The vectors $\xi(t)$ and $\eta(t)$ represent the plant-environment relationship. Given the performance index

$$I(u) = E\{J(u)\} = E\{ \lim_{t_n \to \infty} \frac{1}{t_n - t_o} \int_{t_o}^{t_n} L(z(t),u(t),t)dt\} \tag{3}$$

where L is a known nonlinear, continuous-bounded, nonnegative operator.

The following performance-adaptive, self-organizing solution to the problem is proposed. A specific controller $u_c(t) \in \Omega_u$ designed such that it will sequentially improve a "subgoal" related to the performance index (3) which is to be minimized asymptotically. Such a controller may be constructed as follows:

$$u_c(t) = F\{c^T \phi(z(t),t)\} \in \Omega_u \tag{4}$$

such that $I(u_c)$ is bounded for all controls $u_c(t) \in \Omega_u$, the admissible set; i.e.,

$$0 \leqq I_{min} \leqq I(u_c) \leqq I_{max} < \infty \;\forall\; u_c(t) \in \Omega_u \tag{5}$$

where $\phi(\cdot)$ is a p-dimensional vector of linearly independent, continuous and bounded functions of the measurable outputs properly chosen to span Ω_u; c is a p-vector of adjustable coefficients; and $F(\cdot)$ is an m-dimensional nonlinear bounded operator in Ω_u. Condition (5) implies stability of the system; therefore, I_{max} may be an arbitrary but finite positive number.

Define as a "subgoal" the time average of J(u)

$$J(c_i) = (t_i - t_{i-1})^{-1} \int_{t_{i-1}}^{t_i} L\big(z(t), F\{c_i^T \phi(z(t),t)\}\big)\,dt \qquad (6)$$

$$i=1,2,\ldots,\infty$$

where the length of the interval $T_i = t_i - t_{i-1}$, $i=1,2,\ldots,\infty$ satisfies the conditions

$$T_{i-1} \leqq T_i \quad \forall\, i \text{ and } \lim_{i\to\infty} T_i = \infty\,. \qquad (7)$$

Then c_i^* may be sought through an appropriate search technique applied once at every interval to improve sequentially the value of $J(c_i)$. Suppose that a global random optimization technique is applied for this purpose at every subinterval T_i;

$$c_{i+1} = \begin{array}{l} \rho_i,\ J(c_i) - J(\rho_i) > 2\mu \\ c_i,\ J(c_i) - J(\rho_i) \leqq 2\mu \end{array} \quad \text{at } T_i \qquad (8)$$

where ρ_i is the ith sample value of a vector random variable defined at every subinterval T_i on $\Omega_c = \{c/u_c(t)\varepsilon\Omega_u\}$ with probability distribution function $P(\rho) \neq 0$ for all $\rho\varepsilon\Omega_c$ and μ is an arbitrary positive number. Define the "cost" accrued during the search by

$$V(n) \triangleq \frac{\sum_{i=1}^{n} T_i E_{\xi,\eta}\{J(c_i)\}}{\sum_{i=1}^{n} T_i} \qquad n=1,2,\ldots \qquad (9)$$

where T_i, $i=1,2,\ldots,$ satisfy conditions (7). Define c^* as the value of c that minimizes the performance index (3) with the controller $u_c(t)$ of eq. (4) applied:

$$c^* = \{c/\operatorname*{Min}_c I(u_c) = I_{min}\}\,. \qquad (10)$$

It can easily be shown that after successive applications of c^* for c_i in V(n), i.e.,

$$c_i = c^*, \quad i=1,2,\ldots, \qquad (11)$$

the "cost" is the accumulative performance index over the interval $[t_o, t_n]$:

$$V(n) = \frac{1}{t_n - t_o}\, E\Big(\int_{t_o}^{t_k} L(z(t), F\{c^{*T}\phi(z,t)\}t)\,dt\Big) \qquad (12)$$

and in the limit using eq. (10):

$$\lim_{n\to\infty} V(n) = I(u_c^*) = \operatorname{Min}_c I(u_c) = I_{min} . \tag{13}$$

Then the following theorem and corollary establish the asymptotic optimality of the algorithm in probability, and in the rth mean.

Theorem 1. Given an arbitrary number $\delta > 0$ and the random search (8) with the following properties:

(P1) the function $J(c)$ satisfies (4) and is such that $\Omega_c^* \triangleq \{c | J(c)-J_{min}<\delta, \forall\, \delta, c\varepsilon\Omega_c\}$ has positive measure;

(P2) conditions (7) are satisfied; and

(P3) for any positive numbers ν and n, $\nu \leq n$,

$$\lim_{\nu/n\to 0} \frac{\sum_i^{\nu} T_i}{\sum_{i=1}^{n} T_i} = 0 ,$$

where $\sum_i^{\nu} T_i$ denotes a sum containing ν not necessarily ordered terms of the denominator,

then there exists a number $\mu > 0$ such that as the number of iterations n goes to infinity

$$\lim_{n\to\infty} P(n) \triangleq \lim_{n\to\infty} \operatorname{Prob}\{V(n)-I_{min} \leq 3\delta\} = 1 . \tag{14}$$

Corollary 1. Let Theorem 1 be satisfied, then $V(n)$ converges to $I_{min}+ 3\delta$ in the rth mean for arbitrary $\delta > 0$

$$\lim_{n\to\infty} E\{|V(n)-[I_{min}+3\delta]^{\dagger}|^{r}\}= 0 \; ; \quad 0 < r < n \tag{15}$$

where $[x]^{\dagger} \triangleq 1/2\, [x+|x|]$.

Proofs of Theorem 1 and Corollary 1 are given in the Appendix.

It was found necessary to evaluate $J(c_i)$ at every iteration of algorithm (8) in order to have updated values of the performance index for comparison during the transient period of the system and thus improve the computational results. Since this procedure is done on-line, the additional cost must be included in the accumulative cost function (9) which is redefined by

$$\overline{V}(n) \triangleq \frac{\sum_{i=1}^{n} E_{\xi,\eta}\{T_i J(c_i)+\overline{T}_i J(\rho_i)\}}{\sum_{i=1}^{n} [T_i+\overline{T}_i]} \; ; \; \overline{T}_i = \begin{matrix} T_i, & c_{i+1}=\rho_i \\ 0, & c_{i+1}=c_i \end{matrix} \tag{16}$$

In this case Theorem 2 is given to establish the convergence of $\overline{V}(n)$ to I_{min} in probability and in the rth mean.

Theorem 2. Given an arbitrary $\delta > 0$ and the random search (8) satisfying the properties (P1) to (P3) of Theorem 1 and the following additional property:

(P4)
$$\lim_{n\to\infty} \frac{\sum_{i=1}^{n} \overline{T}_i}{\sum_{i=1}^{n} [T_i+\overline{T}_i]} = 0 ,$$

then there exists a number $\mu > 0$ such that

$$\lim_{n\to\infty} \overline{P}(n) \triangleq \lim_{n\to\infty} \mathrm{Prob}\{\overline{V}(n)-I_{min} \leq 4\delta\} = 1$$

and

$$\lim_{n\to\infty} E\{|\overline{V}(n)-[I_{min}+4\delta]^{\dagger}|^{r}\} = 0, \; 0 < r < \infty$$

The proof of Theorem 2 is also given in the Appendix. A general system with the implementation of the self-organizing algorithm is illustrated in Figure 1.

IV. EXPERIMENTAL RESULTS

The method has been successfully applied to the control of a stochastic fuel regulator problem with a randomly switching delay and has been reported in [1].

New results have been obtained by applying the performance-adaptive SOC algorithm to a linear system. The reason for such a selection was that the output and its derivatives are the obvious choice for ϕ functions of the controller and stability of the system, guaranteeing condition (5), can be maintained by restricting the parameters c to an appropriately computed set Ω_c which will force the eigenvalues of the overall system to have always negative real parts.

The system considered here is given by

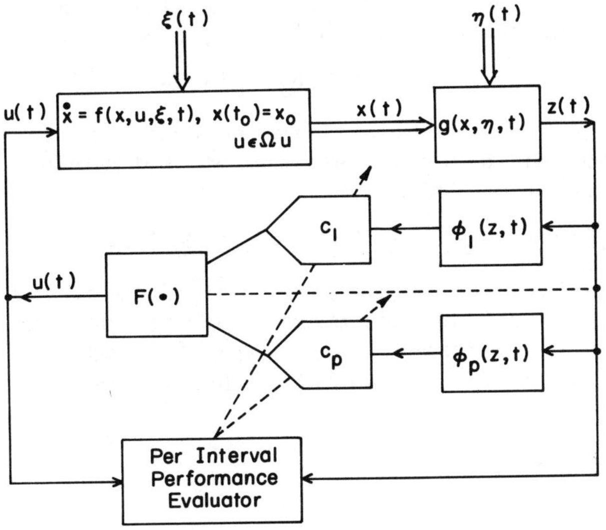

Figure 1. The Performance Adaptive Self-Organizing Control System.

$$\dot{x} = \begin{bmatrix} 0 & 1 \\ -2 & -4 \end{bmatrix} x + \begin{bmatrix} 0 \\ 0.5 \end{bmatrix} u + \begin{bmatrix} 1 \\ 0 \end{bmatrix} w \qquad \begin{matrix} a_1=-2, & b_1=0 \\ a_2=-4, & b_2=0.5 \end{matrix}$$

$$z = (1,0)x + v$$

where

$$w \sim N(0,1)\ , \qquad E\{v(t)w(t-z)\} = 0, \quad -\infty < t < \infty$$

$$v \sim N(0,1)$$

$$x(0) \sim N(0, \begin{bmatrix} 3 & 0 \\ 0 & 2 \end{bmatrix}).$$

The performance index is given by

$$I = E\{\lim_{T\to 0} \frac{1}{T} \int_0^T (z^2+u^2)dt\ .$$

The self-organizing controller is designed to be

$$u(t) = F[c_1\phi_1(z) + c_2\phi_2(z)]$$

where

$$\phi_1(z) = z, \quad \phi_2(z) = \dot{z}\ , \quad F = \left(\frac{d^2}{dt^2} + f_2\frac{d}{dt} + f_1\right)^{-1}$$

and the adjustable parameters are c_1, c_2, f_1 and f_2.

Condition (5) and therefore stability of the system is guaranteed by chosing the set Ω_c such that

$$\Omega_c = \{f_1 < -1,\ f_2 < -2,\ 0 < g_1 < 1,\ g_2 > -1\}\ .$$

This set has been selected to satisfy the Routh criterion for the augmented system and under the assumption that the plant parameters belong to the set

$$\Omega_a = \{a_1 < -1,\ a_2 < -1,\ b_1 = 0,\ 0 < b_2 < 3\}\ .$$

The system is depicted in Figure 2 and the plots of the per-interval performance index $J(c_N)$ and accumulated performance index $V_{(N)}$ in Figure 3 and Figure 4, respectively. $V_{(N)}$ is compared favorably to the optimal solution to the stochastic optimal control problem with known parameters.

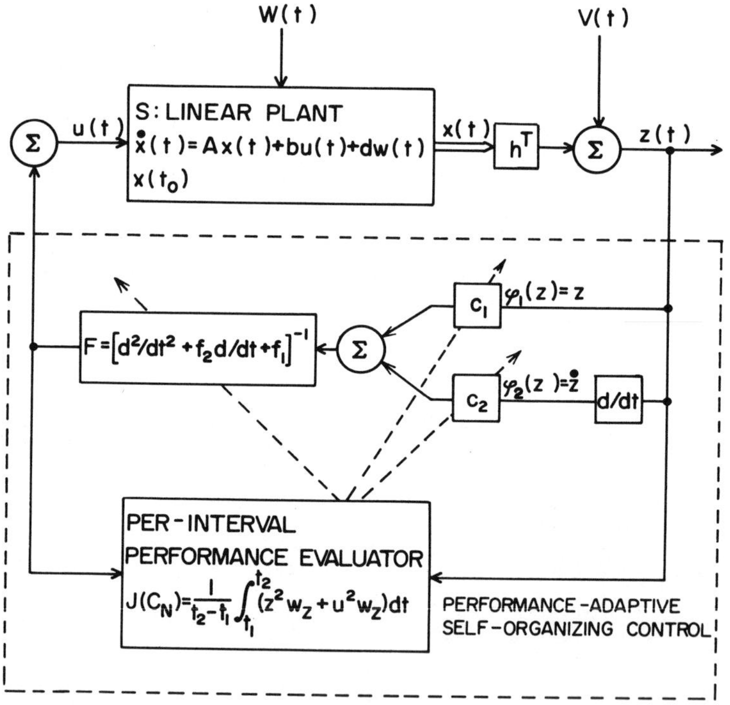

Figure 2. The Performance-Adaptive S.O.C. For a Linear System.

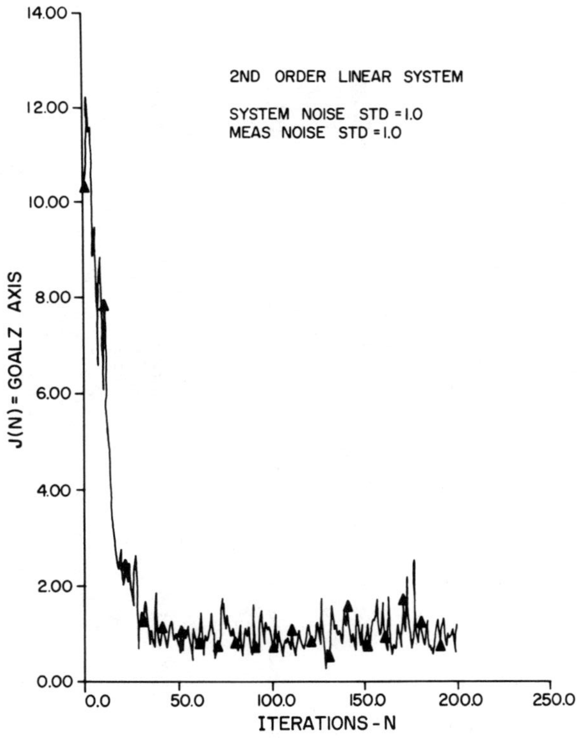

Figure 3. Per-Interval Performance Index $J(C_N)$ Goalz vs. Iterations N.Random Search Technique.

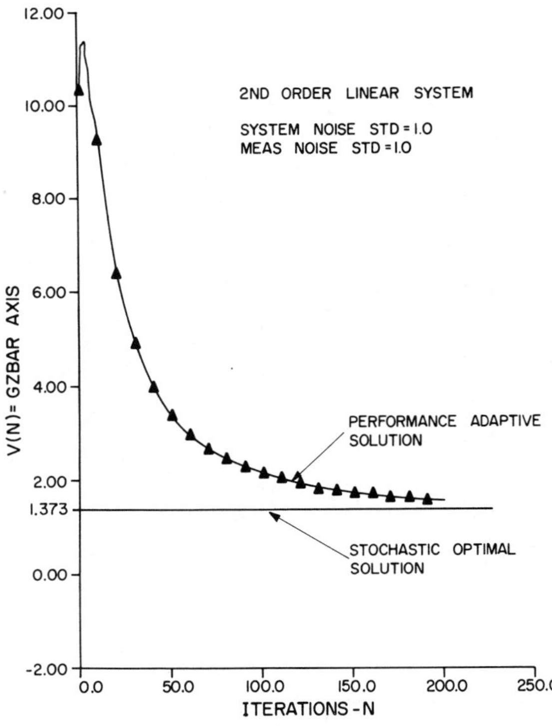

Figure 4. Accumulated Performance Index (V(N) Gzbar vs. Iterations N.Random Search Technique.

V. DISCUSSION

The discipline of Self-Organizing Control was rigorously established in this article and a general solution has been proposed for a class of Performance-Adaptive S.O.C. problems. Similar results have been presented in [2] for the parameter adaptive S.O.C. problem. The advantages of the method proposed here can be summarized as follows:

1. The controller does not explicitly depend on the plant dynamics and noise statistics, but only on input-output measurements so that it can be implemented on-line.

2. A definite relationship has been established between the per interval cost, representing the subgoal, and the overall cost representing the goal in the "behavioral" interpretation of the problem.

3. A global search technique has been used guaranteeing a global minimum of the overall performance index. This algorithm is shown to be more efficient for high dimensional systems than the conventional gradient type methods. The convergence is satisfactory; it can be further improved by using acceleration methods proposed in the literature.

4. The controller is very simple to implement and therefore desirable for practical applications.

The following problems concerning the proposed algorithm have not been resolved yet and they presently are the subject of further investigation.

1. Judicious selection of the set Ω_c to satisfy condition (5) and therefore guarantee stability of the system in the general case where the structure is unknown.

2. Effective choice of the ϕ functions of the controller for completely unknown plants.

3. Effective choice of the growth of the subinterval T_i for optimum convergence rate.

However, the method seems very promising and has given mathematical rigor to heuristic "behavioral" ideas used in learning control problems.

APPENDIX A

Proof of Convergence of Theorem 1

Three Lemmas are pertinent to the proof of convergence of Theorem 1. These proofs are given in [1.2].

Lemma 1. Given a dependent test sequence $\{x_i\}$ where the conditional probabilities of success of each test are bounded from above by the number $\bar{\gamma}$, for any outcome of the remaining tests. Construct an independent test sequence $\{y_i\}$ with probability of success in each test equal to $\bar{\gamma}$. Define $P_x(n_1/n)$ and $P_y(n_1/n)$ as the probabilities of obtaining no less than n_1 successes with n tests of the sequences. Then

$$P_x(n_1/n) \leq P_y(n_1/n)$$

Lemma 2. Given a sequence $\{x_i\}$ of dependent random variables with the conditional distribution function $P(x_i,\lambda)$ of the variable x_i for any fixed values of the variables $x_j, j\neq i$. Construct a sequence $\{y_i\}$ of independent identically distributed (i.i.d) random variables with a distribution function $P(y_i,\lambda) \leq P(x_i,\lambda)$. Define $P_x(\lambda,n)$ and $P_y(\lambda,n)$ the distribution functions of the sums

$$\sum_{i=1}^{n} x_i \quad \text{and} \quad \sum_{i=1}^{n} y_i$$

respectively. Then

$$P_x(\lambda,n) \geq P_y(\lambda,n)$$

Lemma 3. Given an arbitrary $\delta>0$ and a segment of absolutely successful iterations of length n_i, i.e. $E\{J(c_j) - J(\rho_j)\} > \mu$, $j=k,\ldots,k+n_i$, satisfying properties (P1) and (P3). Define

$$P(\bar{n}) \triangleq \text{Prob}[V(n_i) - I_{min} \leq 2\delta;\ n_i > \bar{n}] \qquad \text{(A1)}$$

Then $\lim_{\bar{n}\to\infty} P(\bar{n}) = 1$

The properties of the sequence $\{V(k)\}$ are now investigated. A sufficient condition for a successful iteration, i.e. $J(c_j)-J(\rho_j) > 2\mu$, to be absolutely successful as defined in Lemma 3.is

$$J(c_i) - E\{J(c_i)\} - [J(\rho_i) - E\{J(\rho_i)\}] \leq \mu \qquad \text{(A2)}$$

which occurs with probability $1-\beta$, $0<\beta<1$. β represents the maximum probability of an interruption between absolutely successful iterations by a successful iteration.

Now consider the cost V(n) accrued after n iterations:

$$V(n) = \frac{\sum_{i=1}^{n} T_i E\{J(c_i)\}}{\sum_{i=1}^{n} T_i} = \frac{\sum_{i_1}^{\nu_1} T_i E\{J(c_i)\} + \sum_{i_2}^{\nu_2} T_i E\{J(c_i)\} + \sum_{i}^{n'} T_i E\{J(c_i)\}}{\sum_{i=1}^{n} T_i} \tag{A3}$$

where $\sum_{i_1}^{\nu_1}$ is the sum over ν_1 segments of absolutely successful iterations of length $>\bar{n}$ as defined in Lemma 3, $\sum_{i_2}^{\nu_2}$ is the sum over the $\nu_2 = k - \nu_1$ remaining segments, and $\sum_{i_3}^{n}$ is the sum over n' interruptions. The total length of segments of absolutely successful iterations $\{n_i\}$ is related to the number of interruptions by

$$\sum_{i=1}^{k} n_i = n - n' , \quad n' \geq k-1 \tag{A4}$$

One may rewrite (A3) as:

$$V(n) - I_{min} \leq \frac{1}{\sum_{i=1}^{n} T_i} \left[\sum_{i_1}^{\nu_1} T_i [E\, J(c_i) - I_{min}] + I_{max}\left[\sum_{i_2}^{\nu_2} T_i + \sum_{o}^{n'} T_i\right]\right] \tag{A5}$$

Let F(n',n) be the probability of having n' interruptions in a length n of a sequence $\{c_i\}$ containing successful and unsuccessful iterations. Then from Lemma 1, the definition of β and the binomial distribution it follows that

$$F(n',n) \leq \binom{n}{n'} \beta^{n'} (1-\beta)^{n-n'} \tag{A6}$$

Define the probability of maximum accumulated interruptions $\phi(n,\tau,\beta)$ for some number $\tau>1$ and $\tau\beta< 1$:

$$\phi(n,\tau,\beta) \triangleq \sum_{n'>n\tau\beta}^{\infty} \binom{n}{n'} \beta^{n'}(1-\beta)^{n-n'} \geq \sum_{n'>n\tau\beta}^{\infty} F(n',n) \tag{A7}$$

Then by the central limit theorem for Bernoulli random variables

$$\lim_{n\to\infty} \phi(n,\tau,\beta) = 0 \quad \Rightarrow \lim_{n\to\infty} \sum_{n'>n\tau\beta}^{\infty} F(n',n) = 0 \tag{A8}$$

Now the accumulated cost V(n) for fixed numbers of segments k and interruptions n' is given by

$$k \leq n\tau\beta, \quad n' \leq 2n\tau\beta, \quad m\tau\beta(\bar{n}+2) \leq 1, \text{ for some } m > 1 \tag{A9}$$

It is easily shown that under (A9) at least one segment of length n_i is greater than $\bar{n}$, because otherwise there is a contradiction in the following relation

$$\sum_{i=1}^{k} n_i = n-n' \geq nm\tau\beta(\bar{n}+2)-2n\tau\beta \geq nm\tau\beta\bar{n} \tag{A10}$$

From (A9) it is shown that the quantities ν_2/n and n'/n can be made arbitrarily small

$$\frac{\nu_2}{n} \leq \frac{\bar{n}(k-1)}{n} < \frac{1}{m} \quad , \quad \frac{n'}{n} < \frac{2n\tau\beta}{mn\tau\beta(\bar{n}+2)} < \frac{2}{m(\bar{n}+2)} \tag{A11}$$

Then by property (P3) one may select a number m such that the last expression on the right hand side of (A5) is less than δ.

$$\frac{I_{max}}{\sum_{i=1}^{n} T_i} \left[\sum_{i}^{\nu_2} T_i + \sum_{i}^{n'} T_i \right] \leq \delta \tag{A12}$$

Now define

$$\bar{\theta}_i \triangleq \frac{T_i}{\sum_{i=1}^{n} T_i} \quad , \quad \bar{\xi}_i = E\{J(c_i)\} - I_{min}$$

where $\bar{\xi}_i$ is a random variable defined over $[0,I]$, $I=I_{max}-I_{min}$. Then the first term of the right hand side of (A3) is written as

$$\frac{1}{\sum_{i=1}^{n} T_i} \sum_{i}^{\nu_1} T_i[E\{J(c_i)\}-I_{min}] = \sum_{i}^{\nu_1} \bar{\theta}_i\bar{\xi}_i \tag{A13}$$

From Lemma 3, $0 \leq \bar{\xi}_i \leq \delta$ with probability $P(\bar{n}^*) \leq P(\bar{n})$, $\bar{n}^* < \bar{n}$. Define θ_i and the independent identically distributed random variable ξ_i satisfying conditions of Lemma 2:

$$\theta_i = \frac{\bar{\theta}_i}{\sum_{i}^{\nu_1} \bar{\theta}_i} \quad , \quad \xi_i = \begin{cases} \delta & \text{with probability } P(\bar{n}^*) \\ I & \text{with probability } 1-P(\bar{n}^*) \end{cases} \tag{A14}$$

Then

$$\sum_{i}^{\nu_1} \bar{\theta}_i\bar{\xi}_i \leq \sum_{i}^{\nu_1} \theta_i\xi_i \tag{A15}$$

Define the probabilities for the weighted sum of the sequences $\{\bar{\xi}_i\}$ and $\{\xi_i\}$;

$$\bar{\psi}(\delta,P(\bar{n}^*)) \triangleq \text{Prob}\{ \sum_{i}^{\nu_1} \bar{\theta}_i\bar{\xi}_i \leq 2\delta\} \;;$$

$$\psi(\delta,P(\bar{n}^*)) \triangleq \text{Prob} \sum_{i}^{\nu_1} \theta_i\xi_i \leq 2\delta\}$$

which by using the result of Lemma 2 one may write:

$$\bar{\psi}(\delta,P(\bar{n}^*)) \geq \psi(\delta,P(\bar{n}^*)) \tag{A16}$$

Using the Markov inequality it is easily shown that

$$\text{Prob}\{|\sum_{i}^{\nu_1} \theta_i\xi_i| > 2\delta\} \leq \frac{I}{\delta}(1-P(\bar{n}^*)) \tag{A17}$$

Using the results of Lemma 3 and its definition, $P(\bar{n}^*)$ can be made arbitrarily close to one for a sufficiently large $\bar{n}^*$. This implies that (A17) can be made arbitrarily close to zero and therefore $\psi(\delta,P(\bar{n}^*))$ can be made arbitrarily close to one.

Using now (A5), (A12) and (A16), one establishes that for a fixed number of segments and interruptions

$$P(s/k,n',n) \triangleq \text{Prob}\{V(n)-I_{min} \leq 3\delta/k \text{ segments, } n' \text{ interruptions}\}$$

$$\geq \psi(\delta,P(\bar{n}^*)) \to \text{La},P(\bar{n}^*) \to 1, \text{ as } P(n^*)\to 1 \tag{A18}$$

The overall probability $P(n)$ defined in (14) is analyzed now as follows. Define

$$P(k/n) \triangleq \text{Prob}\{k \text{ segments in } n \text{ iterations}\}$$

$$P(s/k,n) \triangleq \text{Prob}\{V(n)-I_{min} \leq 3\delta/k \text{ segments, } n \text{ iterations}\}$$

$$P(n'/k,n) \triangleq \text{Prob}\{n' \text{ interruptions}/k \text{ segments, } n \text{ iterations}\}$$

Then using (A18)

$$P(n) \triangleq \text{Prob}\{V(n)-I_{min} \leq 3\delta\} = \sum_{k=1}^{\infty} P(k/n)P(s/k,n)$$

$$\geq \sum_{k=1}^{n\tau\beta} P(k/n)P(s/k,n) \geq \sum_{k=1}^{n\tau\beta} P(k/n) \sum_{n'>k-1}^{\infty} P(s/k,n',n)P(n'/k,n)$$

$$\geq \sum_{k=1}^{n\tau\beta} P(k/n) \sum_{n'\geq k-1}^{2n\tau\beta} P(s/k,n',n)P(n'/k,n)$$

$$\geq \psi(\delta,P(\bar{n}^*)) \sum_{k=1}^{n\tau\beta} P(k/n) \sum_{n'\geq k-1}^{2n\tau\beta} P(n'/k,n) \tag{A19}$$

The resulting double sum in (A19) represents the probability of the "event" that among n terms of the sequence there are no more than $n\tau\beta$ segments and $2n\tau\beta$ interruptions. The other possible event is either the number of segments exceeds $n\tau\beta$ or the number of interruptions exceeds $2n\tau\beta$. By (A4) the number of interruptions can be no less than $n\tau\beta$ for both cases. Therefore,

$$\sum_{k=1}^{n\tau\beta} P(k/n) \sum_{n'\geq k-1}^{2n\tau\beta} P(n'/k,n) \geq 1 - \sum_{n'>n\tau\beta}^{\infty} F(n',n) \tag{A20}$$

Therefore substituting back into (A19)

$$P(n) \geq \psi(\delta,P(\bar{n}^*))\,[1-\phi(n,\tau,\beta)] = 1-\gamma;\ \gamma > 0\ . \tag{A21}$$

For a sufficiently large n, $\bar{n}^*$ is made sufficiently large, as from (A8), (A17) and Lemma 3, such that there exists a number μ in (8) for which γ may become arbitrarily small.

Q.E.D.

Proof of Corollary 1. From Loéve Corollary 2, page 164, [3], since $|I_{max}|^r < \infty$ for $0 < r < \infty$ convergence in probability implies convergence in the rth mean.

Proof of Theorem 2. The proof of Theorem 2 is carried on exactly the same way as the proof of Theorem 1 if one replaces equations (A3) and (A5) by

$$\bar{V}(n)-I_{min} = \frac{\sum_{1\,i}^{\nu_1} T_i E\{J(c_i)-I_{min}\}+\sum_{2\,i}^{\nu_2} T_i E\{J(c_i)\}+\sum_{3\,i}^{n'} T_i E\{J(c_i)\}+\sum_{i=1}^{n} \bar{T}_i E\{J(\rho_i)\}}{\sum_{i=1}^{n}(T_i+\bar{T}_i)}$$

$$\leq \frac{1}{\sum_{i=1}^{n}(T_i+\bar{T}_i)} \sum_{1\,i}^{\nu_1} T_i E\{J(c_i)-I_{min}\}+I_{max} \sum_{2\,i}^{\nu_2} T_i+\sum_{i}^{n'} T_i+\sum_{i=1}^{n}\bar{T}_i \tag{A5'}$$

Then by Property (P4) for a sufficiently large n, condition (A12) is replaced by

$$\frac{I_{max}}{\sum_{i=1}^{n}(T_i+\bar{T}_i)} \sum_{2\,i}^{\nu_2} T_i + \sum_{i}^{n'} T_i + \sum_{i=1}^{n}\bar{T}_i \leq 2\delta \tag{A12'}$$

and the proof carries through with V(n) replaced by $\bar{V}(n)$ and 3δ replaced by 4δ. The proof of convergence in the rth mean follows immediately from Corollary 1.

Q.E.D.

REFERENCES

1. G. N. Saridis and H. Gilbert, "Self-Organizing Approach to the Stochastic Fuel Regulator Problem," IEEE Trans. on Systems Science and Cybernetics, Vol. SSC-6, No. 3, July 1970.

2. G. Stein and G. N. Saridis, "A Parameter-Adaptive Control Technique," Automatica, Vol. 5, pp. 731-739, Nov. 1969.

3. Loève, Probability Theory, Van Nostrand, 1963.

ACKNOWLEDGMENT

The author is grateful to the discussors, Professor Asai, Dr. J. Mendel and Professor R. McLaren, for their useful comments to which an attempt has been made to answer as clearly as possible, and to Mr. R. Kitahara for the computational results in the example. This work was partly supported by the NSF Grant GK-24866.

A CONTROL SYSTEM IMPROVING ITS CONTROL DYNAMICS BY LEARNING

Kahei Nakamura

Nagoya University

Nagoya, Japan

1. INTRODUCTION

In this paper, is proposed a learning control system or a real time dynamic optimization system which improves its dynamic character by itself through seeking the optimal or sub-optimal control by an iterative approach. This research aims to develop a learning controller which is applicable to control the plant operating repeatedly in the same environments, such as the start-up control of plant or the batch process control. That is, the objective of this learning controller is to search the best dynamic behavior when it is required to shift from a steady state (arbitrary fixed starting state) to another steady state (given final state) in the repetitive control of the plant, which is to be considered to have an a priori non-linear structure with unknown parameters.

All the signals such as control input, plant output, and desired output are expressed by sequential vectors (composed of time samples of signals), and the learning control process is described as an iterative one in the sequential vector space [1]. Then, the criterion function is the quadratic distance between the current point and the goal in the vector space. The unknown controlled plant which is essentially non-linear can be considered to be linear approximately in each small scale step of the learning process. Repeating the determination of subgoal, the estimation of unknown parameters and the evaluation of the increment of control in each stage of the iterative procedure, the system approaches gradually to the aimed goal and finally reaches to the sub-optimal control which exists in the range with specified small distance from the desired goal. Since the control experience in each stage is utilized in the linealized operation in the succeeding stages, this procedure is worthy to call learning process. This means that the learning controller have learned the way to improve the dynamic (transient) response of the original control system progressively.

2. PROBLEM STATEMENT

Assume that the unknown plant can be simulated by a non-linear system with properly preselected structure such as n-th order differential equation with unknown parameter and operates in the noiseless environment.

Let the desired performance of the control system or the goal point in x-space (sequential vector space of plant output) be x_d, and the pre-fixed starting point be x^o. The criterion function in quadratic form is expressed as $J = || x_d - x ||$. Then the present problem is to search the best control of the admissible control in u-space (sequential vector space of control), causing the plant output to be close to the goal x_d, in other words, is to get the control which minimizes J under specified constraint on control if it exists, through the linearized successive operation in each small-scale stage of the iterative procedure.

3. DESCRIPTION OF SEARCH PRINCIPLE

In this section, the proposed iterative learning approach is explained refering to Figure 1. Initially assume that the arbitrary selected control u^o (for instance step function control with the

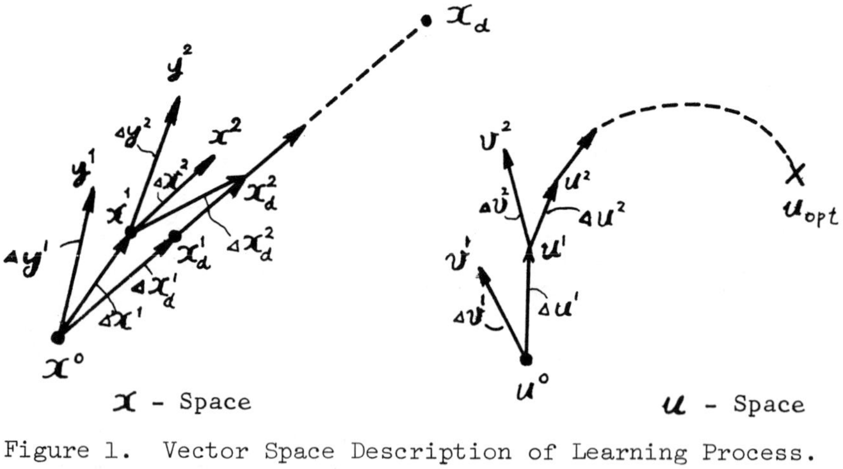

Figure 1. Vector Space Description of Learning Process.

height to cause the desired steady state of plant output) causes the plant output of x^o. Call the association of (u^o,x^o) as the initial situation of the iterative improving approach.

First Stage. The operations in this stage are to set a sub-goal x_d^1 near x^o and to get the control u^1 which causes x_d^1 or its approximation. Since it is reasonable to take the sub-goal on the limited line $\overline{x_0 x_d}$, the following relation is provided:

$$x_d^1 = x^o + \alpha_1 (x_d - x^o) \qquad \alpha_1 < 1 \tag{1}$$

First, select an appropriate small variation of control Δv^1 and observe the plant output y^1 caused by the control $v^1 = u^o + \Delta v^1$. The resultant variation of plant output is $\Delta y^1 = y - x^o$. If the variation of control situation $(\Delta v^1, \Delta y^1)$ exists in the neighborhood of (u^o,x^o) with relevant small distance in the product space $U \times X$, then the minor variation of plant behavior can be regarded as it has occurred under the approximately linearized condition, which is expressed by the relation (2).

$$\Delta y^1 = H^1(x^o)\Delta v^1 \tag{2}$$

where $H^1(x^o)$ is designated Sequential Transfer Matrix or S.T.M. [1] in the first stage, and is evaluated by the procedure described Section 4. Since the evaluated H^1 is available in the neighborhood of x^o, the control variation required to attain the subgoal x_d^1 is derived by the relation (3).

$$\Delta u^1 = [H^1(x^o)]^{-1}\Delta x_d^1 \quad \text{where } \Delta x_d^1 = x_d^1 - x^o \tag{3}$$

By applying the new control $u_1 = u^o + \Delta u^1$, the system reaches at x^1 which is located near x_d^1.

Second Stage. The sub-goal in the second stage x_d^2 is to be selected on the line $x_d^1\, x_d$ as to satisfy the relation (4).

$$x_d^2 = x_d^1 + \alpha_2 (x_d - x_d^1) \qquad \alpha_2 < 1 \tag{4}$$

Since it is reasonable that the $H^1(x^o)$ evaluated in the first stage is still valid approximately in the vicinity of x_d^1, the approximate increment of control required to reach at x_d^2 is calculated by Eq. (5).

$$\Delta u^2 = [H^1(x^o)]^{-1}\Delta x_d^2 \tag{5}$$

where $\quad \Delta x_d^2 = x_d^2 - \Delta x^1$

The usage of $H^1(x^o)$ in Eq. (5) may, of course, cause some considerable error. If the plant output caused by $v^2 = u^1 + \Delta v^2$ is denoted by y^2, the caused output variation is $\Delta y^2 = y^2 - x^1$. The S.T.M. of the second stage $H^2(x^1)$ is calculated by the same way as in the first stage.

That is $$\Delta y^2 = [H^2(x^1)]\Delta v^2 \tag{6}$$

By applying this evaluated $H^2(x^1)$ to

$$\Delta u^2 = [H^2(x^1)]^{-1}\Delta x_d^2 \quad , \tag{7}$$

the control variation Δu^2 to cause the desired variation of output Δx_d^2 can be calculated. By applying to the plant the control $u^2=u^1+\Delta u^2$, the system reaches at x^2 located near x_d^2.

Succeeding Stages. The same procedure as the second stage is repeated in the succeeding stages. Passing through x^3, x^4,... which are located near x_d^3, x_d^4,... respectively, the system approaches the final goal x_d. This iterative process terminates when it enters in the domain with small distance from x_d. Figure 2 is the block diagram for the above iterative procedure.

4. DETERMINATION OF LINEARIZED S.T.M.

A system of n-th degree nonlinear differential equation subjected by a single input is generally expressed by Eq. (8).

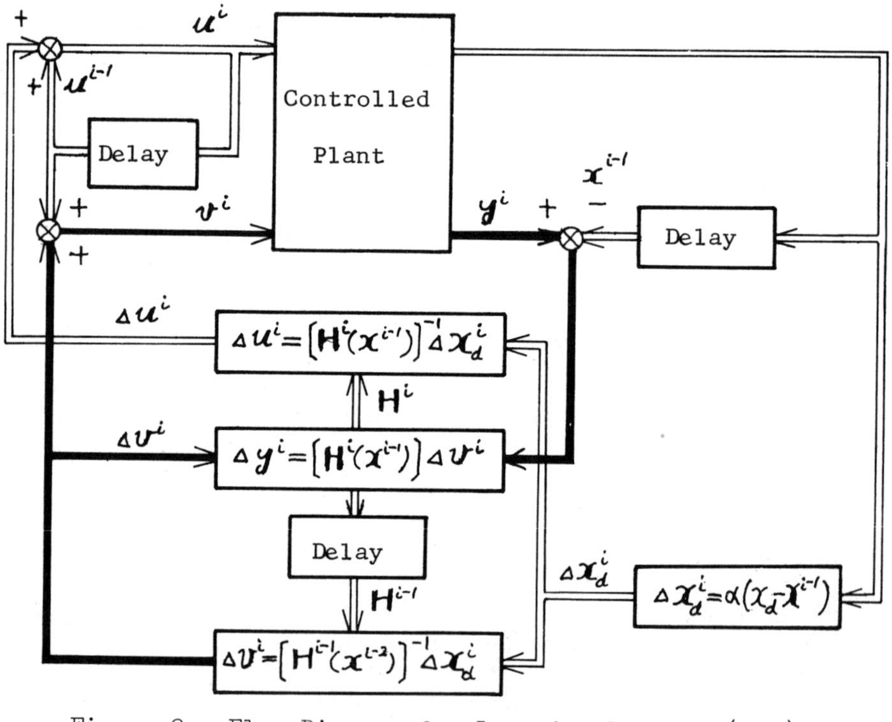

Figure 2. Flow Diagram for Learning Process ($i \geq 2$)

$$f(x,\dot{x},\ldots,\overset{(n)}{x})=g(u) \tag{8}$$

The variational equation of (8) comes to Eq. (9).

$$\Delta\overset{(n)}{x}\frac{\partial}{\partial\overset{(n)}{f}} f(x,x,\ldots,\Delta\overset{(n)}{x})+\Delta\overset{(n-1)}{x}\frac{\partial}{\partial\overset{(n-1)}{f}} f(x,x,\ldots,\overset{(n)}{x})+\ldots$$

$$+\Delta\dot{x}\frac{\partial}{\partial\dot{x}} f(x,\dot{x},\ldots,\overset{(n)}{x})+\Delta x\frac{\partial}{\partial x} f(x,x,\ldots,x\overset{(n)}{x}) = \frac{\partial}{\partial u} g(u)\Delta u$$

$$\Delta\overset{(n)}{x}+\alpha_1(x,\dot{x},\ldots,\overset{(n)}{x})\Delta\overset{(n-1)}{x}+\ldots. \tag{9}$$

$$+\alpha_{n-1}(x,\dot{x},\ldots,\overset{(n)}{x})\Delta\dot{x}+\alpha_n(x,\dot{x},\ldots,\overset{(n)}{x})\Delta x = k(u)\Delta u$$

Now, assume that coefficients α_i's be linear functions of $x,x,\ldots,\overset{(n)}{x}$,

$$\alpha_j(x,\dot{x},\ldots,\overset{(n)}{x}) = c_j(t)+\sum_{\ell=0}^{n}\gamma_{j\ell}\overset{(\ell)}{x}(t) \tag{10}$$

and k(u) be a power-expanded polynominal of u,

$$k(u) = \sum_{\ell=0}^{L} k_\ell u^\ell \tag{11}$$

respectively. Then, Eq. (9) is rewritten as

$$\Delta\overset{(n)}{x}(t)+\sum_{j=1}^{n}\{c_j(t)+\sum_{\ell=0}^{n}\gamma_{j\ell}\overset{(\ell)}{x}(t)\}\Delta\overset{(n-j)}{x}(t) = (\sum_{\ell=0}^{L}k_\ell u^\ell)\Delta u(t) \tag{12}$$

If we select the dimension of sequence vector as a finite number N, and introduce a shift matrix S

$$S = \begin{bmatrix} 0 & 1 & 0 & & 0 \\ 0 & 0 & 1 & & \cdot \\ 0 & 0 & 0 & & \cdot \\ \cdot & \cdot & \cdot & \bullet & \cdot \\ \cdot & \cdot & \cdot & \bullet & 1 \\ 0 & 0 & 0 & \ldots & 0 \end{bmatrix} \text{ N×N matrix} \tag{13}$$

the following relations can be easily derived.

$$\dot{x} = [1-S]x, \tag{14}$$

$$\Delta\dot{x} = [1-S]\Delta x \tag{15}$$

$$\text{sequential vector of } c_j(t)\Delta\overset{(n-j)}{x}(t)=[c_j'R]\Delta\overset{(n-j)}{x} \tag{16}$$

$$\text{sequential vector of } \gamma_{j\ell} x^{(\ell)}(t)\Delta^{(n-j)}_{x}(t) = [\gamma_{j\ell}[1-S]^{\ell}x]'R\Delta^{(n-j)}_{x} \quad (17)$$

$$\text{sequential vector of } u^{\ell}(t)\Delta u(t) = u_{\ell}'R\Delta u \quad (18)$$

where c and u_{ℓ} are sequential vectors of c(t) and $\{u(t)\}^{\ell}$ respectively and

$$R = [e_1e_1' + e_2e_2' + \ldots + e_Ne_N']. \quad \text{e: unit vector} \quad (19)$$

By applying Eq. (14)-Eq. (18) to Eq. (12), the following vector equation is obtained under the assumption that $c_j(t)=c_j$ (constant).

$$\Delta^{(n)}_{x} + \sum_{j=1}^{n} [c_j 1 + \sum_{\ell=1}^{n} \gamma_{j\ell}[1-S]^{\ell}x]'R\Delta^{(n-j)}_{x} = (\sum_{\ell=0}^{L} k_{\ell}u_{\ell}')R\Delta u$$

$$[[1-S]^{n} + \sum_{j=1}^{n} M_j(x,c_j,\gamma_{j\ell})[1-S]^{(n-j)}]\Delta x = K(u,k_{\ell})\, u \quad (20)$$

where $$M_j = [c_j 1 + \sum_{\ell=1}^{n} {}_{j\ell}[1-S]^{\ell}x]'R. \quad (21)$$

M_i, i=1,2,...,n are, in all, N×N diagonal matrices, and their elements are dependent on x (variational base point of output) and jx(n+1) parameters $c_j, \gamma_{j\ell}$, ℓ=1,2,...n. K is also a N×N diagonal matrix and related to u (variational base point of control) and (L+1) parameters k_{ℓ}, ℓ=0,1,2,...,L. Put as

$$H = [[1-S]^{n} + \sum_{j=1}^{n} M_j(x,c_j,\gamma_{j\ell})[1-S]^{(n-j)}]^{-1} \quad (22)$$

A linear equation

$$\Delta x = H\Delta u \quad (23)$$

can be derived.

Eq. (23) is a linearized relation of small-scale stage and the matrix of Eq. (22) is the linearized S.T.M. being valid around the control situation (u,x). The expression of Eq. (23) is not valid when the inverse matrix Eq. (22) does not exist.* However, since the present problem is to calculate of elements of H or c_j's, $r_{j\ell}$'s and k_{ℓ}'s for given (u,x), it is enough to use Eq. (20) containing no inverse matrix, instead of Eq. (23). From Eq. (23), N scalar equations are derived, each of which contains {j(n+1)+L+1}

*The discussion developed here is also correspondent to answer for the question by Professor K. S. Narendra at the Seminar, on the singularity of H.

parameters. If N is selected equal to or larger than number of total parameters, all parameters can be calculated by solving these simultaneous equations.

5. CONVERGENCE OF LEARNING PROCESS

Convergence of the series $x^0, x^1, x^2, \ldots$ is dependent upon the approximation degree of the evaluated H^i, $i=1,2,\ldots$. If H^i is related to the real S.T.M. in i-th stage G^i by

$$G^i = [1-\Delta^i]H^i,$$

then the following expression is true.

$$x^i = x^{i-1}+G^i\Delta u^i = x^{i-1}+G^i[H^i]^{-1}\Delta x_d^i \tag{24}$$

$$= x^{i-1}[1-\Delta^i](x_d^i-x^{i-1}) = \Delta^i(x^{i-1}-x_d^i)+x_d^i \tag{25}$$

While, as for the series $x^0, x_d^1, x_d^2, \ldots$, the following relation is valid.

$$x_d^i = x_d^{i-1}+\alpha_i(x_d-x_d^{i-1}) = x_d+(1-\alpha_i)(x_d^{i-1}-x_d)$$

Iterating this relation, Eq. (26) can be derived.

$$x_d^i = x_d+\prod_{j=1}^{i}(1-\alpha_j)(x^0-x_d) \tag{26}$$

Applying Eq. (26) to Eq. (25), the next relation is derived

$$x^i-x_d = \Delta(x^{i-1}-x_d)+\prod_{j=1}^{i}(1-\alpha_j)(x^0-x_d)$$

From this recurrence relation, Eq. (27) follows.

$$x^i-x_d = [\prod_{m=1}^{i}\Delta^m+\sum_{\ell=0}^{i-1}\prod_{k=0}^{\ell-1}{}^{i-k}[1-\Delta^{i-\ell}]\prod_{j=1}^{i-\ell}(1-\alpha_j)](x^0-x_d) \tag{27}$$

When there is a Δ which satisfies the relation $\|\Delta x\| > \|\Delta x^i\|$, $i=1,2,\ldots$ and it is possible to put $\alpha = \alpha_i$, $i=1,2,\ldots$, the next relation is correct.

$$\|x^i-x_d\| \leq \|[[\Delta]^i+[1-\Delta]\sum_{\ell=0}^{i-1}[\Delta]^\ell(1-\alpha)^{i-\ell}](x^0-x_d)\|$$

$$=\|[\Delta]^i[\Delta+[1-\Delta][1-(1-\alpha)\Delta^{-1}]^{-1}[1-\{(1-\alpha)\Delta^{-1}\}^i]](x^0-x_d)\| \tag{28}$$

Since all components of Δ are usually smaller than 1, the next relation is derived for large number of i.

$$x^i-x_d = \Delta^i[(1-\alpha)\Delta^{-1}+\alpha](x^0-x_d) \tag{29}$$

Then,

$$\lim_{i\to\infty} x^i = x_d \tag{30}$$

For general $\alpha_j < 1$, the convergence of x^i is guaranteed for noise free case.

6. DISCUSSIONS

There are several problems to be discussed.

A. Constraints on Control. The constraint on control is usually norm constraint or saturation constraint. For these cases, the following strategy is promising. Set a hypersphere with specified radius or a hypercube with specified size whose origin is x_d in x-space. When the control trajectory of learning procedure once crosses the hypersphere or hypercube, the best control whose corresponding point in x-space is closest to x_d should be searched on the surface of control constraint sphere or cube, by the relevant searching procedure such as, for example, the optimum gradient method.

B. Effect of Noise*. In this paper, only a noise free case is treated. For a noisy case, the effect of noise contained in plant output may appear on the following two points: One is the effect on evaluation of H, and the other is that on the evaluation of approaching trajectory. The former problem in which measured values of Δx and Δu are contaminated by noise, is easily solved by taking N enough larger than for the noise free case and by solving the simultaneous equations through the least mean square method. The latter problem is the effect of the fluctuation of x and u by noisy signal, then is involved in the case of (C).

C. Modification for Stochastic Environment**. Under stochastic (randomly varying) environment, the initial stationary state, starting point (u_0^0, x^0), passing point (u^i, x^i) and final point (u^N, x^N) on the trajectory may be fluctuated randomly. Then, $J_i = \|x_d - x^i\|$ must be appreciated in stochastic average. So that the approaching process turns reasonably to an algorithm of the stochastic approximation method, research on which is left for the future.

D. Problem on Controllability. Is it always possible to make a system starting at arbitrary x^0 pass through an arbitrary

*This discussion is developed as author's answer to Professor K.S. Narendra's question at the Seminar on a noisy case.

**This is the discussion induced by Professor K.S. Narendra's question at the Seminar.

x_d?* In order to realize a x_d, it must be guaranteed that any x^0 must be transformed to a given x_d, through the system dynamics S (Transforming character of the system), in other words, Transformation $T:x_0 \to x_d$ must involve S. This concerns with controllability problem. For the control of unknown plant, it is of course impossible to predict the controllability of this sense. Therefore, when the desired point is considered to be out of the set which is a transform of x^0 by system dynamics, the author's procedure can only lead the system to an admissible point closest to the desired point.

7. CONCLUSIONS

The learning control scheme offered here aimed to apply to the process control of batch type or the repeating start-stop process. Basic structure of the scheme has been described and the main problem outlined has been discussed. However, there are several problems to be studied in future such as, modifications for stochastic environment, extention to multi-dimensional system, investigation from the view point of Dynamic Programing and so on.

Finally the author expresses his great gratitudes to the researchers in Automatic Control Laboratory, Nagoya University, who have discussed this work.

REFERENCES

1. K. Nakamura, "Sequential Vector Space Description of Descrete Control System", Memoirs of Faculty of Engng., Nagoya University, Vol. 20, no. 1, May 1968.

2. K. Nakamura, Y. Shibata and M. Oda, "A Learning Control System Improving its Dynamic Behavior", Res. Rept. of Aut. Cont. Lab., Nagoya University, Vol. 15, April 1968 (in Japanese).

*This is the problem proposed by Dr. J. M. Mendel at the Seminar.

SELF-LEARNING METHOD FOR TIME-OPTIMAL CONTROL

Hiroshi Tamura

Osaka University

Osaka, Japan

1. ADAPTATION AND LEARNING THROUGH PATTERN IDENTIFICATION

Much work has recently been done in the field of optimal control. The structure of the optimal controller can be determined mathematically on the basis of Pontryagn's maximum principle, if the dynamic characteristics of the plant are known.

The practical design procedure of the optimal controller has also been studied by many people.

In general, the plant dynamics fluctuate due to variations of surrounding circumstances. Therefore we need to continuously adapt the control function to follow the change of the dynamic performance. So far, the optimal controller has been designed under the assumption that the dynamic characteristic of the given plant is definitely known. Hence the design procedure is invalid when the dynamic characteristic is unknown or varies randomly with respect to time.

An adaptive control method is developed to implement optimal control under such situations. In case of adaptive control, the following three procedures are, in general, necessary in order to realize the optimal control:

(1) to identify the process parameters;
(2) to design quickly the optimal controller on the basis of information obtained in the item (1);
(3) to correct the structure of the optimal controller on the basis of design data obtained in item (2).

Because of difficulties in the identification and the high-speed optimization, it is difficult to apply the above procedure to the process under operation. Thus it has been an important and long-pending problem for control engineers to find the control system which adapts its control action to the varying dynamic plant.

We have developed some methods to modify control function to adapt the change in dynamic characteristic of the plant, not by using identification and optimization principles, but by using pattern identification techniques. Our method is derived from the fact that the state of the plant shows distinctive behavioral patterns depending on whether the control function is optimized or not. And modification of the control function is made directly from the detection of those patterns. Thus we neither use a priori information about parameters of plant characteristics, nor identify them. Besides, this method is likewise applicable both to linear and non-linear systems.

An adaptive control system having some means to decide the direction to modify the control function in itself, may be called self-adaptive. On the other hand, when the direction of modification is only given by a superviser outside of the system, the system may be called forced-adaptive. Widrow proposed a forced-adaptive control system by introducing a learning machine as a controller. His system may learn to control a plant at the minimum time, if appropriate instructions are given by the superviser.

But we have to differentiate a system with learning machines from a learning control system. Learning control system is an improved adaptive system which can memorize the optimal control function once established through adaptation and can immediately execute optimal control without adaptive search when a once experienced situation takes place again.

A learning control system, therefore, should have memory facilities to store pairs of experienced situations and the results of adaptation. Moreover, it should have the capability to relate a certain control function with the present situation.

This paper will show that, by making use of the above mentioned pattern identification techniques, self-learning control can be accomplished for time-optimal control of unknown plant dynamics.

2. DESCRIPTION OF SYSTEM

Figure 1 shows the structure of the system to be considered in this paper. This system has two loops, namely, the feedback loop and the adaptive loop. In this figure, the symbol R represents an input signal or the desired state of the plant. The symbol X, E and u denote state of the plant, the error and the control signal,

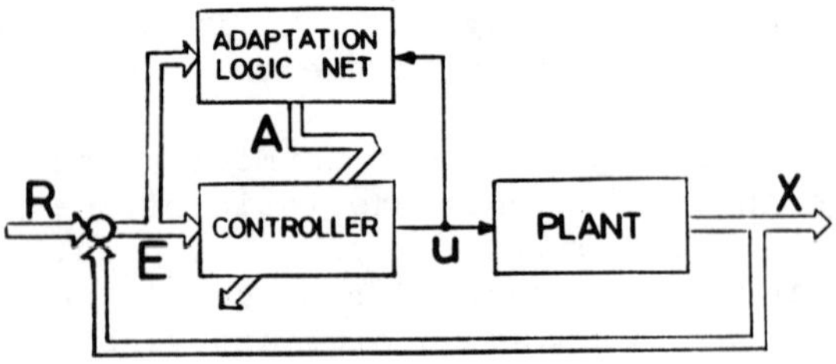

Figure 1. An Adaptive Control System.

respectively. The control function of the controller is corrected by the adaptive control signal A which is produced by the adaptation logic element.

The plant, is assumed to be of nth order, and the state $X=(X_1, X_2, \ldots, X_n)$ of the plant is described by the following system of equations:

$$\begin{aligned} \dot{X}_i &= X_{i+1} \qquad (i=1, 2, \ldots, n-1) \\ \dot{X}_n &= h(X) + u(X,R) \qquad (h(0)=0) \end{aligned} \tag{1}$$

Further we assume that the control variable is constrained such that

$$|u| \leq 1. \tag{2}$$

The optimal control theory states that the normal time optimal control can be implemented by switching u from 1 to -1 or vice versa. A control of this kind is usually called bang-bang control, and such switchings of u can be made by a switching surface set up in the state space.

Let the state space be represented regarding the error E=R-X, so that the target state always remains at the origin of the space. Then the control is given by the following switching logic:

$$u = \Phi[G(E)] \tag{3}$$

where

$$\Phi[\xi] = \begin{cases} 1 & \text{for } \xi \geqq 0 \\ -1 & \text{for } \xi < 0 \end{cases} \tag{4}$$

G(E) is called a switching function and G(E)=0 a switching surface.

In addition to above assumptions, we assume that the desired state can be reached from any initial state by switching over the control signal u at most (n-1) times. We designate the region of the state space satisfying the above assumption as the controllable region.

The controllable region defined above is unlimited in most of practical problems. The above assumption, however, may look like a constraint in case of a conservative process, the controllable region for which is shown in Figure 2. Even in this case, reference input R being bounded as $|R|<1$ under normal control situation, the maximum error due to sudden change of reference input is bounded as $|E|\leq 2$. Thus, under normal control situations, the actual state of the plant is bounded in the controllable region.

In order that the switching of u is such that the control time from any initial state to the terminal is minimum, we have to determine the optimal switching function G(E), which is yet unknown. The switching function G(E) is assumed to be able to approximated within necessary precision by appropriate choice of the approximate function $f_j(E)$ and corresponding weight W_j in approximate switching function $\hat{g}(E)$.

$$\hat{g}(E) = \sum_j W_j f_j(E) \tag{5}$$

where $f_j(E)$'s are linearly independent functions of E.

3. SELF-ADAPTATION LOGIC

When plant dynamics are unknown, weights W_i's are not possible to determine analytically and some adaptive method is necessary. The author shows that the pattern of state transition or trajectory in relation to temporal switching surface could be sufficient information to decide the direction of adaptive modification of W_i's.

A very important property should be noticed on the optimal switching surface of the minimum time problem. That is, a trajectory

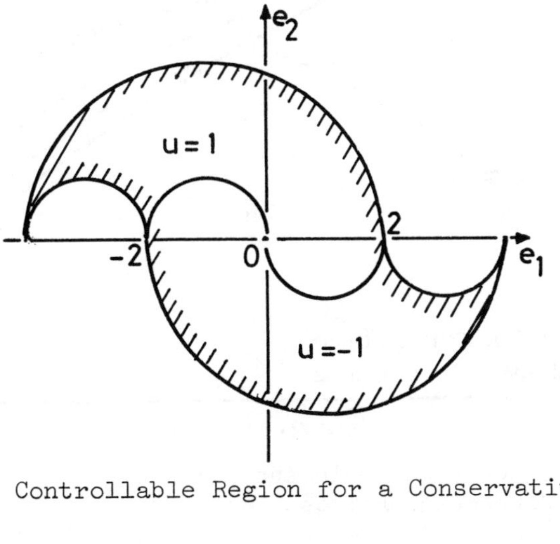

Figure 2. Controllable Region for a Conservative Process

starting from an arbitrary state point in the controllable region is on the optimal switching surface itself, after the first switching have done. Taking this property into consideration, it can be concluded that the switching surface is not correctly set at each point where those patterns of state transition as shown in the Figure 3 are observed.

Figure 3 shows switching curves of a 2nd order plant which are not yet optimized and some typical patterns of state transition. Every time such a pattern of state transition is observed, the weights of the Eq. (2) are subjected to the following correction:

$$W_j = W_j' \pm \delta \cdot f_j(E) \qquad (j=1,2,\ldots) \tag{6}$$

where W_j' denotes previous value and the signs + and - correspond to the downward and upward arrows in Figure 2 respectively. Besides

$$\delta = \frac{\alpha}{\sqrt{\Sigma f_j^2(E)}} \qquad (\alpha>0), \tag{7}$$

α is called a correction coefficient and usually decreased with the increase in adaptation steps.

This adaptation logic yields sometimes non-effective corrections marked with o as well as effective ones marked with ⊙. However,

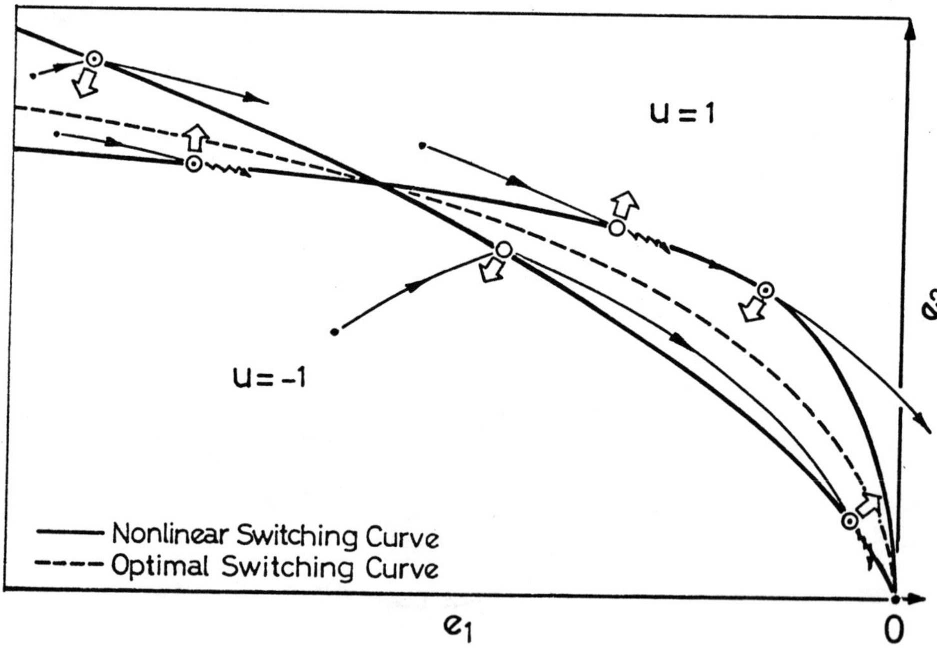

Figure 3. The Adaptation Logic and Effective (⊙) and Non-effective (o) Correction Points.

an effective correction takes place at a point nearer to the origin always after such non-effective correction, as far as the optimal switching curve for a given plant is convex upward in the second quadrant. This holds true in most systems of practical importance. Speed and convergence of this logic are subject to shape of the optimal switching curve, type of $f_j(E)$, input patterns and its distribution. Besides the stability of the feedback loop in the process of adaptation is an interesting thema left for further studies. But computer experiments applying this self-adaption logic to the various control situation have proved satisfactory results with regard to convergence and stability.

4. ADAPTATION PROCESS

Figure 4 shows the result of application of this method to a 2nd order plant

$$\begin{aligned} \dot{X}_1 &= X_2 \\ \dot{X}_2 &= -X_2 + u \end{aligned} \tag{8}$$

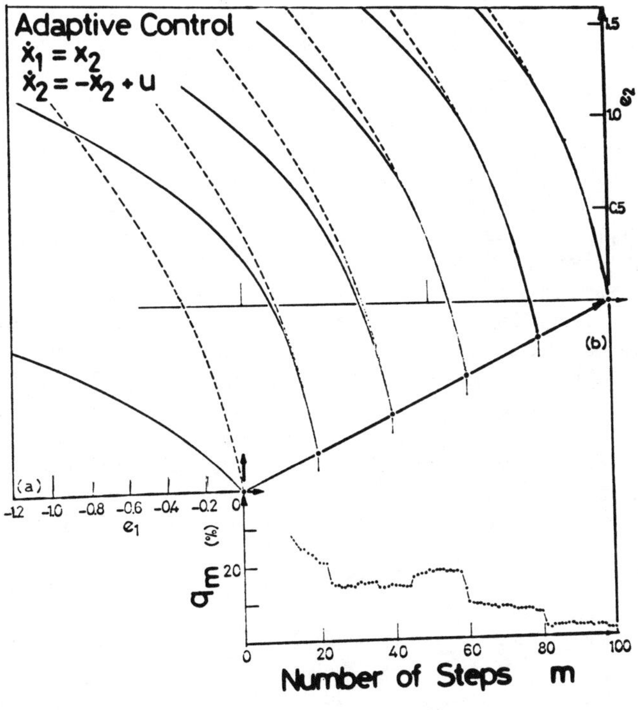

Figure 4. Adaptation Process to a Second Order Plant.

The particular form of switching function was

$$\hat{g}(E) = w_1 e_1 + w_2 e_2 + w_3 e_2^3 \tag{9}$$

and correction coefficient was $\alpha=0.3$. In the figure, the solid curves denote the nonlinear switching curves and the dotted curves denote the optimal ones at each adaptation step. The switching curve was initially set at the position far from the optimal one. Then it converged to the optimal one step by step and showed good coincidence after about 60 steps (Figure 4a).

Here each step begins with the setting of an initial state and ends with the arrival of the terminal state, the origin.

15 initial states are randomly chosen from the state space and applied cyclically. The control quality q_m at the m-th step is defined as

$$q_m = \frac{100}{m-k+1} \sum_{j=k}^{m} \frac{t_j - t_{oj}}{t_{oj}}, \tag{10}$$

$$k = \begin{cases} 1 & \text{for } m \leq 14 \\ m - 14 & \text{for } m \geq 15 \end{cases}$$

where t_j is the actual control time at j-th step and t_{oj} is the possible minimum time for the same initial state. The time course (Figure 4b) of q_m shows good improvement of control quality. The resultant switching curve at the 100th step is

$$e_1 + 0.217 e_2 + 0.085 e_2^3 = 0 \tag{11}$$

Adaptation logic of this type have been examined also for approximate function $f_j(E)$ other than polynomial. For second order plant, piecewise linear functions are quite effective as well, and some modified adaptation logic have been proposed [2,3]. For higher order plant, it seems much easier to use polynomial functions than piecewise linear functions. The optimal switching surface of higher order plant is a continuous surface which is composed of two smooth surfaces, and at the connecting edges of these two surfaces, it is not smooth. Since this edge is a part of final trajectory, the accuracy of approximation around it is very critical to the control quality. Good approximation around the connecting edge is difficult to attain when a set of polynomial function is used. We used two smooth surfaces and certain logic to connect them. Thus polynomial function method came out successful also for higher order plants.

5. LEARNING SYSTEM SUBJECTED TO VARIOUS INPUTS

The adaptive controller described in 4. can also adapt to variations of input patterns as well as to those of the plant dynamics. For instance, it constructs the corresponding optimal switching curve when ramp inputs of a constant velocity are imposed, and thereafter when the input turns to step inputs, it can reconstruct the optimal switching curve for the step inputs. Besides it operates stably against the variations of the input patterns.

The above-mentioned system is adaptive but not learning, in point that it destroys the once established control function and reconstructs for the new input pattern. Now the system is expected to be improved so that it may be possible to adapt to the new input pattern without destroying the past results and come to execute optimal control as soon as the same input pattern as before is observed again. Such learning control may be particularly useful for the plant of 0-type since the optimal switching curve has to be specified separately for each magnitude of step inputs.

Let Y denote the pattern which represents input R(t). Defining the new state space as a direct product of E and Y, and setting up a switching surface in the state space, a learning control system is realized by applying the above-mentioned adaptation logic. The coordinates Y serve as a kind of the memory storage.

Results of learning process are shown in Figure 5. Here three kinds of magnitudes of the step input [Y = (y); y = 0.5, 0, -0.5] are imposed to the conservative oscillatory plant

$$\begin{aligned} \dot{X}_1 &= X_2 \\ \dot{X}_2 &= -X_1 + u \end{aligned} \qquad (12)$$

Self-learning process can be read from the time course of the control quality. Since adaptation to a certain environment causes to forget the established experience for other environments at earlier stage of learning, the controller cannot follow the sudden changes of the environment. It is, however, proved that the controller comes to be able to utilize, as the learning process proceeds, the result of adaptation for the input patterns experienced earlier.

6. SELF-LEARNING IN VARYING ENVIRONMENT

Plants are surrounded by various environmental conditions, so their dynamics have, in general, a trend to vary under the influences of the environment. Sailing ships have varying dynamics depending on the wind and the sea wave, and lifts have to be controlled by the adequate control function according to the weight. In such cases, it becomes very important to find the control function depending on environmental patterns. Therefore learning control has great advantages in such a situation.

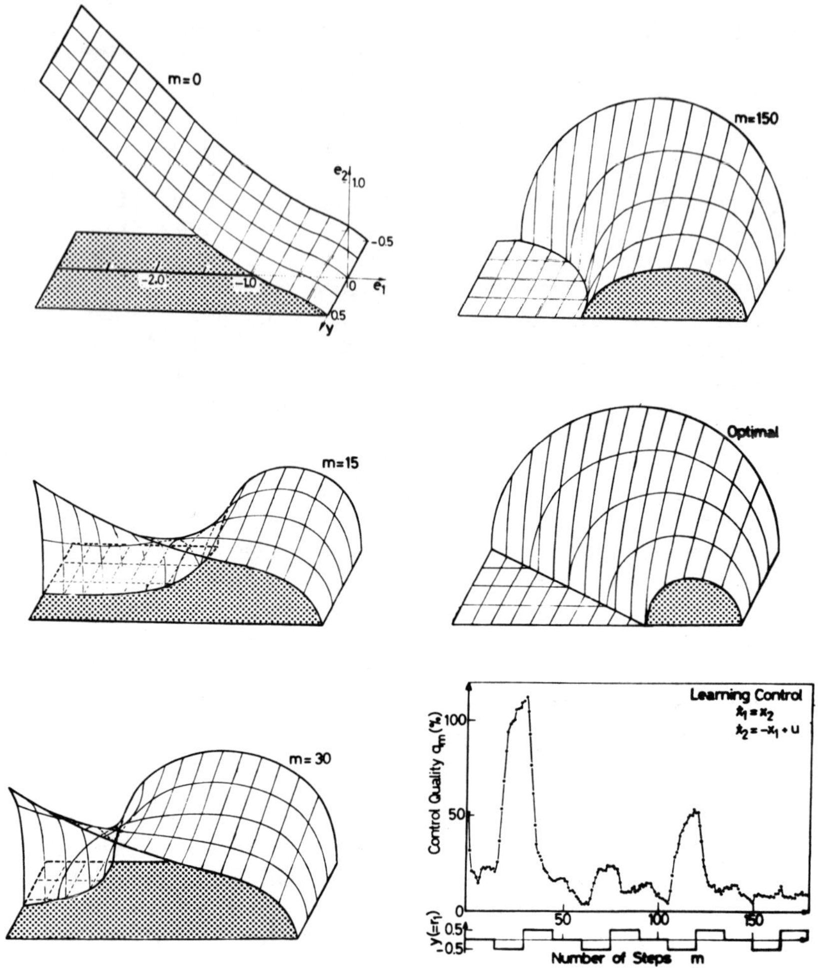

Figure 5. Self-learning Process for Periodically Varying Input.

Consider a learning control system drawn in Figure 5. Z is a environmental pattern which influences the plant dynamics. Let us assume that environmental patterns can be observed by the controller. Considering the influences of the environment, the 2nd equation of Eq. (2) is rewritten as follows:

$$\dot{x}_n = h_Z(X) + u(X, R). \tag{13}$$

The function $h_Z(X)$ is unknown but only assumed to relate with Z in one-to-one manner. The augmented state space introduced in 4. is given here as a direct product E × Y × Z.

This scheme has been applied to the case where the environmental pattern is simply represented by a scalar z which influences linearly the plant dynamics. The results are shown in Figure 6. The plant is described by

$$\dot{x}_1 = \dot{x}_2$$
$$\dot{x}_2 = -zx_2 + u \tag{14}$$

Under random occurrences of the three environmental patterns [z=0, 1, 2] the switching surface changes adaptively for step inputs and at the 120th step it highly approximates the optimal switching surface.

Moreover the switching surface for the unexperienced environment [for example z=0.5, 1.5] is simultaneously constructed and becomes well approximating through adaptation to other environments although it experiences only three kinds. In other words, the proposed learning control system can estimate control for the unexperienced environment from the learned results by interpolation.

7. PATTERN-SELECTIVE LEARNING CONTROL

Adaptive control and learning control which are implemented by the controller having self-adaptation capability derived from the patterns of state transition of the plant have been described. Though the simple situations for 2nd order plants have been considered, the proposed method can easily be extended to more complicated situations of higher order plants.

In spite of versatility of this method, it is not always efficient, because of complexity which the switching surfaces have, to include all the possible variables of input and environmental patterns in the state space. It may be expedient to provide the

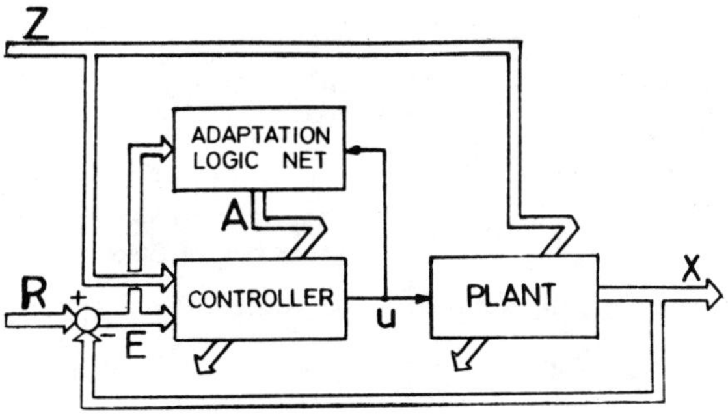

Figure 6. A Learning Control System in the Varying Environment.

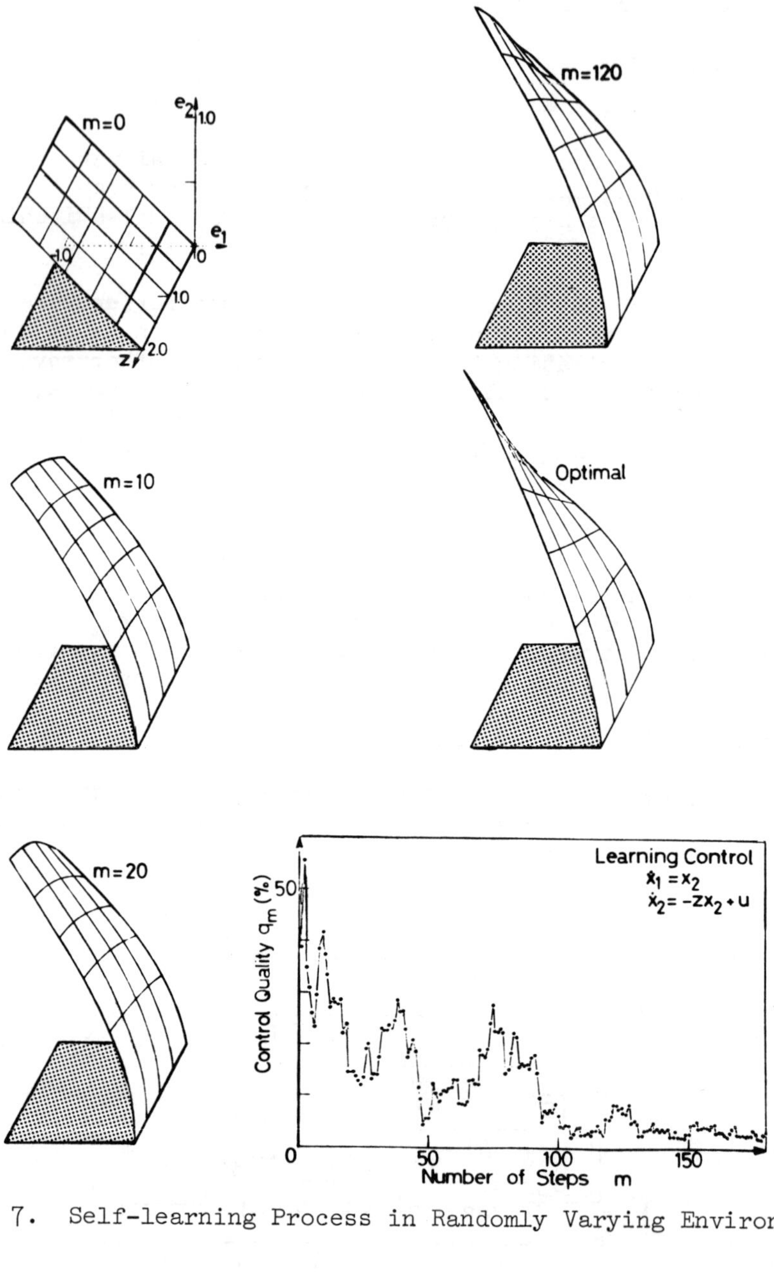

Figure 7. Self-learning Process in Randomly Varying Environment.

different controllers for the extremely different classes of input and environmental patterns.

Figure 8 illustrates one of such systems. Each controller (selector controller) contains a selector in order to recognize the input or environmental patterns which the controller should take charge of. While the selector can be designed a priori, learning mechanism could also be attached to it so that it may recognize patterns flexibly. If the patterns which the selectors can accept do not overlap, only one of the controllers is at work at each instant and adaptive correction is given by this controller.

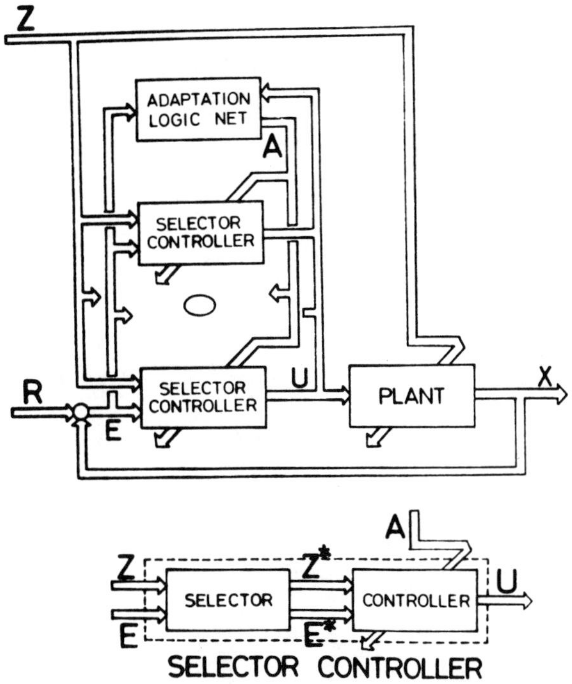

Figure 8. A Pattern-selective Learning Control System.

REFERENCES

1. B. Widrow and F. W. Smith, "Pattern-recognizing Control Systems," Computer and Inform. Sci. 1, 288-317, (1964).

2. H. Tamura and T. Kurokawa, "Adaptive Construction of Time Optimal Switching Functions," J. Inst. Elect. Engrs. Japan 88-8 1531-1540, (1968).

3. H. Tamura and T. Kurokawa, "Adaptive Pattern Classifiers Applied to Time Optimal Control," IFAC Symp. on Technical and Biological Problems in Control, Yerevan (USSR), (1968).

4. T. Kurokawa and H. Tamura, "A Learning Method for the Construction of Optimal Switching Surfaces," Trans. IECE Japan, 54-C, 904-911, (1970).

LEARNING CONTROL VIA ASSOCIATIVE RETRIEVAL AND INFERENCE

Julius T. Tou

University of Florida

Gainesville, Florida, U.S.A.

INTRODUCTION

One of man's greatest assets is his ability to acquire information from his environment and to make intelligent decisions on the basis of this information. It would be extremely desirable if man-made systems could be built with such learning capability. We understand that learning is not a mathematical operation, such as multiplication, differentiation and integration; nor is it an engineering process, such as computation, identification and control. Rather, it is a psychology term, which describes a complex dispositional property of behavior. Consequently, the mechanization of a learning process is by no means a trivial task. This paper introduces a new approach to learning control by making use of associative retrieval and inductive inference.

Before discussing machine learning, we will have a brief review of some of the major empirical findings in human learning. From psychologists' point-of-view, human learning is concerned with six problem areas; namely, capacity, practice, motivation, understanding, transfer, and forgetting. Learning ability of a human being is dependent upon a level of intelligence. An individual can transcend his immediate level of intelligence by learning. Intelligence implies knowing how to ask relevant questions, interpret the answers, discover and correct mistakes, recognize patterns, recall and use past experience. Intelligence is a property of behavior, of decision-making, of doing, and of solving problems. In human

---

This work was supported in part by the National Science Foundation under Grant No. GK 2786.

learning, small pieces of information are grouped into larger ones with higher information content. Repetition of situations does not in itself modify behavior. It may be harmful if correct responses are being made. The number of reinforcements appears to be more important than the amount of reinforcement. Learning proceeds most rapidly when every correct response is reinforced. Distributed practice is superior to lumped practice if the information is to be retained for a long period of time. Prompt reward is more effective than delayed reward. Learning under the control of reward is usually preferable to learning under the control of punishment. Understanding relations between parts and wholes is of value in learning. Transfer to new tasks will be better if the learner can discover relationships, formulas or algorithms for himself. Learning may be impaired when it is followed closely by another activity, especially somewhat resembling the first.

It would be extremely interesting if we could incorporate these empirical findings of human learning in the design of machine learning. However, the present design techniques and mathematical theories do not seem to provide us with a tool to imbed these findings in artificial intelligent systems. We may use these empirical findings as guidelines and for generating motivations in the design of artificial intelligence. Some aspects of the findings, such as distributed practice and reinforcement, reward and punishment, generation of algorithms, understanding relations between parts and wholes, are worth consideration in the design of learning systems.

The major control tasks involved in the operation of an automated processing plant are the low-level local control and the high-level global supervisory control. The former is carried out by electronic and electromechanical devices, and the latter is usually performed by human operators. Mechanization of high-level global supervisory control is primarily concerned with the design of learning and artificial intelligence for the system.

The role of human operators of an automated processing plant can be viewed as that of an extremum controller equipped with recognition and learning capabilities. Under normal operating conditions, they take regular observations on the plant and, on the basis of past experience, select a suitable control strategy so that some optimum output from the plant is generated. The principal tasks and actions of the operator can be described as follows:

(1) The operator recognized the features characterizing the present situation of plant from observed data.

(2) He will try to retrieve from his memory whether he has encountered the same or similar situation in the past.

(3) If the present situation has occurred before, he knows what the control strategy should be and he will apply it to the process.

(4) In case it is a new situation, he will learn what the control strategy for this situation should be and apply it.

(5) He will try to memorize this situation and the corresponding control strategy for future use.

When we replace the human operator by some intelligent controller, it is evident that the controller must perform the tasks of optimization, recognition, storage, retrieval, and learning. The controller will recognize the features characterizing the situation of the plant, generate or retrieve a control strategy for future use. Recognition, storage, retrieval and inference are essential steps in the learning process. It is the purpose of this paper to discuss the retrieval and inference aspects in the design of learning control.

FUNCTIONAL REQUIREMENTS FOR MACHINE LEARNING

The functional requirements of a learning control system involve the concepts of <u>search</u>, <u>recognition</u>, <u>storage</u>, <u>retrieval</u>, and <u>inference</u>. The search concept is exemplified by goal-seeking behavior for system self-optimization and may be implemented by a gradient-type hill-climbing procedure or through random search strategies. Recognition of the features or patterns characterizing a control situation and the subsequent association with a learned control strategy stored in some form of memory involve the latter four concepts. Many ingenious configurations have been proposed which meet some of these functional requirements. The theory of operation of these proposed models and systems has traditionally drawn on Markov processes, statistical decision theory, automata theory, hill-climbing techniques, information theory, and pattern recognition. This paper attempts to develop a new approach to the design of learning control making use of information retrieval and inference concepts.

Learning may be defined as the acquisition of skill to perform meaningful self-modification and to improve performance on the basis of past experience. We consider a learning controller to be a mechanization of a control law which relates the observed state of the plant, the input signals and the environment in such a way that a control signal for improving the plant operation is generated. The controller can thus be viewed as a device which matches or associates the state of the plant, the input signals, and the environmental parameters with a certain control signal; or more generally, the controller must associate a control situation with a control strategy. In this context, the learning controller may be considered as consisting of three major units: the recognition unit, the storage unit, and the retrieval unit, as shown in Figure 1.

The recognition unit senses and records the situation parameters of the plant, which include the state variables, the input signals and the environment parameters, and converts the recorded situation parameters into a control feature via encoding. During the learning and training period, the control strategy for a specified control situation is determined and it is stored along with the corresponding control feature. During plant operation, the retrieval unit selects an appropriate control strategy from the storage unit when a certain control situation arises. If the control situation is not clearly defined, the learning controller has to apply such techniques as inductive inference to retrieve the desired control strategy. A technique for the realization of the inference process is proposed in this paper. The structure of the learning controller allows the evaluation of its control strategies for a given control feature relative to a prescribed performance criterion, and constantly up-dates the contents of the storage unit.

LEARNING CONTROL AS AN INFORMATION RETRIEVAL PROCESS

The preceding discussions point out that information retrieval is an essential process in a learning control system. A retrieval scheme is used to identify the control strategy which is relevant to an inquiry consisting of a specification of pertinent control situations or features. Let the control feature vector $\underline{y}$ be an n-dimensional vector denoted by

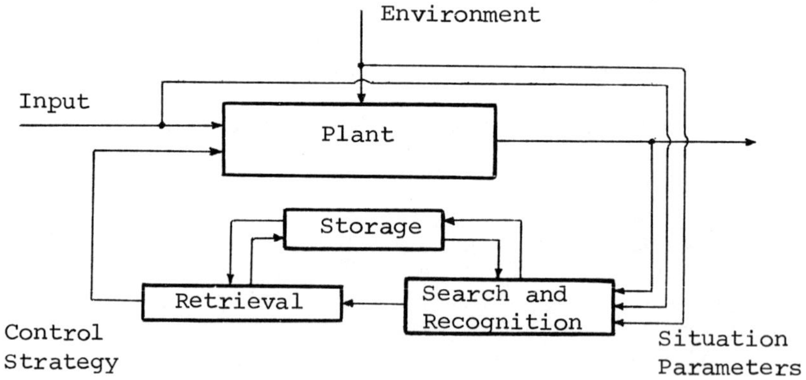

Figure 1. General Structure of a Learning Controller

$$\underline{y} = \text{col } (y_1, y_2, \ldots, y_i, \ldots, y_n) \tag{1}$$

where y_i denotes the ith feature parameter which may be a meter reading describing the performance measure, the input signal, or the environmental parameter. The control feature vector may be converted to a binary vector, if we quantize the feature parameters and express each measurement in a binary number. A group of y_k's will be assigned to a feature parameter. Determination of the control feature vector from observed data is a problem of encoding.

Let the control strategy vector $\underline{z}$ be an m-dimensional vector denoted by

$$\underline{z} = \text{col } (z_1, z_2, \ldots, z_j, \ldots, z_m) \tag{2}$$

where z_j is the value assigned to the jth control strategy by the retrieval scheme in response to inquiry $\underline{y}$ from the recognition unit. The inquiry is characterized by the control feature vector. In the proposed retrieval scheme, an observed feature vector $\underline{y}$ will assign a value to each of the m control strategies. The control strategy with the highest score is considered the most relevant to the feature vector $\underline{y}$ and will be selected accordingly. Here we have assumed that the plant can be controlled by m admissible strategies.

Corresponding to an observed control feature vector there is a relevant or optimum control strategy. These two quantities are interrelated in a very intricated manner. As a first approximation, the relationship between control strategies and control features may be expressed in a matrix, called the strategy-feature matrix $\underline{C}$, given by

$$\underline{C} = \begin{bmatrix} c_{11} & c_{12} & \cdots & c_{1n} \\ c_{21} & c_{22} & \cdots & c_{2n} \\ \vdots & \vdots & & \vdots \\ c_{m1} & c_{m2} & \cdots & c_{mn} \end{bmatrix} \tag{3}$$

In the design of the learning control system, this matrix is determined during the training period. This matrix may be written in the form of a binary matrix through proper encoding. A simple retrieval scheme can be implemented by making use of the relationship

$$\underline{z} = \underline{C}\,\underline{y} \tag{4}$$

This approach will result in a simple retrieval scheme for learning control. A more sophisticated process can be designed by including strategy and feature associations. A possible method consists of assuming that the retrieval transformation is linear and that a linear relation exists between control strategy and control feature vectors. This will lead to a linear associative retrieval scheme for learning control.

ASSOCIATIVE RETRIEVAL FOR LEARNING CONTROL

In eq. (4), the control strategy vector is given as a linear transformation of the control feature vector. If we will take into consideration the current measurement as well as the past experience, we will express the linear transformation as

$$\underline{z} = \underline{C}\ \underline{v} \tag{5}$$

where $\underline{v}$ is an identifier vector, which is given by

$$\underline{v} = \underline{y} + f(\underline{z},\underline{C}) \tag{6}$$

In eq. (6), $f(\underline{z},\underline{C})$ is a function of the control strategy vector and the learned strategy-feature matrix. This function provides an expression which characterizes the past experience. Equation (5) reduces to eq. (4) when the experience function $f(\underline{z},\underline{C})$ is zero; that is, when the past experience is ignored. The feature vector $\underline{y}$ may be regarded as a first-order identifier vector. If we choose

$$f(\underline{z},\underline{C}) = \lambda\ \underline{C}'\ \underline{z} \tag{7}$$

with $0 < \lambda \leq 1$, we have a linear associative retrieval scheme.

Combining eqs. (6) and (7), we obtain

$$\underline{v} = \underline{y} + \lambda\ \underline{C}'\ \underline{z} \tag{8}$$

Substituting eq. (8) into eq. (5) yields

$$\underline{z} = \lambda\ \underline{C}\ \underline{C}'\underline{z} + \underline{C}\ \underline{y} \tag{9}$$

which reduces to

$$\underline{z} = (\underline{I} - \lambda\ \underline{C}\ \underline{C}')^{-1}\ \underline{C}\ \underline{y} \tag{10}$$

Upon expanding the inverse matrix term, we have

$$\underline{z} = [\underline{I} + (\lambda\ \underline{C}\ \underline{C}') + (\lambda\ \underline{C}\ \underline{C}')^2 + \dots + (\lambda\ \underline{C}\ \underline{C}')^k + \dots]\ \underline{C}\ \underline{y} \tag{11}$$

where the term $(\underline{C}\ \underline{C}')^k$ represents a measure of the kth order strategy-strategy associations derived by matrix multiplication from the strategy-feature matrix. These association terms may be used to register past experience for the learning control system. It is noted that if all the strategy associations derived from the description matrix $\underline{C}$ are dropped, eq. (11) reduces to eq. (4). In eq. (11), the coefficient λ is introduced to secure fast convergence of the series expansion.

INDUCTIVE INFERENCE IN LEARNING CONTROL

In the case of real-world systems, the observed control situations are often poorly defined and the feature parameters are only partially known. Under these circumstances, the learning controller will not be able to retrieve the appropriate control strategy. To solve this problem, we will make use of inductive inference in the retrieval process. Since the measured feature parameters are either incomplete or inaccurate, they may not be adequate for retrieving the desired control strategy. In the design of the learning controller, we introduce the concept of generating an augmented feature set via inference. The inferred feature set will then facilitate the retrieval of the most relevant control strategy in response to the observed data.

By proper coding the strategy-feature matrix $\underline{C}$ may be expressed as a 0-1 matrix. From the strategy-feature matrix, we may determine the association between the ith and the jth feature, which is characterized by the association coefficient a_{ij}. As a simple example, a_{ij} may be given by the ratio of the number of strategies associated with both y_i and y_j to the number of strategies associated with either y_i or y_j or $y_i \cap y_j$. Such an association coefficient is given by

$$a_{ij} = \frac{N_{ij}}{N_i + N_j - N_{ij}} \tag{12}$$

In eq. (12), the N's denote the numbers of control strategies explained above. These association coefficients form the feature association matrix

$$\underline{A} = \begin{bmatrix} 1 & a_{12} & a_{13} & \cdots & a_{1n} \\ a_{21} & 1 & a_{23} & \cdots & a_{2n} \\ \vdots & \vdots & \vdots & & \vdots \\ a_{n1} & a_{n2} & a_{n3} & \cdots & 1 \end{bmatrix} \tag{13}$$

From the feature association matrix we may create a feature profile for each feature parameter, which consists of a set of closely associated feature parameters. For instance, a reasonable definition of the feature profile is

$$P(y_j) = \{y_j, y_k \ni a_{jk} \geq \theta \quad \forall k \neq j\} \tag{14}$$

where θ is a pre-specified threshold. Feature parameter y_k is said to be closely associated with feature parameter y_j if $a_{jk} \geq \theta$.

Such a feature profile can be easily derived from the feature association matrix. The main motivation to introduce feature profile is that the feature set will cover the actual feature parameters which may be different from the measure feature parameters or even missing from the measurements. The feature profile is then used to generate an inferred feature set from the incomplete or inaccurate measurements. The inferred feature set will be used as an inquiry for retrieving the desired control strategy.

To illustrate the generation of the inferred feature set, we assume that the observed feature parameters are y_i, y_j, and y_k. Then the augmented set is obtained by substituting the feature parameters by their respective profiles. Thus,

$$Q = P(y_i) \cap P(y_j) \cap P(y_k)$$
$$= \{y_i, y_\ell \ni a_{i\ell} > \theta \;\forall\; \ell \neq i;\; y_j, y_p \quad a_{ip} > \theta \;\forall\; p \neq i;$$
$$y_k, y_q \quad a_{iq} > \theta \;\forall\; q \neq i\} \tag{15}$$

By repeating the process, we derive the inferred feature set as

$$\hat{Q} = \{y_\alpha, y_\beta, y_\gamma, \ldots\} \tag{16}$$

The inferred feature set is then used to determine the significance of every feature parameter y_i, $i=1,2,\ldots,n$. The significance is measured by the feature weight. A simple way to calculate the feature weight is to find the sum of the association coefficient with respect to the particular feature under consideration. For instance,

$$w(y_i) = \sum_{y_k \in \hat{Q}} a_{ik} \tag{17}$$

By following the above procedure, the retrieval of the desired control strategy may be completed by calculating the relevance number for each strategy in response to the observed control situation. The relevance vector $\underline{R}\ (z_i)$ can be readily determined from the feature weight vector $\underline{W}\ (y_i)$ and the strategy-feature matrix $\underline{C}$.

$$\underline{R}(a_i) = \underline{C}\ \underline{W}(\underline{y}_i) \tag{18}$$

The control strategy with the largest relevance number will be selected.

CONCLUSIONS

This paper reviews the functional requirements for learning in automatic control and introduces the associative retrieval and inference approach to the design of learning controller for a plant with unknown or partially known system dynamics. The inference aspects are discussed along the line of generation of the inferred feature set, which plays an important role in retrieving the desired control strategy.

Major efforts in learning control system studies have been devoted to the search and recognition aspects. Little work has been done on the storage, retrieval, and inference aspects. It is hoped that this paper will stimulate research activities in these aspects which are much needed in the design of full-fledged learning control systems.

REFERENCES

1. E. Hilgard, Theories of Learning, Appleton-Century-Crofts, Inc., New York, 1956.
2. B. Berelson and G. Steiner, Human Behavior: An Inventory of Scientific Findings, Harcourt, Brace and Lould, New York, 1964.
3. J. T. Tou and J. D. Hill, "Steps Toward Learning Control," Proceedings of the 1966 JACC, Seattle, Washington.
4. J. T. Tou, Optimum Design of Digital Control Systems, Academic Press, New York, 1963.
5. J. T. Tou, Modern Control Theory, McGraw-Hill Book Company, New York, 1964.
6. K. Nakamura and M. Oda, "Fundamental Principle and Behavior of Learntrols," in Computer and Information Science - II, edited by J. T. Tou, Academic Press, New York, 1967.
7. K. S. Fu, "Learning Control Systems," in Advances in Information Systems Science (Vol. 1), edited by J. T. Tou, Plenum Press, New York, 1968.
8. Ya. Z. Tsypkin, "Learning Systems," in Advances in Information Systems Science (Vol. 2), edited by J. T. Tou, Plenum Press, New York, 1969.
9. M. E. Senko, "Information Storage and Retrieval Systems," in Advances in Information Systems Science (Vol. 2), edited by J. T. Tou, Plenum Press, New York, 1969.
10. V. A. Kovalevsky, "Pattern Recognition: Heuristics or Science," in Advances in Information Systems Science (Vol. 3), edited by J. T. Tou, Plenum Press, New York, 1970.

STATISTICAL DECISION METHOD IN LEARNING CONTROL SYSTEMS

Shigeru Eiho and Bunji Kondo

Kyoto University

Kyoto, Japan

INTRODUCTION

Statistical decision is an effective method to control an unknown system with the aid of incomplete or slight information (e.g., a few past control experiences about the system). The statistical decision method itself has no learning property. But adequate treatment of each element of decision theory (e.g., probability function, loss function, etc.) makes it possible for the statistical decision method to acquire learning property.

Most published works on the statistical decision method assume that the probability distribution of the parameters of the system is known fully or partially beforehand [1,2]. This paper deals with a case where we have slight knowledge about the statistical property of the system [4].

1. STATEMENT OF THE PROBLEM

Consider the usual optimizing control system with one control variable (Figure 1). Optimal control input which brings the P.I. (performance index) to the maximum point, is a function of disturbance z. There may, however, be unobservable random disturbances. The optimal value of x may, therefore, be thought of as a random variable with a certain probability distribution function.

Assuming that z is a discrete variable, the optimal control input can be determined by the statistical decision method as follows. Optimal decision $d_o(z_t)$ when observed value of z is z_t, is $d(z_t)$ which minimizes $L(z_t)$:

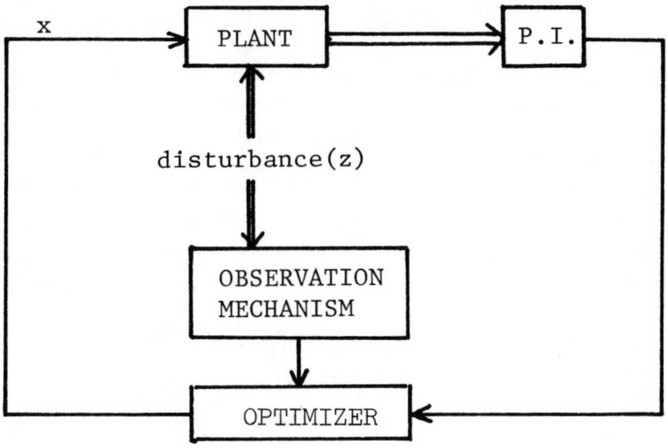

Figure 1. Optimizing Control System.

$$L(z_t) = \int f(x,d(z_t))\, P(x/z_t)\, dx \quad (1)$$

where $f(x,d(z_t))$ represents the loss associated with decision $d(z_t)$ when the optimal input $x_o=x$, and $P(x/z)$ is the conditional probability density function. However, $P(x/z)$ in many cases is not known beforehand. We can use the past control experiences as the pair (x_o,z_t) of optimal input and observed disturbance.

The problem now is how to evaluate d when $z=z_t$ is observed. Here we assume that z_t is statistically independent of z_r $(r \neq t)$. Then d_o can be evaluated only by using the past control experiences $(x_{o1},z_t),\ldots,$ and (x_{oN},z_t). For the sake of simplicity, subscript t and the variable z are omitted in the following.

Let us state the problem again. Assume X to be a random variable which has $p(x)$ as its probability density function, U to be the control input or decision variable, and $f(x,u)$ to be the loss function when $X = x$ and $U = u$. The optimal u, then, is obtainable from such u as minimizes the following $L(u)$:

$$L(u) = \int f(x,u)p(x)\, dx\ . \quad (2)$$

But $L(u)$ is incalcuable because $p(x)$ is unknown. Assume, then, that samples of X, $x_1,x_2,\ldots,$ and x_n are given. The problem now is how to fix u when $f(x,u)$ and $x_1,\ldots,$ and x_n are given.

There are two ways to solve this problem: (a) calculate eq. (2) by estimating $p(x)$ (section 2); and (b) take the minimax strategy (sections 3-5).

2. ESTIMATING PROBABILITY FUNCTION

There are a number of methods to estimate p(x). Since we know nothing about p(x), we use the experimental cumulative probability distribution function (ECPDF) [3] and the mth order polynomial of x to fit it to the representative points of ECPDF.

Assume that (x_i, y_i)'s are representative points of ECPDF given by samples where i=1,2,...,N. The problem here is how to fit the following mth order polynomial of x to these N points:

$$y = a_o + a_1 x + a_2 x^2 + \dots + a_m x^m = F(x) . \qquad (3)$$

The problem of fitting a curve to points is soluble, as a rule, by the least-square method. But since a probability function has a non-decreasing property, we use the linear programming method to solve this problem. Assume that W_i is the error at the ith points. Make the sum of W_i as amall as possible. The linear programming problem, then, is as follows:

$$\text{constraints:} \quad |y_i - F(x_i)| \leqq W_i \ , \quad i=1,\dots,N \qquad (4)$$

$$\text{i.e.:} \quad \left.\begin{array}{l} a_o+a_1x_i+a_2x_i^2+\dots+a_mx_i^m+W_i \geqq y_i \\ a_o+a_1x_i+a_2x_i^2+\dots+a_mx_i^m-W_i \leqq y_i \end{array}\right\} \qquad (4')$$

$$a_1 + 2a_2x_j^2+\dots+ma_mx_j^{m-1} \geqq 0$$

$$j = N+1,\dots,N+K \qquad (5)$$

$$\text{objective function:} \quad z = W_1 + W_2 + \dots + W_N \ \rightarrow \min . \qquad (6)$$

The non-decreasing property is checked at K points, as is done in eq. (5). It is proved through examples that the linear programming method excels the usual least-square method.

Figure 2 is the example of probability distribution function estimated by linear programming method, and Figure 3 is the example of probability function estimated by least-square method. The representative points of ECPDF are shown by small circles (7 points) in these figures. Constraint (5) is made at points 0.1, 0.2,..., and 1.0 (11 points).

Note that, with adequate experience of control, it is possible to make an optimal decision by the method which this paper suggests. The reason is that the more samples we use, the closer ECPDF gets to the true probability function.

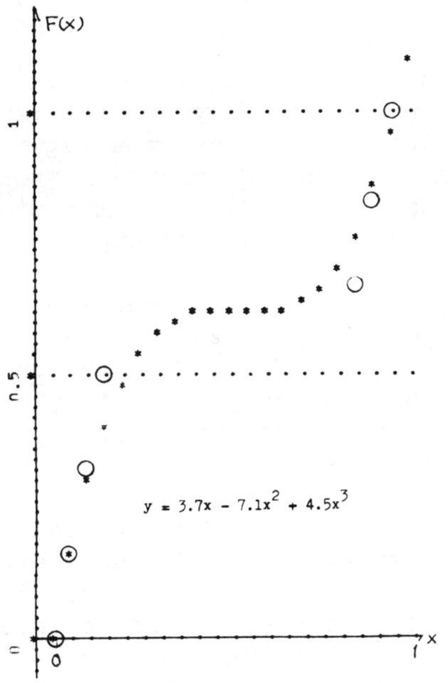

Figure 2. Probability Distribution Estimated by Linear Programming Method.

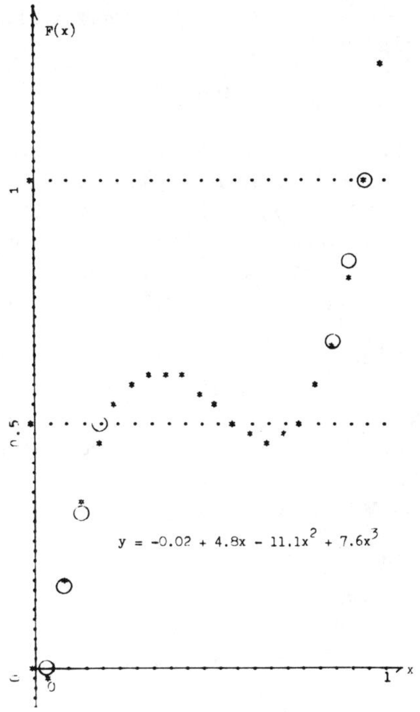

Figure 3. Probability Distribution Estimated by Least-Square Method

3. MINIMAX STRATEGY WITH RESTRICTED STRATEGY OF NATURE

Assume that U can take only discrete values. Eq. (2) can then be expressed as follows:

$$L_i = \int f_i(x)p(x)\,dx = \int f_i(x)\,dF(x)\ . \qquad (7)$$

The problem now is to choose an i which minimizes this quantity. Assuming that probability distribution function F(x) is an mth order polynomial, as in eq. (3), then eq. (7) can be expressed as follows:

$$L_i = a_1 c_{i1} + a_2 c_{i2} + \ldots + a_m c_{im} \qquad (8)$$

where c_{ij} is:

$$c_{ij} = \int jx^{j-1} f_i(x)\,dx\ . \qquad (9)$$

Use the concept of the confidence interval. The confidence interval $[\alpha_i, \beta_i]$ is obtainable from samples $(x_1, \ldots, x_n)$ and confidence level a%. The probability that $X \leq x_i$, $F(x_i)$, satisfies the following inequality with confidence level a %:

$$\alpha_i \leqq F(x_i) \leqq \beta_i\ . \qquad (10)$$

Now about the minimax strategy by which to make an optimal decision:

$$L_{iopt} = \min_{i} \sup_{F} L_i \tag{11}$$

where we maximize L_i under the condition given in eqs. (3) and (10). Namely, we restrict the strategy of nature to satisfy the eq. (10) (and eq. (3)). Therefore, sup L_i is obtainable by calculating $a_o, a_1, \ldots, a_m$ by linear programming with the use of constraints of eqs. (5) and (10) and objective function of eq. (8) to be maximized. The optimal i_o is obtainable by comparing the values of sup L_i.

The confidence interval $[\alpha_i, \beta_i]$ decreases with an increase in the amount of data. This means that the sub-space in which $a_o, a_1, \ldots, a_m$ can be chosen freely shrinks gradually until, finally, the value of $a_o, a_1, \ldots, a_m$ are determinable only by data. The final stage implies that our learning of the statistical property of an unknown system is complete.

The minimax strategy is a very conservative policy. We can moderate its conservativeness by manipulating the confidence level, although it is difficult to tell what percentage of confidence level is desirable. A low confidence level reduces the confidence interval and curtails the range of probability as estimation by eq. (10) sharply. This puts the above minimax strategy on a par with the empirical Bayes solution, an optimistic strategy if used when the amount of data at hand is limited.

Figures 4 and 5 are examples. The confidence intervals (eq. (10)) are used at 11 points; 0.0, 0.1,..., 0.9 and 1.0. The samples of x are 0.21, 0.22, 0.23, 0.71, and 0.72. The order of the polynomial is 4(m=4).

$$y = A0 + A1\cdot x + A2\cdot x^2 + A3\cdot x^3 + A4\cdot x^4 .$$

The loss function $f_i(x)$ is:

$$f_i(x) = (x-u_i)^2 \qquad i=1,2,3,4,5$$

where $u_1=0.2$, $u_2=0.4$, $u_3=0.6$, $u_4=0.8$, $u_5=1.0$. We can see from these figures that small confidence level make almost the same probability function; that is, the strategy of nature is restricted.

4. MINIMAX STRATEGY WITH RANDOMIZED DECISION FUNCTIONS

Take a case where X is a discrete random variable. Use the following symbols:

X: discrete random variable having k different possible values: $x_1, x_2, \ldots, x_k$;

d: decision, assuming q different possible decisions: $d_1, \ldots, d_q$;

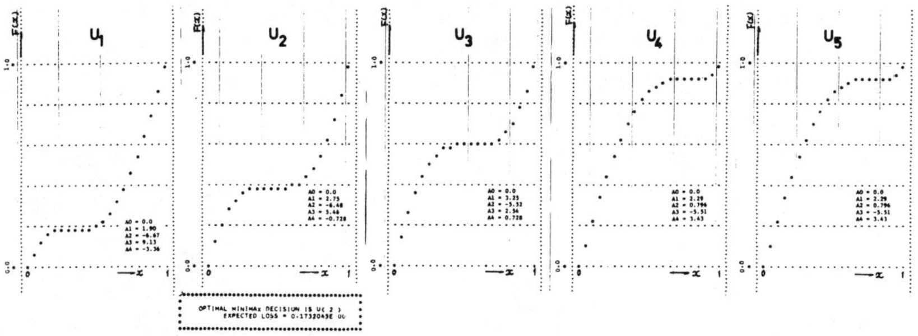

Figure 4. Estimated Probability Function (Confidence Level=0.90)

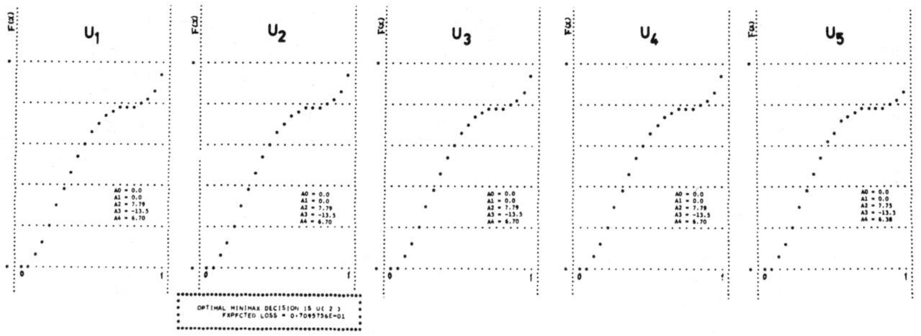

Figure 5. Estimated Probability Function (Confidence Level=0.40)

m_i: number of past observations for which $X = x_i$, $i=1,\dots,k$;
n: sample size: $n=m_1+m_2+\dots+m_k$;
$\underline{m}=(m_1,\dots,m_k)^T$: vector expression of sample (data);
M: set of $\underline{m}$, whose element is expressed by $\underline{m}_p$;
L_{ij}: loss associated with the decision d_j when $X=x_i$;
$L=[L_{ij}]$: loss matrix;
$P(x_i) = \theta_i$: probability of $X=x_i$, $i=1,2,\dots,k$;
$\underline{\theta} = (\theta_1,\theta_2,\dots,\theta_k)^T$: vector expression of probability of X.

The problem is which $d_1,d_2,\dots,$ or d_q is to be chosen when $\underline{m}$ is given. To solve this problem, let us define the randomized decision function. This function means the function of data $\underline{m}_p$ and that its value is probability vector $\underline{\pi}_p = (\pi^1_{\underline{m}_p},\dots,\pi^q_{\underline{m}_p})$, where

$$\sum_{j=1}^{q} \pi^j_{\underline{m}_p} = 1, \quad p=1,\dots,s \ (= {}_{k+n-1}C_n) .$$

The expected loss (risk when decision $d=d_j$ is used) is:

$$r(j,\underline{\theta}) = \theta_1 L_{1j} + \theta_2 L_{2j} + \dots + \theta_k L_{kj} . \tag{12}$$

The expected risk when decision function D is used is:

$$R(D,\underline{\theta}) = \sum_{p=1}^{s} \sum_{j=1}^{q} r(j,\underline{\theta})\ \pi^{j}_{\underline{m}_p}\ P(\underline{m}_p/\underline{\theta})\ , \tag{13}$$

$P(\underline{m}/\underline{\theta})$ being the probability with which $\underline{m}$ materializes when the probability distribution function of X is $\underline{\theta}$; i.e.,

$$P(\underline{m}/\underline{\theta}) = \theta_1^{m_1}\ \theta_2^{m_2}\ \ldots\ \theta_k^{m_k} \cdot n!/(m_1!\ m_2!\ \ldots\ m_k!)\ . \tag{14}$$

With $\underline{\theta}$ known, the optimal decision function is obtainable by choosing a d_j which minimizes $r(j,\underline{\theta})$. This decision function is called "a simple Bayes solution." With the a priori probability density function of $\underline{\theta}$ known, Bayes decision function may be used as the optimal decision function [2].

It happens occasionally in conventional systems that the a priori density does not exist or is unknown if it exists. The value of $\underline{\theta}$ even in this case can be estimated if there are enough past data to draw on; that is, if sample size n is large. Then the simply Bayes solution is obtainable. With insufficient data, however, it is advisable to use the minimax decision function, D_m, as defined by the following inequality [5]:

$$\max_{\underline{\theta}} R(D_m,\underline{\theta}) \leqq \max_{\underline{\theta}} R(D,\underline{\theta}) \qquad D:\ \text{arbitrary d.f.} \tag{15}$$

Consider the zero-sum two-person game $\Gamma(\underline{\theta},\underline{P},L)$ where $\underline{\theta}$ and $\underline{P}$ are mixed strategies which players I and II take, respectively, and L is a payoff matrix for player I. The following theory of minimax decision function is obtainable by using the theory of games [6].

Theorem 1 [4]. Assume that D_m is a minimax decision function and also assume that $R(D_m,\underline{\theta})$ takes its maximum value at $\underline{\theta} = \underline{\theta}_m$. Then

$$R(D_m,\underline{\theta}_m) = \min_{\underline{P}} \max_{\underline{\theta}} R(D,\underline{\theta}) = \min_{\underline{P}} \max_{\underline{\theta}} \underline{\theta}^T L \underline{P} = v \tag{16}$$

Especially $R(D_m,\underline{\theta})$ has its maximum value at $\underline{\theta} = \underline{\theta}^*$, where $\underline{\theta}^*$ and $\underline{P}^*$ are the maximin and minimax solution of game Γ, and v is the value of game Γ.

The following theorem is also obtainable by using the theory of game and the above theorem 1.

Theorem 2. A minimax decision function satisfies the following equation:

$$\sum_{p=1}^{s} \pi^{j}_{\underline{m}_p}\ P(\underline{m}_p/\underline{\theta}^*) = P^*_j \tag{17}$$

where $\underline{\theta}^*$ and $\underline{P}^*=(P_1^*,P_2^*,\ldots,P_q^*)$ are the maximin and minimax solution of game Γ and

$$\sum_{j=1}^{q} \pi^j_{\underline{m}_p} = 1 \qquad p=1,2,\ldots,s \tag{18}$$

There can be many minimax decision functions which satisfy eq. (17). We deal, therefore, with optimal minimax decision function D_{om} as defined by the following inequality:

$$\int_{\underline{\theta}} R(D_{om},\underline{\theta})d\underline{\theta} \leqq \int_{\underline{\theta}} R(D_m,\underline{\theta})d\underline{\theta} \equiv S(D_m) \tag{19}$$

D_m: arbitrary minimax d.f.

Quantity S(D) is calculable as follows:

$$S(D) = \int_{\underline{\theta}} R\, d\underline{\theta} = \sum_{j=1}^{q} \sum_{p=1}^{s} \pi^j_{\underline{m}_p} K(\underline{m}_p,j) \tag{20}$$

where

$$K(\underline{m},j) = [n!/(n+k)!] \sum_{i=1}^{k} L_{ij}(m_i+1) \tag{21}$$

Generally, many solutions satisfy eq. (17). Some are not minimax decision functions. Many examples show that a decision function which satisfies the condition (17) and minimizes S(D) is a minimax decision function and, therefore, an optimal minimax decision function. The following conjecture seems to be borne out by considering the following: minimizing S(D) acts as bringing $R(D,\underline{\theta})$ down as a whole. Moreover, $R(D,\underline{\theta})$ is, if D is fixed, the (n+1)th order polynomial function of $\underline{\theta}$, and therefore, the curve of $R(D,\underline{\theta})$ does not undulate violently.

Conjecture

The decision function that minimizes the following S(D) under the conditions (22) and (23) is the optimal minimax decision function:

$$\sum_{p=1}^{s} P(\underline{m}_p/\underline{\theta}^*)\, \pi^j_{\underline{m}_p} = P_j^* \qquad j=1,\ldots,q \tag{22}$$

where $\pi^j_{\underline{m}_p} \geqq 0$ for $\underline{m}_p \in M$, and $j=1,\ldots,q$, and

$$\sum_{j=1}^{q} \pi^j_{\underline{m}_p} = 1 \qquad \underline{m}_p \in M \tag{23}$$

$$S(D) = \sum_{j=1}^{q} \sum_{p=1}^{s} \pi^j_{\underline{m}_p} K(\underline{m}_p,j) \tag{24}$$

where $\underline{\theta}^*$ and $\underline{P}^*$ are the maximin and minimax solutions of game Γ.

Note that eqs. (22) and (23) are linear constraints on $\pi^j_{\underline{m}_p}$ and $S(D)$ is also a linear function of $\pi^j_{\underline{m}_p}$. Therefore, the optimal minimax decision function is obtainable by the linear programming method.

The proof of this conjecture is very difficult, and we have only obtained it when $k = q = 2$. If game Γ has more than two minimax and/or maximin solutions, the above conjecture must be modified. (The modification is omitted here.)

5. CONVERGENCE PROBLEM OF OPTIMAL MINIMAX DECISION FUNCTION

Before discussing the convergence problem of optimal minimax decision functions, let us note that the decision function obtained by minimizing $S(D)$ of eq. (24) with no constraints agrees with Bayes decision function concerning the uniform distribution as the a priori probability density of $\underline{\theta}$. Therefore, this decision function gets close to optimal decision function if $n \to \infty$ [7].

Assume that matrix $\tilde{L}$ is made by deleting the ith row of L, which corresponds to $\theta_i^* = 0$ in the solution of game Γ. Assuming $\theta_i^* = 0$, then $P(\underline{m}'/\underline{\theta}^*) = 0$ for data $\underline{m}'$ in which $m_i \neq 0$. Thus there are no constraints on such data $\underline{m}'$. We can, therefore, discuss the convergence problem by using $\tilde{L}$.

First, take a case where $P_j^* \neq 0$ for all j. By the law of large numbers, $Pr(|m_i/n - \theta_i^*|) \to 0$, as $n \to \infty$. Therefore, the constraining condition (22) binds only data $\underline{m}_p$'s which satisfy the inequality $\|\underline{m}_p/n - \underline{\theta}^*\| < \varepsilon$. Moreover, assuming $\underline{\theta} = \underline{\theta}^*$, the expected losses corresponding to various decision functions are the same.

Finally, the expected loss resulting from the optimal minimax decision function in this case converges uniformly on the expected loss resulting from an optimal decision.

Take a case where $p^*_{j_o} = 0$ at some j_o. Then, in the light of eq. (22), $\pi^{j_o}_{\underline{m}} = 0$ for all $\underline{m}$. If there is such i_o as $\min_i L_{i_o j} = L_{ij}$ in matrix $\tilde{L}$, then the optimal decision is $\pi^{j_o}_{\underline{m}} = 1$ at $\theta_{i_o} = 1$. It is advisable, in such cases, to relax the constraint of minimax condition (22) by the following inequality:

$$\left| \sum_{p=1}^{s} P(\underline{m}_p/\underline{\theta}^*)\, \pi^j_{\underline{m}_p} - P_j^* \right| \leq \gamma(n) \tag{25}$$

where $\gamma(n)$ is a nondecreasing function of n. Then the expected loss resulting from this relaxed minimax decision function converges on the expected loss resulting from an optimal decision.

CONCLUSIONS

This paper has dealt with the question of how to use the statistical decision method where the statistical property of the system are not known. The learning property is included in the mechanism of estimating the probability function by the data (section 2) or of making the confidence interval (section 3). The third method (section 4) has learning property included in the mechanism of minimizing S(D). By using a relaxed minimax decision function, we can make an optimal decision for every loss matrix when we have gained enough control experience.

APPENDIX - PROOF OF THEOREM 2

We prove here the special case of Theorem 2 where game Γ has only one pair of minimax and maximin solution (and k = q). The minimax and maximin solution of game Γ in this case are shown as follows [6]:

$$\underline{P}^* = (\text{adj } L)\, I_k \,/\, (I_k^T (\text{adj } L)\, I_k^T) \qquad \text{(A-1)}$$

$$\underline{\theta}^* = (\text{adj } L)^T\, I_k \,/\, (I_k^T (\text{adj } L)\, I_k^T) \qquad \text{(A-2)}$$

From Theorem 1, $R(D,\underline{\theta})$ has its maximum value at $\underline{\theta} = \underline{\theta}^*$. Therefore, the next equation is necessary:

$$\left.\frac{dR}{d\theta_i}\right|_{\underline{\theta}=\underline{\theta}^*} = 0 \qquad i=2,\ldots,K \qquad \text{(A-3)}$$

Calculating eq. (A-3) by substituting eq. (13), we get:

$$\frac{dR}{d\theta_i} = \sum_{j=1}^{q} r(j,\underline{\theta}) \frac{d}{d\theta_i} P(j,\underline{\theta},D) + \sum_{j=1}^{q} P(j,\underline{\theta},D) \frac{d}{d\theta_i} r(j,\underline{\theta}) \qquad \text{(A-4)}$$

where

$$P(j,\underline{\theta},D) = \sum_{p=1}^{s} \pi^j_{\underline{m}_p} P(\underline{m}_p/\underline{\theta}) \qquad \text{(A-5)}$$

If the game has only one maximin solution and one minimax solution, the theory of games leads the following equation:

$$r(j,\underline{\theta}^*) = v \qquad \text{for all } j \qquad \text{(A-6)}$$

where v is the value of game Γ. Considering the fact that:

$$P(1,\underline{\theta},D) + P(2,\underline{\theta},D) + \ldots + P(q,\underline{\theta},D) = 1\ , \qquad \text{(A-7)}$$

the first term of eq. (A-4) vanishes. Then

$$\left.\frac{dR}{d\theta_i}\right|_{\underline{\theta}=\underline{\theta}^*} = \sum_{j=1}^{q} P(j,\underline{\theta}^*,D)\,(L_{ij}-L_{1j}) = 0 \ . \tag{A-8}$$

The solution of eq. (A-8) is:

$$\underline{P} = (\text{adj } L)\, I_k \,/\, (I_k^T(\text{adj } L)\, I_k) \tag{A-9}$$

where I_k is a column vector having k elements which are all 1, and

$$\underline{P} = (P(1,\underline{\theta}^*,D),\ P(2,\underline{\theta}^*,D),\ \ldots,\ P(q,\underline{\theta}^*,D))^T \tag{A-10}$$

This coincides with the minimax solution of game Γ; i.e.,

$$P_j^* = P(j,\underline{\theta}^*,D) = \sum_{p=1}^{s} \pi^j_{\underline{m}_p}\ P(\underline{m}_p/\underline{\theta}^*) \tag{A-11}$$

Q.E.D.

The proof of general cases is similar to the above special case, though it is omitted here.

REFERENCES

1. Y. Sawaragi, Y. Sunahara, and T. Nakamizo, Statistical Decision Theory in Adaptive Control Systems, Academic Press, 1967.
2. T. T. Lin and S. S. Yau, "Bayesian Approach to the Optimization of Adaptive Systems," IEEE Trans. on Systems Science and Cybernetics, Nov. 1967.
3. Z. Bubnicki, "Least Interval Pattern Recognition and Its Application to Control Systems," IFAC IV-th Congress, 1969.
4. B. Kondo and S. Eiho, "Statistical Min-Max Decision Methods and Their Applications to Learning Control," IFAC IV-th Congress, 1969.
5. A. Wald, Statistical Decision Functions, Wiley, 1950.
6. J. C. C. McKinsey, Introduction to the Theory of Games, McGraw-Hill, 1952.
7. M. Aoki, "On Some Convergence Questions in Bayesian Optimization Problems," IEEE Trans. on Automatic Control, April 1965.

A CONTINUOUS-VALUED LEARNING CONTROLLER FOR THE GLOBAL OPTIMIZATION OF STOCHASTIC CONTROL SYSTEMS

R. W. McLaren

University of Missouri

Columbia, Missouri, U.S.A.

INTRODUCTION

The optimization of stochastic control systems has received increased attention in recent years [1,2,3,4]. The stochastic nature of a control situation can arise from a variety of sources such as measurement noise, parameter drift, or other, possibly external disturbances. Such random properties represent a description of inaccurately known disturbances, signals, or system changes. If there exists much a priori information or assumptions concerning plant dynamics and/or statistical properties, then a stochastic controller selecting control policies on the basis of a state estimation procedure may be appropriate. With known statistical properties, the estimation is amenable to a Kalman filter or a Bayesian approach. However, the lack of sufficient a priori assumptions concerning the stochastic plant dynamics requires a learning controller to interact with the stochastic plant to gain information for improving performance. Several algorithms for the learning controller for seeking optimal control policies have been considered elsewhere [5,6,7]. One such algorithm which has received considerable attention is stochastic approximation; it is a stochastic form of a more general class of "gradient" or hill-climbing search algorithms. Such algorithms, however, tend to have local optimization properties. Hence, although often providing rapid convergence, such an algorithm can "miss" an extremum while seeking a local minimum or maximum.

This paper describes the application of a particular learning control algorithm to several configurations of a class of digital stochastic control systems. The learning controller applys the algorithm to the unknown stochastic plant to search for global optimal control policies over a continuum of values while extremizing a performance measure.

A CLASS OF LEARNING CONTROL SYSTEMS

The interaction of a stochastic plant with the learning controller is shown in Figure 1. Referring to this figure, discrete-time quantities are U(k), the control vector, X(k), the state vector for the plant, and the instantaneous performance, Z(k), a scalar, k=0,1,.... Stochastic properties can arise from various sources; here, measurement and system parameter noise are emphasized. Only additive measurement noise is shown in Figure 1. Referring to this diagram, the learning controller first forms a normalized, instantaneous performance evaluation, Z(k), $0 \leq Z(k) \leq 1$. The measurement, Y(k), may also be available to allow the controller to form a state estimate, $\hat{X}(k)$. Generally, Z(k) is a continuous function of Y(k), Y(k-1), and U(k-1), k=1,2,.... Then, the learning controller uses Z(k) to select U(k) in an iterative learning algorithm. The instantaneous performance, Z(k), serves as a "reaction" of the stochastic plant to U(k-1). The dashed lines in Figure 1 show different configurations for the class of stochastic control problems. The particular configuration which applys depends on (1) the mode of operation of the learning controller, (2) the amount and form of information available to the controller concerning the operation (performance) of the plant, and (3) the assumed structure of the plant. Four configurations are considered.

1. Performance evaluation only. In this configuration, the only information available to the learning controller is the

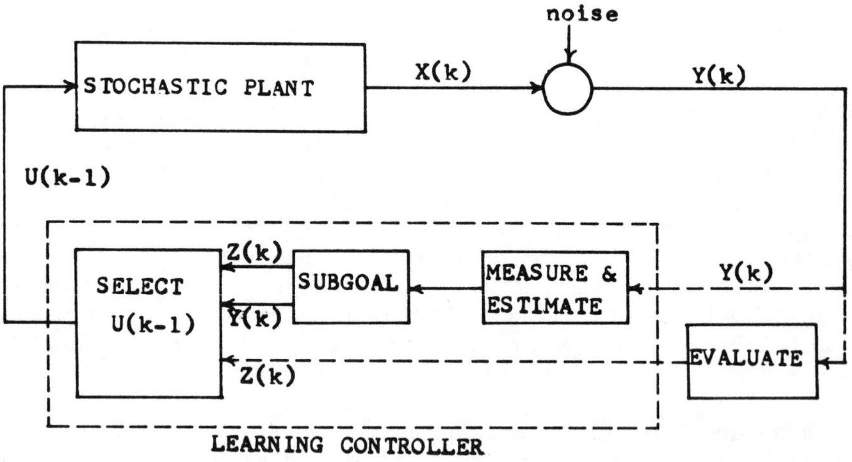

Figure 1. Illustration of the Interaction of the Learning Controller with the Stochastic Plant.

instantaneous performance, Z(k). This can be considered as a reaction to the choice of a control policy selected by the controller. In Figure 1, this is represented by the external evaluation, which, in general, is an unknown function of Y(k) and U(k-1). The controller seeks, on-line, a control which is optimal averaged over all states.

2. State measurement available. In this configuration, the Y(k) measurements are available to the controller along with its own selection of U(k-1). Then, the learning controller has a state measurement along with a performance evaluation, Z(k), provided externally or calculated from Y(k) and U(k-1). Then, using Z(k), the learning controller seeks on-line an optimal control on a per-state basis. As with model 1, the controller seeks on-line optimal controls to yield long-term improvement in performance.

3. Optimal finite control sequence. A more general form of a performance index would be function of each X(k) (and possibly, U(k-1)), k=1,2,..., for a given X(0)=Y(0). Hence, given X(0), this learning controller configuration will utilize Y(k) and Z(k) to seek an optimal finite control sequence, U*(0),...,U*(N-1), where N is the length of the sequence. Such an approach depends on the repeatability of sequences of states rather than that for individual states for seeking optimal. This leads to off-line optimization, the system being repeatedly reinitialized at a given X(0).

4. A growing stochastic automaton for the plant. In the previous three configurations, a fixed-state stochastic automaton structure is imposed on the plant. In this configuration, a model for the stochastic plant bridges the gap between a finite-state model and an analog system, with the state defined over a continuum of values. The approach is to use a growing automaton structure for the plant, while the controller uses Z(k) to seek an optimal control as with model 1, except over an increasing number of states.

DESIRED PERFORMANCE AND LEARNING

It is generally desired to cause the plant to behave in a prescribed (optimal) fashion; this refers to the behavior or sequences of values for U(k-1) and X(k) for k=1,...,N, for some N (or infinite N). The behavior could be described by a fairly general cost functional,

$$J_N = J[U(k-1), X(k),\ k=1,\ldots,N;\ \underline{\theta} \,|\, X(0)], \tag{1}$$

where N is a terminal time, $\underline{\theta}$ is a set of generally unknown parameters, and X(0) is a specified initial state. This function summarizes the behavior in that it would reflect the actual behavior in terms of U(k-1) and X(k) compared with that desired. It will be assumed that it is desirable to minimize J_N. Furthermore, because of the stochastic nature of the system, X(k) would be replaced by Y(k), the measurement of X(k). The resulting stochastic nature of J_N would make it more meaningful to minimize the expected value of J_N over the random variations. This becomes even more meaningful when one considers that, in general, the selection of U(k-1) is based on a randomized strategy.

If the plant dynamics were fairly well-defined, then one could, at least theoretically, using perhaps a variational technique, determine an optimal control sequence for a given X(0) and $\underline{\theta}$. This may, of course, involve estimating X(k), given Y(k) or a sequence of Y(k)'s; in a linear system this situation would allow the use of a Kalman filter. However, it is generally assumed that the dynamics are unknown as well as the characteristics of the random sources. To achieve optimization under these conditions, it is necessary to use a self-optimizing or learning controller. Rather than estimating dynamics and/or statistical characteristics, the controller is to sequentially select values of U (control policies) to extremize a performance index such as $E[J_N]$. This selection will be based on performance information and possibly a noisy measurement of X(k), depending on the model or configuration. The self-optimizing or learning process involves gaining information from the environment (the stochastic plant in this case) through repeated selection and evaluation of control policies in given learning situations or conditions. For this reason, the basic form for J_N will need to be modified in formulating a related subgoal consistent with a particular system configuration, the information available to the controller, and the level of the plant/controller interaction. For example, in J_N, N generally may have little meaning due to the required repetition during the learning process. Consistent with the controller configurations under consideration, the desired performance subgoal can be expressed in terms of the following learning behavior:

(1) determine a control policy U=U* (or a sequence of such policies to minimize

$$P(K|U) = \frac{1}{K} \sum_{k=1}^{K} E[Z(k)|\ U(k-1)] \ , \qquad (2)$$

for sufficiently large K; and

(2) drive the time average of E[Z(k)] monotonically toward a minimum value.

It is emphasized the U(k) is rather general; it can be simply a level of a step function or it could represent a set of parameter values determining a waveform input to a continuous time system. The measured output, y(t), for $(k-1)T < t < kT$, could then be compared with a desired behavior to form Z(k) as, say, an average mean square of the difference. Hence, the controller would be exercising sample control. The performance $Z(k)$ would then depend on $U(k-1)$, $X(k-1)$, and system dynamics.

THE LEARNING CONTROL ALGORITHM

Let the control vector, U, assume a continuum of values over a bounded region, R_U, in E^n. Let U^* be the optimal value of U, in general. The self-optimizing or learning algorithm to be applied here seeks a control (or control sequence) which is arbitrarily close to U^*. Now, define a uniform probability distribution of level $L(k)$ over R_U. Given $X(0)$, control policy $U = U_1$ is randomly selected and assigned probability $p_1(1) = \alpha[1-Z(1)]$, while the uniform level changes to $1-p_1(1)$. The parameter α represents a learning parameter to be selected by the designer. Let $Pr[U(k)=U_i]=p_i(k)$ and let $L(k)$ be the level of the uniform probability distribution over R_U at step k. After $U(0)=U_1$ is selected and yields "reaction" $Z(1)$, set

$$p_1(1) = (1-\alpha)[1-Z(0)], \quad L(1) = \alpha + (1-\alpha)Z(0) \tag{3}$$

where $0 < \alpha < 1$. A new value of U will be generated when $L(k)$ occurs. After this step, say at step k, let $s=s(k)$ be the number of values of U in existence at step k. Then, (a) if $U(k)=U_j$, $j=1,2,\ldots,s(k)$, set

$$\begin{aligned} p_j(k+1) &= \alpha p_j(k) + (1-\alpha)[1-Z(k+1)] \\ p_i(k+1) &= \alpha p_i(k) + (1-\alpha)\, Z(k+1)/s \\ L(k+1) &= \alpha L(k) + (1-\alpha)\, Z(k+1)/s \ , \end{aligned} \tag{4}$$

where $i=1,2,\ldots,s(k)$, $i \neq j$, and $0 < \alpha < 1$; (b) if $U(k) = U_{s+1}$, a new control policy, then eq. (4) above applies with $j=s(k)+1$, while $s(k)$ is replaced with $s(k)+1$. This particular algorithm is discussed further in [8] while a discrete-valued one is described in [9]. Remembering that $Z(k)$ represents an instantaneous performance index, define the conditional average value of $Z(k)$ as $m(U)=E_Z(Z|U)$. Let $m(U)$ possess bounded partial derivatives and assume that the conditional probability density function $p(Z|U)$ is stationary for a given U. As $k\to\infty$, the expected values of the probabilities of the generated or existing values of U tend to approach limit values which fall in reverse magnitude order as the corresponding conditional expected performance values. That is, if $m_i=E[Z|U_i]$ and $E_v(k)=E[p_v(k)]$, $v=1,\ldots,s=s(k)$, then,

$$E_i(k) > E_j(k) \leftrightarrow m_j > m_i \ . \tag{5}$$

Specifically, as k increases, the value of U generated arbitrarily close (based on performance) to U* is associated with min $[m_i=m(U_i), i=1,\ldots,s(k)]$, over existing values of U at step k. This value of U will thus have the largest expected probability. This ordering of probabilities can serve as the basis for selecting the best policy. The expected probabilities can be estimated using time averages. In addition, the expected performance measure has the form,

$$E[Z(k)] = \sum_{v=1}^{s} E_v(k)m_v + E[L(k)]\bar{m} , \tag{6}$$

where $\bar{m}$ is the average of m(U) over U and $m_v=m$ for $U=U_v$. Now, E[Z(k)] tends to converge monotonically toward a minimum value. Using results from [8], as s(k) increases, $s(k) \sim s(k)+1$, and thus $\sum_i E_i(k)$ approaches 1. One then has

$$\begin{aligned} E_i(k+1) &\sim \alpha E_i(k)+(1-\alpha)[(1-m_i)E_i(k) + \sum_{\substack{v=1\\ \neq i}}^{s} \frac{m_v}{s} E_v(k)] \\ E[L(k+1)] &\sim \alpha E[L(k)] + (1-\alpha) \sum_{j=1}^{s} \frac{m_j}{s} E_j(k) , \quad i=1,\ldots,s, \end{aligned} \tag{7}$$

where s=s(k). If $E_i(k)=E_j(k)$, $i \neq j$, then $E_i(k+1) > E_j(k+1)$ if and only if $m_i < m_j$. As k increases, s(k) increases; the $E_i(k)$ decrease with k in such a fashion that eq. (5) holds. Also, from eq. (7) above, $E[L(k)] \to 0$, decreasing faster if the average of Z(k) is low due to many m_j values being relatively low (corresponding to many desirable values of U already having been generated). Under such conditions, the probability of generating new policies should be reduced more rapidly. The expected probability of a new value of U being generated is given by

$$E_{s+1}(k+1) = (1-\alpha)(1-\bar{m}) E[L(k)] ; \tag{8}$$

this reflects a larger value when the average value of m, $\bar{m}$, over U is smaller. As time progresses, the "best" U value (minimum m) is sought from among existing values of U while the set of probabilities of existing values of U depends on the "quality" of the existing set of U values. If the existing values are poor, then the probability of generating additional values to try will be higher. Eventually, values of U located in an arbitrarily small region containing the optimal U, U*, will be generated and then selected using time average estimates of probabilities.

Furthermore, from eq. (7) and using reference [8],

$$\Delta E_i(k+1) = E_i(k+1)-E_i(k) \sim (1-m_i)E_i(k)+ \sum_{v=1\neq i}^{s} \frac{m_v}{s} E_v(k) . \tag{9}$$

From this equation, $\Delta E_i(k) \to 0$ as k increases, indicating that the performance index tends to decrease monotonically toward a minimum value over time.

The parameter α represents a learning parameter which determines the rate and variation in learning. This corresponds to mean and variance of $\Delta Z(k)$. There is a trade-off between higher learning rates for smaller values of α and the resulting greater magnitude of fluctuation in $Z(k)$.

STRUCTURE OF THE PLANT

For purposes of analysis and comparison of different configurations considered here, a fairly general structure for the plant will be assumed for the first three models or configurations.

Let the state space $X(k)$ be partitioned into disjoint regions, R_i, $i=1,\ldots,r$; if $X(k) \in R_i$, let state $Q(k)=q_i$, any i. Let $Pr[Q(k)=q_i] = a_i(k)$, all i. Then the stochastic plant can be described as a stochastic automaton (Q,U,Y,f,g). Q is the set of internal states (partition of X), $q_1,\ldots,q_r$, U is the set of control policies, $U_1,\ldots,U_{s(k)}$, and f is a stochastic mapping from (U,Q) to Q. This mapping is defined by a matrix of state transition probabilities, $t_{ij}(U)$, where $t_{ij}(U)$ is the transition probability from state q_j to state q_i for $U(k)=U$. Y is the set of output states, $y_1,\ldots,y_r$ and g is a stochastic output mapping defined by $b_{ij} = Pr[Y=y_i|Q=q_j]$, all i,j pairs. To complete the description, let $a_i(0) = 1/r$, all i.

MODEL I: PERFORMANCE EVALUATION ONLY

Suppose that the only information available to the learning controller is the instantaneous performance measure, $Z(k)$. Let a cost, $c_{ij}(U)$, be associated with transition $t_{ij}(U)$. Assume that $c_{ij}(U)$ is unknown but is continuous in U and its partial derivatives. Given $U(k-1)$, the conditional expected value of $Z(k)$ is expressed by

$$E[Z(k)|U(k-1)] = \sum_{i=1}^{r} \sum_{j=1}^{r} c_{ij}(U)\, t_{ij}(U)\, a_j(k-1) . \tag{10}$$

Each $a_j(k)$ is defined recursively using the $t_{ij}(U)$'s, $U = U(k-1)$, and $a_j(0) = 1/r$, all j. As the plant passes through an unknown sequence of states, the learning controller, using $Z(k)$, applies a single learning algorithm on-line to seek the control $U = U^*$ which would minimize $m_i(k) = m(U,k) = E[Z(k)|U(k-1)]$, where $Q(k) = q_i$, $i=1,\ldots,r$, $k=1,\ldots$. No one control is optimal for all states. From eqs. (7) and (9), the sign and magnitude of $\Delta E_i(k)$ as k increases reflect an average effectiveness of the selected values of U over a sequence of states. The most effective U value minimizes m(U) over all states, $q_1,\ldots,q_r$. With p(U,k) as the p.d.f. of U(k), E[Z(k)] is the average of E[Z(k)|U(k-1)] over U(k-1).

Suppose that the distribution of U(k-1) were constant over k; then E[Z(k)] would vary as the plant changed from state to state, the state distribution, $\{a_j(k-1)\}$, changing with k. Furthermore, let the $a_j(k)$'s converge monotonically toward final values (for any given U). Then as the learning controller searches for the optimal U, the $E_i(k)$ converge monotonically. This implies that p(U,k) converges monotonically. Then

$$E[Z(k)] = \int E[Z(k)|U(k-1)=U]\, p(U,k)\, dU \tag{11}$$

converges monotonically toward a minimum as desired. If the $a_j(k)$'s simply converge (for a given U), E[Z(k)] will converge toward a minimum. A subgoal which is appropriate for this on-line learning would be to seek a U = U* to minimize the time average of E[Z(k)], as in eq. (2), after many iterations. It is noted that U(k-1) could represent a waveform to be applied between (k-1)T and kT; the performance, Z(k), would then depend on the system dynamics as well as on other factors. This approach would require some modification of the form of eq. (10) for the performance in terms of cost.

MODEL II: STATE MEASUREMENT AVAILABLE

Suppose now that the learning controller is able to measure X(k), resulting in Y(k). An instantaneous performance measure, Z(k), is still available, either directly from the environment or is calculated by the controller from Y(k) and U(k-1). Now, using the probabilities $b_{ij} = \Pr[Y=y_i|Q=q_j]$, all i,j pairs, the effect of noise on X(k) can be described by

$$v_{ij}(k) = \Pr[Q(k)=q_i|Y(k)=y_j] = \frac{b_{ji}a_i(k)}{d_j(k)}, \tag{12}$$

where $d_j(k) = \Pr[Y(k)=y_j] = \sum_i b_{ji}a_i(k)$. Suppose first that there is no measurement noise; then the controller uses Z(k) and X(k) to determine a control policy U_i that will minimize $E[Z(k_i)|U_i]$, and thus its time average for large k_i for state q_i, $i=1,\ldots,r$, where k_i represents the number of times that state q_i has been entered. The p.d.f. $p(U,k_i)$ for state q_i tends to converge for any i. Here it is also desired to drive the time average of E[Z(k)], over all states, monotonically toward a minimum value. The desired performance can be achieved by applying or assigning the basic learning algorithm on-line to each state independently. With this rule, the optimal U_i, U_i^*, can be determined for each state as it is repeatedly re-entered. When the plant enters state q_i, only the algorithm assigned to that state applies; all others remain unchanged. Compared with Model I, the optimization here is per state, instead of averaged over all states, yielding better performance. To examine this, suppose that for state q_i, $U = U_i^*$ is optimal, while say $U = U^*$ is the optimal control averaged over all states (Model I). Then using eq. (10) to find the difference in conditional expected values of Z(k), using subscripts to indicate Models I and II,

$$E_1[Z(k)|U(k-1)]-E_2[Z(k)|U(k-1)] = \sum_{i=1}^{r} \sum_{j=1}^{r} c_{ij}(U^*)t_{ij}(U^*)a_j(k-1) - \sum_{i=1}^{r} \sum_{j=1}^{r} c_{ij}(U_j^*)t_{ij}(U_j^*)a_j(k-1) . \quad (13)$$

This can then be rewritten as,

$$\sum_{i=1}^{r} \sum_{j=1}^{r} [c_{ij}(U^*)t_{ij}(U^*) - c_{ij}(U_j^*)t_{ij}(U_j^*)] \; a_j(k-1) . \quad (14)$$

Now, for any given state, q_j, the optimality of U_j^* means, by definition,

$$M_j = \sum_{i=1}^{r} c_{ij}(U_j^*)t_{ij}(U_j^*) \quad (15)$$

is a minimum over U. Hence, as long as $U^* \neq U_j^*$ for at least one value of j, the quantity in eq. (14) will be strictly positive. Thus, the performance for Model II will be better.

Suppose now that there are only noisy state measurements available. Let $v_{ii}(k)$ be the maximum of $v_{ij}(k)$ over j for any i. If U_t^* is the optimal U for state q_t using Y(k), then using eq. (10) with $a_t(k-1) = 1$,

$$E[Z(k)|U_t^*] \; v_{it}(k) > E[Z(k)|U_t^*] . \quad (16)$$

Generally, as the measurements become noisier, $v_{ii}(k)$ decreases, increasing the average cost thus yielding worse performance.

Now, the initial partitioning of state space into disjoint regions represented by $q_1, q_2, \ldots, q_r$ may not, in general, be efficient in the sense that some states (regions) may never be entered, while others might require a finer partition for "closer" control. An alternate approach involves generating the states on-line, using Y(k) (or X(k)) as the center of a hyperspherical region of radius R to define a state. If Y(k) falls outside of all states at a given step, a new state is formed. If Y(k) falls inside of two or more existing states, it is assigned to the state whose center is closest.

This is a more efficient representation. The representation and performance depend to some extent on the radius R. For a single radius, there is a trade-off between having only a "rough" description of the plant for large R and having an unwieldly number of states for small R; a compromise could be arrived at through computer simulation. Another approach is to make R_i dependent on the number of times that state q_i is entered; the higher that this number is, the lower that R_i will be.

MODEL III: OPTIMAL FINITE CONTROL SEQUENCE

In this model, the instantaneous performance, Z(k), is available to the learning controller. A knowledge or measurement of the state X(k) or Q(k) of the plant may or may not be available. The previous two models dealt with on-line optimization to determine a control policy to minimize a long-term average performance measure. If U(k-1) represents an instantaneous control vector (rather than a waveform, etc.), then the optimal control tends to depend on the steady-state characteristics of the plant and not on its dynamics as it approaches steady-state. The reason for this rests on two factors: (1) the on-line nature of the plant/controller arrangement and (2) a restriction to determine a single control policy for a basically variable (dynamic) system. The on-line nature restricts in that once a particular control policy has been selected, applied, and evaluated, there is no going back to try again using a different policy. The iterative nature of the algorithm is based on the idea that at each step, a new learning situation is presented; applying the algorithm leads toward average performance--a smoothing process. This is particularly true for Model I. In Model II, the limiting knowledge is an optimal policy for each state; hence, for any sequence of states, an optimal sequence of corresponding optimal controls could be generated. This does have some measure of generality; however, these controls do not depend on the state-to-state transition characteristics (dynamics), for a specific plant. For on-line optimization to involve the plant dynamics, more knowledge concerning such dynamics and/or statistical characteristics would need to be available. Hence, there is a trade-off between the "degree" of involvement of the dynamics in an on-line optimization situation and the a priori assumptions. Here, the relative lack of knowledge concerning characteristics of the plant will be assumed. To involve dependence on the dynamics under such assumptions will lead to an off-line learning situation.

The learning controller is to determine or seek the optimal control sequence, $S_N^* = [U(0),\ldots,U(N-1)]^*$, so as to minimize,

$$P(S_N) = \frac{1}{N} \sum_{k=1}^{N} E[Z(k)|U(k-1)] \ , \tag{17}$$

for a given Q(0) or X(0), where $N=1,\ldots,M < \infty$. Per-state optimization to achieve long-term low average performance as in Model II is not sufficient to determine a finite optimal control sequence as considered here. Rather, the learning controller must repeat control sequences, reinitializing at X(0) in an off-line mode. It is desired that the controller seek a control sequence S_N^* to achieve the following behavior. Letting $P[S_N|X(0)] = P(S_N)$ for a given X(0), it is desired to determine S_N so that,

(a) $P[S_N^*|X(0)]$ = minimum over S_N , and

$$\text{(b)} \quad \frac{1}{kN} \sum_{v=1}^{kN} E[Z(v)] \rightarrow \text{minimum} . \tag{18}$$

These two conditions are related in the sense that for large k, the optimal sequence sought will result in the term of eq. (18a) dominating the time average expressed in eq. (18b) as S_N^* is chosen more frequently. In general, the N-step performance measure could be a function of all U(k-1) and X(k). If the X(k) are not accessible, then the desired result in eq. (18) is over the unknown states. In this model, the learning controller selects and applies a sequence or vector of N values of U(k-1), $k=1,\ldots,N$. Based on Σ Z(k) over N steps (for the N controls), S_N is generated from a uniform distribution defined over a continuum of sets of N controls. As the number of repetition increases (with reinitialization), sequences of controls closer to the optimal one will acquire higher expected probabilities while the average expected performance decreases. This is similar to using an Nth order stochastic automaton model for the controller. Let Q(0) be specified. If Q(1) is known, the learning controller can then seek an optimal (N-1) control sequence, S_{N-1}^* for every known Q(1). The length of the desired control sequence decreases as the Q(k), $k < N$ become known. In this case, more control sequences, such as U(N-1),...,U(k) given X(0), X(1), ...,X(k), $k < N$, would need to be stored (with the corresponding performances), but performance will be improved because of this added information.

MODEL IV: GROWING STOCHASTIC AUTOMATON FOR PLANT

In this model, the structure of the plant is reconsidered. In the previous models, there is state-space quantization of the original continuous-valued X(k). This approximation would improve as the number of states increases, but memory requirements and learning rate degrade. The continuous-valued (growing) stochastic automaton model previously considered can be applied to the plant structure itself. The $t_{ij}(U)$ transition probabilities now depend on k; the resulting $t_{ij}(U,k)$ are iteratively related through the global optimization algorithm, based on Z(k). This plant representation is more efficient in terms of the number of states and their significance in terms of performance. Generally, it is expected that performance will improve. Additional investigation of this model is required to establish its significance and how to meaningfully change the $t_{ij}(U,k)$'s so as to bring about the desired behavior.

COMPUTER SIMULATION

In order to demonstrate the applicability of the learning control algorithm to a class of stochastic control systems, a high-speed digital computer was used to simulate a basic second-order plant having additive measurement noise. The instantaneous performance was set up as a quadratic function of Y(k) and U(k-1) with X(0) = 0. The learning controller selected U(k-1), while state-space was allowed to be continuous. The control policy, U(k-1) was chosen from a continuum of values over a bounded region, using the algorithm to minimize an average performance over all states as in Model I. The results indicate that the "better" controls have higher time average probabilities and that the time average of Z(k) tended to decrease toward a minimum. In addition, an increase in α decreased the learning rate and smoothed the variations in the time average of Z(k). A typical result based on an ensemble study of 200 systems is shown in Figure 2.

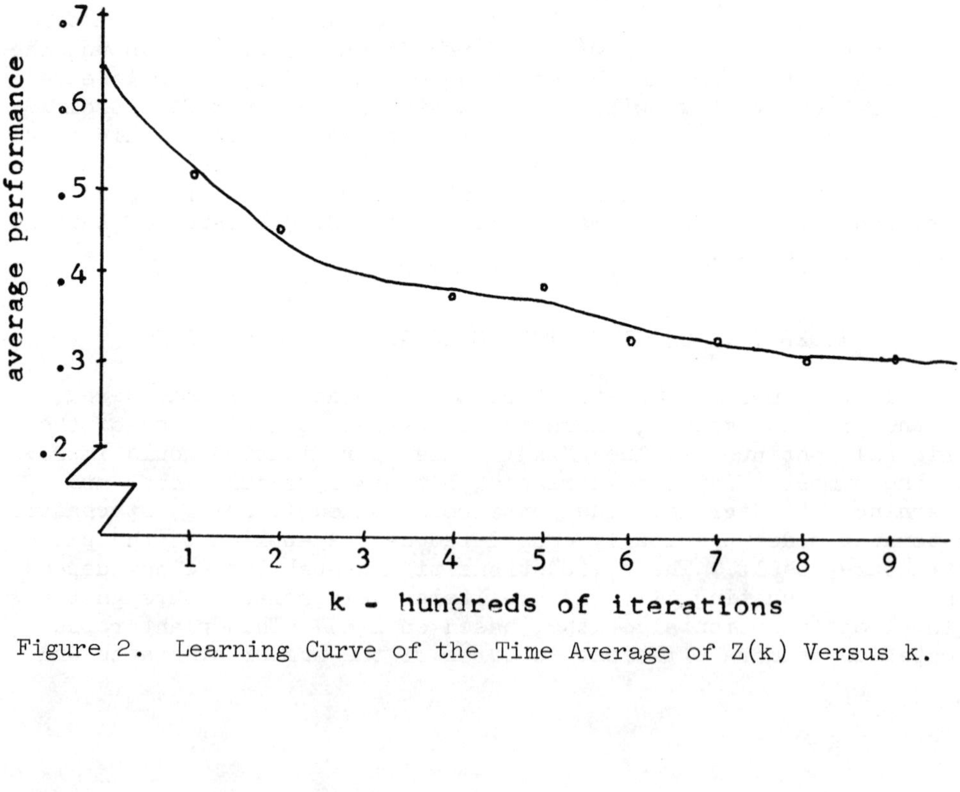

Figure 2. Learning Curve of the Time Average of Z(k) Versus k.

DISCUSSION AND CONCLUSIONS

Figure 2 reveals (1) a tendency toward relatively slow convergence and (2) a leveling off of the average performance at a fairly high value. However, with respect to (1), it should be remembered that the algorithm is a global one; in the author's view, it is an efficient one compared with, say, an exhaustive search which, in general, would be time-consuming. The global search algorithm considered here represents, perhaps, a sacrifice in terms of an increased number of required iterations compared with a local optimization technique. Moreover, it is observed that this algorithm also gives an ordering (over control policies) according to their relative (expected) performances. With respect to (2), it should be remembered that as the expected performance decreases, the optimal control policy is achieving a higher average probability. Thus, one could accelerate convergence by eliminating controls having low average probabilities of occurrence and continuing, starting with the higher average probability controls that remain.

A particular learning control algorithm has been applied to several configurations of a class of stochastic control systems. Two configurations represent on-line methods; basically, they seek a single steady-state control. The third model seeks, off-line, a dynamic optimal control sequence, while the last model involves a growing automaton model for the plant. The mode of application and some resulting properties for these configurations have been discussed. Results indicate that the learning controller (1) can determine a control arbitrarily close to the globally optimal one over a continuum of values and (2) causes the average performance measure to decrease toward a minimum value.

REFERENCES

1. H. F. Karreman, Stochastic Optimization and Control, John Wiley and Sons, New York, 1968.

2. R. S. Bucy and P. D. Joseph, Filtering for Stochastic Processes With Applications to Guidance, John Wiley and Sons, New York, 1968.

3. T. Kailath, et.al., Stochastic Problems in Control, Symposium held at 1968 JACC, University of Michigan, June 1968.

4. J. Meditch, Stochastic Optimal Linear Estimation and Control, McGraw-Hill, New York, 1969.

5. J. S. Riordon, "An Adaptive Automaton Controller for Discrete-Time Markov Processes," in Preprints of the 1968 JACC, Univ. of Michigan, June 1968.

6. K. S. Fu and J. M. Mendel, Adaptive, Learning, and Pattern Recognition Systems, Academic Press, New York, 1970.

7. Proceedings of the U.S.-Japan Seminar on Learning Process in Control Systems, held August 18-20, 1970, in Nagoya, Japan.

8. R. W. McLaren, "Application of a Continuous-Valued Control Algorithm to the On-Line Global Optimization of Stochastic Control Systems," Proceedings of the National Electronics Conference, Chicago, Ill., 1969.

9. K. S. Fu and R. W. McLaren, "An Application of Stochastic Automata to the Synthesis of Learning Systems," Purdue Univ. Tech. Rept. No. EE 65-17, School of Electrical Engineering, Purdue Univ., Lafayette, Ind., Sept. 1965.

ON VARIABLE-STRUCTURE STOCHASTIC AUTOMATA

R. Viswanathan and Kumpati S. Narendra

Yale University

New Haven, Connecticut, U.S.A.

I. INTRODUCTION

A stochastic automaton with a variable structure (SAVS) may be described by the sextuple $\{X, \Phi, \alpha, p, U, G\}$ where $X=\{x_o,x_1,\ldots,x_k\}$ is the input set, $\Phi = \{\phi_1,\phi_2,\ldots,\phi_s\}$ is the internal state set, $\alpha = \{\alpha_1,\alpha_2,\ldots,\alpha_r\}$ with $r \leq s$ is the output or action set, p is the state probability vector (i.e., at stage or time instant n, $p(n) = (p_1(n),p_2(n),\ldots,p_s(n))$ governing the choice of the state, U is an updating scheme which generates $p(n+1)$ from $p(n)$ and $G:\Phi \to \alpha$ is the output function. In general, G may be a stochastic function. In this paper it is assumed that G is deterministic and one-to-one (i.e., $r = s$), $k = 1$ and $s < \infty$.

Figure 1 depicts a closed loop structure consisting of an SAVS and a stochastic environment. The environment responds to the automaton actions by producing a penalty ($x=1$) or non-penalty ($x=0$) at the input of the automaton. A penalty probability set $\{c_1,c_2,\ldots, c_r\}$ is used to characterize the environment; c_i is the probability that the environment penalizes the action α_i. Only stationary environments are considered in this paper so that all the c_i are constant. Further, it is assumed that $c_i \in (0,1)$ $(i=1,2,\ldots,r)$, the set $\{c_i\}$ has a unique minimum and the actual values of c_i are not known. The expected value of penalty, $M(n)$, for an SAVS functioning in a random environment is defined as follows:

$$M(n) = \sum_{i=1}^{r} \overline{p_i(n)}\ c_i \qquad (1.1)$$

where $\overline{p_i(n)}$ refers to the expected value of $p_i(n)$. The asymptotic average penalty is denoted by M:

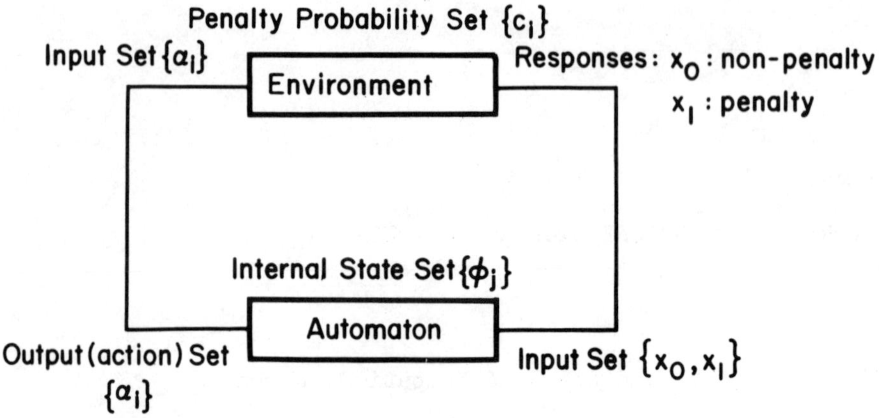

Figure 1. Automaton-Environment Feedback Configuration

$$M = \lim_{n\to\infty} M(n) \tag{1.2}$$

Expediency and Optimality [1]. An SAVS functioning in a stationary random environment is said to be:

(i) expedient, if

$$M < M_0 \triangleq \frac{1}{r} \sum_{i=1}^{r} c_i \tag{1.3}$$

or as a sufficient condition if

$$\lim_{n\to\infty} \overline{p_i(n)} > \lim_{n\to\infty} \overline{p_j(n)} \quad \text{whenever } c_i < c_j \qquad i,j = 1,2,\ldots,r,\ i \neq j \tag{1.4}$$

and

(ii) optimal, if

$$M = c_k \tag{1.5}$$

where the index k is defined by

$$c_k = \operatorname*{Min}_i c_i \tag{1.6}$$

or equivalently if

$$\left.\begin{aligned} &\lim_{n\to\infty} \overline{p_k(n)} = 1 \\ &\lim_{n\to\infty} \overline{p_{i\neq k}(n)} = 0, \; i = 1,2,\ldots,r \end{aligned}\right\} \tag{1.7}$$

The closer the value of M [satisfying (1.3)] to c_k the more expedient is the automaton. The degree of expediency can be defined similarly so that it is maximum when $M = c_k$ and zero when $M = M_0$. Expediency assures a performance better than that obtained by choosing the automaton actions with equal probabilities. However, expediency allows in the limit, finite and perhaps significant probabilities of occurrence of poorer actions (those having higher penalty probabilities). Furthermore, as the probabilities $p_i(n)$ $(i=1,2,\ldots,r)$ are bounded below and above by 0 and 1 respectively, the optimal convergence defined in (1.7) results in zero variance for the limiting state probability vector. Hence optimality is more desirable than expedient behavior.

A meaningful way of constructing an SAVS would be to enhance better actions and suppress poorer ones. At each stage, the stochastic automaton changes its structure depending upon the response from the environment, by varying its transition probabilities [2,7] or equivalently its total probabilities [3-7]. Only total probability updating is considered here, for convenience. The structural variation in this case can in general be given as follows:

$$p_i(n+1) = p_i(n) + f_i^n[p_1(n),p_2(n),\ldots,p_r(n)]$$
$$i = 1,2,\ldots,r; \; n = 1,2,3,\ldots \tag{1.8}$$

where $f_i^n[\cdot]$ is an updating term for p_i at the n-th stage of operation of the automaton-environment combination. To preserve probability measure, it is necessary that the following condition be satisfied for all n:

$$\sum_{i=1}^{r} f_i^n[p_1(n),p_2(n),\ldots,p_r(n)] = 0 \tag{1.9}$$

Let $\alpha(n) = \alpha_i$. Clearly, the action α_i is penalized if $f_i^n[\cdot] < 0$ and rewarded if $f_i^n[\cdot] > 0$. The updating scheme by which the automaton changes its structure is said to be a reward-penalty scheme if it rewards an action that causes the environment to produce a non-penalty output and penalizes an action that makes the environment respond with a penalty. Similarly reward-reward, reward-inaction, inaction-penalty and penalty-penalty updating schemes can be defined.

Linear and Nonlinear Schemes. An updating scheme is
(i) linear, if for each $i \in \{1,2,\ldots,r\}$ the updating term $f_i^n[\cdot]$ depends linearly on $p_k(n)$ $(k = 1,2,\ldots,r)$

and

(ii) nonlinear, if it is not linear.

An updating scheme is said to yield expedient or optimal convergence depending upon whether the stochastic automaton which uses it is expedient or optimal. For an updating scheme the expected change in $p_i(n)$ at time n, denoted by $\Delta p_i(n)$ and referred to as the expected step size in $p_i(n)$, is defined as

$$\Delta p_i(n) = \overline{p_i(n+1)|p(n)} - p_i(n) \tag{1.10}$$

In this paper several expedient and optimal updating schemes are discussed. A necessary condition for optimal condition is stated and it is shown that optimal convergence cannot be obtained if a constant linear penalty updating is used. Several optimal nonlinear schemes are discussed and a method of comparing the various schemes is proposed.

As the labeling of c_i is arbitrary, in what follows it is assumed, for convenience, that

$$c_1 < c_2 \leq c_3 \leq \cdots \leq c_r \tag{1.11}$$

II. A NECESSARY CONDITION FOR OPTIMAL CONVERGENCE

The class $\mathcal{S}$ of updating schemes to be considered is defined below. An updating scheme belonging to $\mathcal{S}$ is described by (1.8) where $f_i^n[p_1(n),\ldots,p_r(n)]$ $(i = 1,2,\ldots,r)$ satisfy (1.9) and the following conditions:

(i) they are continuous in all the arguments

(ii) when the action $\alpha(n) = \alpha_i$,

$$f_i^n[p_1(n),p_2(n),\ldots,p_r(n)] = \begin{cases} (f_i)_+[p_1(n),p_2(n),\ldots,p_r(n)] & \text{if the environment responds with a non-penalty} \\ (f_i)_-[p_1(n),p_2(n),\ldots,p_r(n)] & \text{otherwise} \end{cases} \quad \ldots (2.1)$$

and (iii) the step size factors in the updating functions are such that the updated probabilities lie within [0,1] at each stage 'n'.

A necessary condition for optimal convergence of an updating scheme belonging to $\mathcal{S}$ is stated in the following theorem.

Theorem 2.1. A necessary condition for an updating scheme belonging to the class $\mathcal{S}$ to be optimal is that its updating functions satisfy the following:

$$(f_i)_{\pm}[p_1(n),p_2(n),\ldots,p_r(n)]\Big|_{p_i(n)=1} = 0, \quad i = 1,2,\ldots,r \qquad (2.2)$$

The proof may be briefly outlined as follows. There is a finite probability that α_j corresponds to the minimum penalty probability since no prior knowledge about the environment is assumed. If $p_j(n) = 1$ and $(f_j)_x[p_1(n), p_2(n),\ldots, p_r(n)] = \lambda \neq 0$ ($x = +$ or $-$) then with probability $1, p_j(n+1) = 1 + \lambda$ which is either greater than 1 or less than 1 depending on the sign of λ. However the former is impossible and the latter leads to a contradiction to the hypothesis that the updating scheme is optimal.

A sufficient condition for the existence of the limit, $\lim_{n\to\infty} \overline{p_1(n)}$, is that

$$\Delta p_1(n) \geq 0 \quad \cdot \quad p_1(n) \in [0,1]$$

equality holding only when $p_1(n)$ equals 0 or 1; in this case $\{p_1(n)|n = 1,2,\ldots\}$ is a Semi-Martingale [8] with a bounded non-negative sequence of random variables and a well-known Semi-Martingale convergence theorem applies ([8], p. 324, Theorem 4.1s).

III. LINEAR SCHEMES

It is convenient to describe a linear scheme in the following form [9]:

$$p_k(n+1) = p_k(n) + \beta(n)[\Lambda_k(n)-p_k(n)] \qquad (3.1)$$

$$k = 1,2,\ldots,r$$

In view of (1.9) it is necessary that

$$\sum_{k=1}^{r} \Lambda_k(n) = 1 \qquad (3.2)$$

For linear reward-penalty scheme L_{R-P}, $\beta(n)$ and $\Lambda_k(n)$ ($k = 1,2,\ldots,r$) are defined in (3.3) for the case when $\alpha(n) = \alpha_i$. If the environment responds with a

$$\left.\begin{aligned} &\text{(i) \underline{non-penalty}, set} \\ &\qquad \Lambda_i(n) = 1 \\ &\qquad \Lambda_{j\neq i}(n) = 0 \\ &\qquad \beta(n) = (1-a),\ 0 < a < 1 \\ &\text{(ii) \underline{penalty}, set} \\ &\qquad \Lambda_i(n) = 0 \\ &\qquad \Lambda_{j\neq i}(n) = \frac{1}{r-1} \\ &\qquad \beta(n) = (1-b),\ 0 < b < 1 \end{aligned}\right\} \tag{3.3}$$

When $b = a$, the updating scheme given by (3.1) and (3.3) with $r = 2$ reduces to the one considered by Varshavskii and Vorontsova [2] and gives

$$M = \frac{2c_1c_2}{c_1+c_2} \tag{3.4}$$

which corresponds to an expedient convergence. For $r > 2$ it can be shown [4] that L_{R-P} is expedient with

$$M = \frac{r}{\sum_{i=1}^{r} \frac{1}{c_i}} \tag{3.5}$$

A constant linear penalty updating can be either of the form

$$p_i(n+1) = p_i(n) - \beta[1 - p_i(n)],\ 0 < \beta < 1$$

or of the form

$$p_i(n+1) = p_i(n) - \beta p_i(n),\ 0 < \beta < 1$$

The former is not admissible as it renders $p_i(n+1)$ to be negative when $p_i(n)$ is close to zero and the latter does not satisfy the necessary condition for optimality given in (2.2). Hence a constant linear penalty updating leads to non-optimality.

IV. NONLINEAR SCHEMES

For the class of nonlinear updating schemes considered in this section, it is assumed that $r=2$ and $c_1 < c_2$ as before. Let $\alpha(n) = \alpha_i$.

Then (i) non-penalty: $p_i(n+1) = p_i(n) + f_+[p_i(n)]$

(ii) penalty: $p_i(n+1) = p_i(n) + f_-[p_i(n)]$ (4.1)

where $f_+[\cdot]$ and $f_-[\cdot]$ are the nonlinear updating terms to be specified.

4.1 Conditionally Optimal Schemes

(a) β-Model. For the β-model of the mathematical learning theory [10, 4] we have

$$\left.\begin{aligned} f_+(p) &= \frac{(1-a)p(1-p)}{p + a(1-p)} \\ f_-(p) &= -\frac{(1-a)p(1-p)}{ap + (1-p)} \\ \text{where} \quad & 0 < a < 1 \end{aligned}\right\} \qquad (4.2)$$

It can be shown ([11], p. 239, Theorem 4) that the β-model given in (4.1) and (4.2) is optimal if

$$c_1 < 1/2 < c_2 \qquad (4.3)$$

As a straightforward extension, let

$$\left.\begin{aligned} f_+(p) &= \frac{(1-a)p(1-p)}{p + a(1-p)}, \quad 0 < a < 1 \\ f_-(p) &= -\frac{(1-b)p(1-p)}{bp + (1-p)}, \quad 0 < b < 1 \\ \text{with} \quad & b = a^k, \ k \geq 1 \end{aligned}\right\} \qquad (4.4)$$

Using another result from [11] (p. 240, Theorem 5) it is found that the β-model as given in (4.1) and (4.4) is optimal provided that

$$c_1 < \frac{1}{k+1} < c_2 \qquad (4.5)$$

Thus the latter scheme yields optimality for a wider range of c_1 and c_2.

(b) The $N_{R-P}^{(1)}$ Scheme. For the $N_{R-P}^{(1)}$ scheme,

$$\left.\begin{aligned} f_+(p) &= ap^\alpha(1-p)^\alpha \\ f_-(p) &= -bp^\alpha(1-p)^\alpha \\ & \alpha \geq 1 \\ \text{where} \quad & 0 < a, b < 1 \quad \text{when } \alpha = 1 \\ & 0 < a, b < \frac{(2\alpha-1)^{2\alpha-1}}{\alpha^\alpha(\alpha-1)^{\alpha-1}} \quad \text{when } \alpha > 1 \end{aligned}\right\} \qquad (4.6)$$

Varshavskii and Vorontsova [2] considered the case when b = a and showed that optimal convergence is obtained when (4.3) holds. A simple extension of their approach is to let

$$b = ka, \; 0 \leq k \leq 1 \tag{4.7}$$

It is not difficult to show that the updating scheme given by (4.1), (4.6) and (4.7) is optimal when c_1 and c_2 satisfy (4.5). Hence when prior knowledge of the magnitudes of the penalty probabilities c_i is available, optimal convergence can be achieved by choosing an appropriate k.

In general, as prior information about the environment is not known these schemes, though optimal, are not very useful. In the following section we consider optimal nonlinear schemes which assume no knowledge of c_1 and c_2.

4.2 Unconditionally Optimal Schemes

Conditions on $f_{\pm}(p)$ in (4.1) for unconditional optimality are stated in the following theorem [5]:

Theorem 4.1. The nonlinear updating scheme described in (4.1) is optimally convergent if

$$\text{(i)} \quad f_{\pm}(p)\Big|_{p=1 \text{ or } 0} = 0 \tag{4.8}$$

$$\text{(ii)} \quad p\, f_{\pm}(p) = (1-p) f_{\pm}(1-p) \tag{4.9}$$

$$\text{(iii)} \quad [f_{+}(p) - f_{-}(p)] > 0 \;\forall\, p \in (0,1) \tag{4.10}$$

$$\text{(iv)} \quad \int_0^1 dy \exp\left(-\int_{x_o}^{y} \frac{2a(z)}{b(z)}\,dz\right) = \infty \;\forall\, x_o \in (0,1) \tag{4.11}$$

while

$$\int_x^1 dy \exp\left(-\int_{x_o}^{y} \frac{2a(z)}{b(z)}\,dz\right) < \infty \;\forall\, x, x_o \in (0,1) \tag{4.12}$$

where

$$a(z) = E[p_1(n+1) - p_1(n) \mid p_1(n) = z] \tag{4.13}$$

$$b(z) = E[(p_1(n+1) - p_1(n))^2 \mid p_1(n) = z] \tag{4.14}$$

and (v) $\forall\, p \in (0,1)$

$$0 \leq p + f_{\pm}(p), \quad p + f_{\pm}(1-p) \leq 1 . \tag{4.15}$$

It is observed that (4.8) ensures that the necessary condition for optimality given by theorem 2.1 is satisfied. (4.9) implies that $xf_{\pm}(x)$ is even-symmetric about 1/2. The condition (4.15) on the step size limits the updated probabilities to lie in [0,1].

(a) <u>The $N_{R-P}^{(2)}$ Scheme</u>. For the $N_{R-P}^{(2)}$ scheme [5], let

$$\left.\begin{aligned} & f_+(p) = A_1 p^\alpha (1-p)^{\alpha+1} \\ & f_-(p) = -A_2 p^\alpha (1-p)^{\alpha+1} \\ & \alpha \geq 2 \\ & 0 < A_1 \leq 2^{2\alpha} \\ & 0 < A_2 \leq \frac{(2\alpha)^{2\alpha}}{(\alpha+1)^{\alpha+1}(\alpha-1)^{\alpha-1}} \end{aligned}\right\} \tag{4.16}$$

By theorem 4.1, $N_{R-P}^{(2)}$ is optimal. Furthermore, all the variations of $N_{R-P}^{(2)}$, namely, $N_{R-R}^{(2)}$ $(A_1 > 0,\ A_2 < 0)$, $N_{R-I}^{(2)}$ $(A_2 = 0,\ A_1 > 0)$, $N_{I-P}^{(2)}$ $(A_1 = 0,\ A_2 > 0)$ and $N_{P-P}^{(2)}$ $(A_1 < 0,\ A_2 > 0)$ are optimally convergent.

(b) <u>The $N_{R-P}^{(3)}$ Scheme</u>. The $N_{R-P}^{(3)}$ scheme is defined by (4.1) with

$$f_{\pm}(p) = \pm B \sum_{\alpha=2}^{\infty} p^\alpha (1-p)^{\alpha+1} = \pm \frac{Bp^2(1-p)^3}{1-p(1-p)}, \quad 0 < B < 3 \tag{4.17}$$

$N_{R-P}^{(3)}$ is optimal in view of theorem 4.1.

(c) <u>The $N_{R-P}^{(4)}$ Scheme</u>. For the $N_{R-P}^{(4)}$ scheme,

$$f_{\pm}(p) = \pm \frac{C}{p} (\sin \pi p)^\beta, \quad \beta \geq 3, \quad 0 < C < 1/4 \tag{4.18}$$

Application of theorem 4.1 establishes the optimal convergence of the $N_{R-P}^{(4)}$ scheme.

V. FURTHER COMMENTS

The updating schemes discussed in the preceding sections can be compared on the basis of optimality, speed of convergence and variance of probability vector $p(n)$. For instance, as mentioned in section I, optimally convergent schemes are theoretically preferable to expedient schemes in view of the fact that the limiting probability vector has zero variance. From a practical standpoint, however, expedient convergence may be quite satisfactory provided M is sufficiently close to $\underset{i}{\text{Min}}\ \{c_i\}$.

For optimal schemes, the expected step size $\Delta p_1(n)$ may be used as a criterion for comparison. For example, a scheme A may be defined to be better than a scheme B if $\Delta p_1(n)$ for A is larger than $\Delta p_1(n)$ for B whenever $p_1^A(n) = p_1^B(n)$. Using such a criterion it can be shown that the $N_{R-P}^{(2)}$ scheme is better than the $N_{R-I}^{(2)}$, $N_{R-R}^{(2)}$, $N_{I-P}^{(2)}$, and $N_{P-P}^{(2)}$ schemes.

In some cases it may be desirable to combine two or more optimal schemes in the updating procedure. A hybrid scheme obtained in such a fashion by combining two schemes U_1 and U_2 and denoted by $H[U_1,U_2]$ can be defined as one which switches between U_1 and U_2 in an arbitrary manner; i.e., $H[U_1,U_2] = U_1$ for some values of n (stage number) and U_2 for others. In order to completely define a hybrid scheme it is thus necessary to specify a switching policy; i.e., how $H[U_1,U_2]$ switches between U_1 and U_2. Often it is possible to construct a hybrid scheme in which the switching from one constituent scheme to another occurs on the basis of observations on $p_i(n)$, thus involving feedback. If U_1 and U_2 are both optimal with $\Delta p_1(n) > 0$ for all $p_1(n) \in (0,1)$, then it is clear that $H[U_1,U_2]$ will also be optimal. Optimal hybrid schemes designed to improve the expected step size $\Delta p_1(n)$ have been considered in [7].

REFERENCES

1. M. L. Tsetlin, "On the Behavior of Finite Automata in Random Media," Automatika i Telemekhanika, Vol. 22, pp. 1345-1354, Oct. 1961.
2. V. I. Varshavskii and I. P. Vorontsova, "On the Behavior of Stochastic Automata with a Variable Structure," Automatika i Telemekhanika, Vol. 24, pp. 353-360, March 1963.
3. K. S. Fu and G. J. McMurtry, "A Study of Stochastic Automata as Models of Adaptive and Learning Controllers," School of Elec. Engr., Purdue University, Lafayette, Ind., Rept. TR-EE 65-8, 1965.
4. B. Chandrasekaran and D. W. C. Shen, "On Expediency and Convergence in Variable-Structure Automata," IEEE Trans. on Systems Science and Cybernetics, Vol. 4, pp. 52-60, March 1968.
5. I. P. Vorontsova, "Algorithms for Changing Stochastic Automata Transition Probabilities," Problemy Peredachi Informatsii, Vol. 1, No. 3, pp. 122-126, 1965.
6. I. J. Shapiro and K. S. Narendra, "Use of Stochastic Automata for Parameter Self-Optimization with Multimodal Performance Criteria," IEEE Trans. on Systems Science and Cybernetics, Vol. 5, pp. 352-360, Oct. 1969.

7. R. Viswanathan and K. S. Narendra, "Expedient and Optimal Variable-Structure Stochastic Automata," Dunham Lab., Yale University, New Haven, Conn., Tech. Rept. CT-31, April 1970.

8. J. L. Doob, Stochastic Processes, John Wiley and Sons, Inc., New York, 1953.

9. R. R. Bush and F. Mosteller, Stochastic Models for Learning, John Wiley and Sons, Inc., New York, 1955.

10. R. D. Luce, Individual Choice Behavior, John Wiley and Sons, Inc., New York, 1959.

11. J. Lamperti and P. Suppes, "Some Asymptotic Properties of Luce's Beta Learning Model," Psychometrika, 25, pp. 233-241, 1960.

ACKNOWLEDGMENT

This project was supported by the National Science Foundation under grant GK 20580.

A CRITICAL REVIEW OF LEARNING CONTROL RESEARCH

K. S. Fu

Purdue University

Lafayette, Indiana, U.S.A.

I. INTRODUCTION

In designing an optimal control system, if the a priori information required is unknown or incompletely known, one possible approach is to design a controller which is capable of estimating the unknown information during its operation and determining the optimal control action on the basis of the estimated information. If the estimated information gradually approaches the true information as time proceeds, then the controller designed will approach the optimal controller; and, consequently, the performance of the control system is gradually improved. Because of the gradual improvement of performance due to the improvement of the estimated unknown information, this class of control systems has been called learning control systems. Design techniques proposed for learning control systems include: (1) trainable controllers using pattern classifiers, (2) reinforcement learning algorithms, (3) Bayesian estimation, (4) stochastic approximation, and (5) stochastic automata models. A survey of these techniques can be found in [1]. A general formulation using stochastic approximation has been treated extensively in [2,3]. Practical applications include spacecraft control systems, the control of valve actuators, power systems, and production processes. In addition, several nonlinear learning algorithms have recently been proposed.

II. PROBLEMS FOR FURTHER INVESTIGATIONS IN LEARNING CONTROL

Regardless of the fact that preliminary results have been quite promising, learning control is still a new area of research. The following problems have not been effectively solved.

(1) Improvement of learning rate and the stopping rule. The rate of learning of existing learning algorithms is considered rather slow, particularly for fast-response systems. It may be improved either by appropriate utilization of a priori knowledge or by developing new and faster learning algorithms. Furthermore, most of the existing algorithms have been proved to be convergent asymptotically. In practice, with a finite time of operation, the time available for learning may be rather limited. Consequently, the quality of a learning algorithm at a finite number of iterations becomes important. If the tolerable or satisfactory performance of the system can be specified, an appropriate stopping rule is required so that the learning time will not be longer than necessary.

(2) Learning in nonstationary environment. Most of the existing learning algorithms are valid only in a stationary environment (estimation of stationary parameters). Because of the plant dynamics involved or possibly nonstationary environmental disturbance, algorithms for learning in nonstationary environment definitely need to be developed. Applications of nonlinear and two-step linear learning algorithms, and the use of pattern recognition techniques to detect changes in the environment have been suggested [4].

(3) Hierarchical structure of learning. In learning control systems where several different levels of learning processes are involved, the convergence of the overall system learning requires special attention. In most cases, performances of learning processes at different levels are closely related. The overall system learning may not be convergent even though the convergence of the learning process at each level has been guaranteed. In the case of one learning algorithm being imbedded within another one, it is sometimes very easy to draw the conclusion that the overall learning is accelerated. However, in this case, the complexity of the system for two learning algorithms and the actual learning rate in real time should also be considered.

(4) Fuzzy set approach to learning control. The concept of fuzzy set [5] has recently been applied to the design of learning control systems [6,7]. The notion of fuzziness has been motivated by practical problems in the real physical world. A basic concept which makes it possible to treat fuzziness in a quantitative manner is that of a fuzzy set. Since the problem of learning control was primarily generated from practical situations (control under incomplete a priori information) it should be interesting to see further studies of the fuzzy set approach to learning control.

(5) Control systems with human controller. One good example of learning control systems is a control system with a human

controller. It is expected that a study of the adaptive and learning behavior of human controllers will provide some clues to the synthesis of learning control systems. The modelling of human controllers in manual adaptive control systems has been studied by a number of authors [8]. A block diagram is given in Figure 1 [9]. For a fixed type of plants, the Force Program Calculator usually performs a parameter-adjustment, model-matching process based on a prespecified performance criterion. If, in addition to the changes of plant parameters, the type of plants also changes (for example, a change from a first-order plant to a second-order plant), the Pattern Recognizer will recognize the change. Consequently, a change of control strategy will be initiated. The adjustment of control parameters will be performed according to the new strategy. So far, proposed models of human adaptive controllers are satisfactory only for narrowly defined simple tasks. Although a certain type of adaptive and learning behavior can be demonstrated by these models, their "intelligence" level can be considered to be only very low. The pattern recognition model introduced, although satisfactory for very simple situations, is rather primitive. Only very limited study has been devoted to the model of decision-making and learning processes of human controllers, particularly in the cases where performance of various different control tasks is required [10,11]. Feedback signals are usually measured directly from the plant output rather than from the environment. This will certainly be insufficient if more complex and unknown environments are encountered.

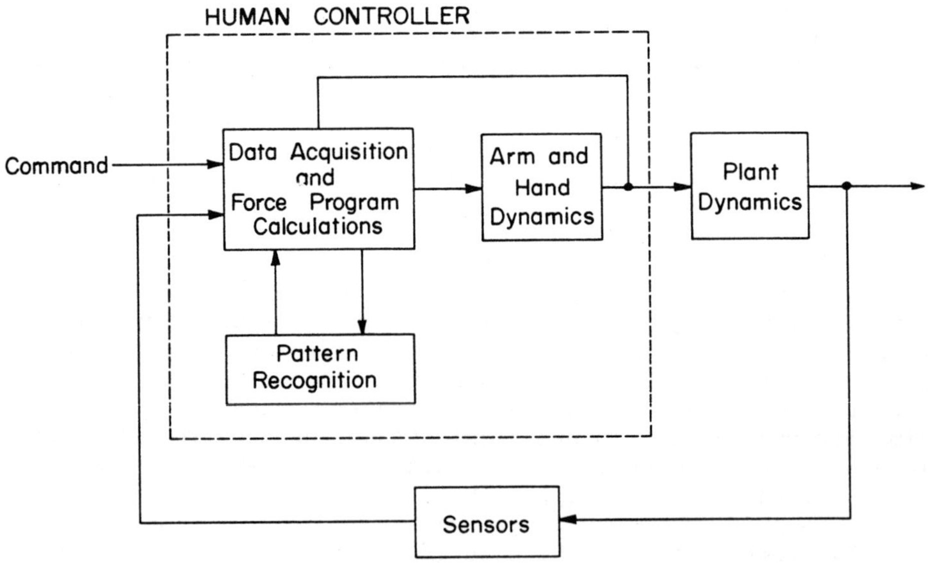

Figure 1. Pattern Recognizing Model for a Manual Control System

(6) Structure learning versus parameter learning. Most of the learning algorithms developed so far are concerned with learning (or updating) of parameters in a specified control law (or function). In other words, the structure of the controller is more or less preselected. It is anticipated that in the case of complex and/or nonstationary (unknown) environments, a flexible structure of the controller would be preferred. Information received by the learning controller could be used not only to update the parameter values in a given function (structure), but to learn the "best" controller structure. Unfortunately, not many results have yet been available.

Remarks. Sometimes a complete optimal design may result in a very complex or low learning rate controller. In order to obtain relatively simple but practical solutions, suboptimal (but satisfactory) and heuristic approaches may be preferred [12-15]. Also, if the problem formulation is so complicated that we have not been able to obtain algorithmic solutions so far, heuristic but workable solutions should be acceptable with respect to the practical problem-solving. As can be seen in Section III, solutions obtained for integrated robot systems are primarily from heuristic approaches. However, we probably should keep in mind the fact that usually very few general conclusions can be drawn from a special system design by heuristic methods.

III. FROM LEARNING CONTROL TO "INTELLIGENT CONTROL" SYSTEMS

If we naturally extend our goal to the design of automatic control systems which behave more and more intelligently, then learning control systems research is only a preliminary step toward a more general concept of "intelligent" control systems. Perhaps a convenient explanation is that the area of intelligent control systems describes the activities in the intersection of "automatic control systems" and "artificial intelligence." Many research activities in artificial intelligence are concerned with control problems. On the other hand, control engineers interested in adaptive and learning control research have attempted to design more and more human-like controllers. This overlap of interest between the two areas has created a new area of interest to engineers. Intelligent control systems can be classified into two types--those with human supervision and those without human supervision.

(1) Control systems with man-machine controller. A study has recently been initiated for the control systems in which a man-machine combination is employed as the controller. Presumably, relatively simple control tasks can always be performed by the machine (computer) controller only. In more complicated control tasks, human supervision, acting as a teacher and a controller, is necessary. Or, in some cases (e.g., for remote manipulation tasks),

machines are needed to help the human controller in planning and executing the tasks. One example is the human-supervised, computer-controlled, remotely manipulated system shown in Figure 2 [16]. The remote loop, closed through the computer at the distant site, represents the remote device acting as a machine controller with a short-range goal. The supervisory loop, closed through the human controller, represents his functions of intermittently setting goals and supervising the learning for the remote device, and, if necessary, even guiding the effectors directly. The local loop is independent of the remote device, and signifies the human controller's use of his computer as an aid to model the remote system so that he may predict its behavior and hence improve his supervision.

A state space formulation of the remote manipulation problem has been proposed [17]. A state vector is defined, containing not only variables which describe the manipulative device, but also important parameters (features) of the environment, possibly including locations of relevant objects and obstacles. This vector, suitably quantized, spans a discrete state space which contains many different plant-environment control situations. A manipulation task is defined as a new state which the man-computer controller wishes the plant-environment to occupy. The state transitions are accomplished by commands--elementary motions of the manipulator. The goal for the controller is to determine, for given costs of each state transition, the optimal (minimum cost) path from the present state to the desired state. A method similar to dynamic programming has been employed to determine the optimal path.

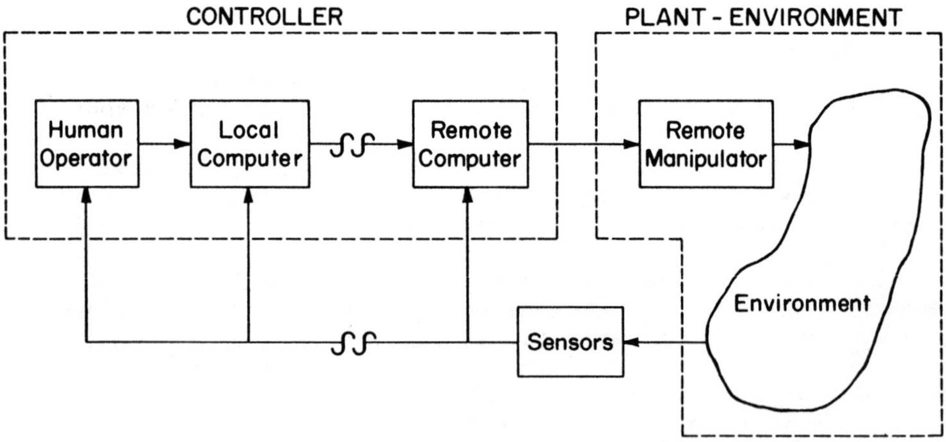

Figure 2. Supervisory Controlled Remote Manipulator

(2) Autonomous robot systems. In the use of a man-machine combination as a controller, decisions requiring higher intelligence can be performed by the human controller; for example, recognition of complex environmental situations, setting subgoals for the machine controller, etc. On the other hand, activities requiring relatively lower intelligence such as data collection, routing decisions and on-line computations, can usually be carried out by the machine controller. In designing an intelligent control system, our purpose is to try to transfer as much as possible, the human operator's intelligence relevant to the specified tasks to the machine controller. In addition to the study of man-machine controllers (or supervised controllers), a more direct and ambitious study is to design a nonsupervised machine controller for an intelligent control system.

One project in intelligent control systems research is the autonomous robot, which is intended to maneuver and manipulate unaided in a distant environment [18-20]. A block diagram of the SRI (Stanford Research Institute) robot system is shown in Figure 3. If the robot system is viewed as a computer-controller system in a complex environment, the controller should perform at least the following three major functions: (i) problem solving, (ii) modelling, and (iii) perception. The block "Reflexive Responses" in Figure 3 is similar to a conventional controller which executes

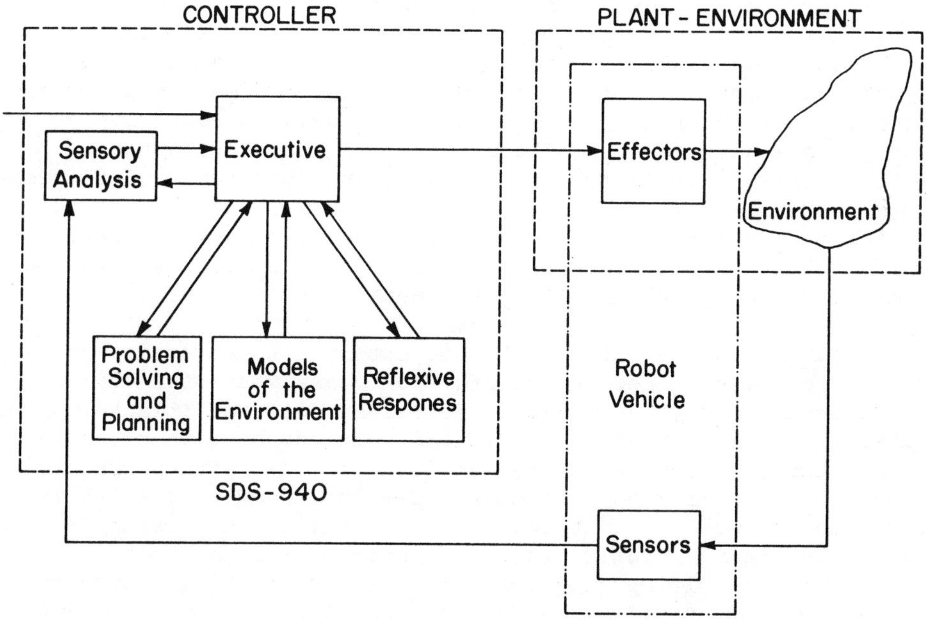

Figure 3. SRI Robot System

preprogrammed solutions for simple and more-or-less predictable environmental situations. The robot vehicle is propelled by two stepping-motors independently driving a wheel on either side of the vehicle. It carries a vidicon TV camera and optical range-finder in a movable head. A typical task performed by the system requires the vehicle to rearrange (by pushing) simple objects in its environment. The system accomplishes the specified tasks by performing a sequence of elementary actions, such as wheel motions and camera readings. For efficiency, a task should first be analyzed into a sequence of actions calculated to have the desired effect. This process is often called problem solving and planning because it is accomplished before the robot begins to act. Scene analysis and object recognition are required to provide the robot with a description of its environment. The knowledge of the effects of its actions constitutes a type of model of the environment. The problem-solving process utilizes the information stored in the model to calculate what sequence of actions will cause the controlled process (plant-environment) to be in the desired state. As the environment changes, the model must be updated to record these changes. The addition and the updating of information about the model is a learning process and is similar to the process introduced in many learning control systems.

Simple environments can sometimes be expressed in terms of a small number of quantized plant-environmental situations (states). In these cases, for relatively well-defined tasks, the method of dynamic programming has been suggested to determine the sequence of actions (the optimal control policy) required to reach the desired state [23]. Sophisticated pattern recognition techniques must be used for the analysis of complex environments. There are no optimal algorithms currently available which enable a robot system to accomplish a wide variety of tasks in a complex environment. Heuristic problem-solving processes have been proposed for determination of the "best" sequence of actions; however, because of the heuristic procedures used, such control policies are generally suboptimal. With appropriate learning processes introduced, the performance of the problem-solving process can be improved.

Remarks. It is noted that some attempts to determine optimal control policy have been successful for cases where the environment can be meaningfully quantized to represent different plant-environmental (control) situations or states. In essence, the function of sensory analysis or pattern recognition has been either greatly simplified or performed by a human operator. For complex environments, the number of quantized environmental situations may become very large making identification of environmental situations difficult and the resulting computation scheme infeasible. Furthermore, if the environment is unknown or only partially known a priori, a prespecified meaningful quantization for a variety of tasks can hardly be obtained. On-line pattern recognition is necessary to identify (classify) various significant and meaningful plant-environmental situations. The optimal control policy for a given task (if one exists) will then be determined on the basis of the classified plant-environmental situations. However, in order to perform various different tasks, an on-line task-dependent decision-making process must be implemented. It should be interesting to study on-line decision-making processes in conjunction with pattern recognition for a variety of control tasks in complex environments.

REFERENCES

1. K. S. Fu, "Learning Control Systems--Review and Outlook," IEEE Trans. on Automatic Control, Vol. AC-15, No. 2, April 1970.

2. Ya. Z. Tsypkin, "Adaptation and Learning in Automatic Systems," (English Translation), Academic Press, 1971.

3. Ya. Z. Tsypkin, "Learning Control Systems," Automatika i Telemekhanika, No. 4, April 1970.

4. L. R. Cockrell and K. S. Fu, "On Search Techniques in Switching Environments," Proc. IEEE 9th Symposium on Adaptive Processes, Dec. 7-9, 1970, Austin, Texas.

5. L. A. Zadeh, "Fuzzy Sets and Systems," Symposium on System Theory, Polytechnic Institute of Brooklyn, April 20-22, 1965.

6. W. G. Wee and K. S. Fu, "A Formulation of Fuzzy Automata and Its Application as a Model of Learning Systems," IEEE Trans. on Systems Science and Cybernetics, Vol. SSC-5, July 1969.

7. K. Asai and S. Katajima, "Learning Control Using Fuzzy Automata," US-Japan Seminar on Learning Processes in Control Systems, Nagoya, Japan, August 1970.

8. L. R. Young, "On Adaptive Manual Control," IEEE Trans. on Man-Machine Systems, Vol. MMS-10, No. 4, Dec. 1969.

9. D. W. Gilstad and K. S. Fu, "A Two-Dimensional, Pattern Recognizing, Adaptive Model of a Human Controller," Proc. 6th Annual Conference on Manual Control, April 1970.

10. A. E. Preyss and J. L. Meiry, "Stochastic Modeling of Human Learning Behavior," USC-NASA Conference on Manual Control, NASA SP-144, March 1967.

11. R. E. Thomas and J. T. Tou, "Evaluation of Heuristics by Human Operators in Control Systems," IEEE Trans. on Systems Science and Cybernetics, Vol. SSC-4, No. 1, March 1968.

12. D. Michie and R. A. Chambers, "Boxes: an Experiment in Adaptive Control," Machine Intelligence, Vol. 2, edited by E. Dale and D. Michie, 1967.

13. A. G. Ivakhnenko, "Heuristic Self-Organization in Problems of Engineering Cybernetics," 4th IFAC Congress, Warsaw, Poland, June 16-21, 1969.

14. K. Nakamura and M. Oda, "Heuristics and Learning Control," US-Japan Seminar on Learning Process in Control Systems, Nagoya, Japan, August 1970.

15. J. E. Doran, "Planning and Generalization in an Automaton/Environment System," Machine Intelligence, Vol. 4, edited by B. Meltzer and D. Michie, 1969.

16. T. B. Sheridan and W. R. Ferrell, "Human Control of Remote Computer Manipulators," Proc. International Joint Conference on Artificial Intelligence, May 1969.

17. D. E. Whitney, "State Space Models of Remote Manipulation Tasks," Proc. International Joint Conference on Artificial Intelligence, May 1969.

18. A. Freedy, F. Hull and J. Lyman, "The Application of A Theoretical Learning Model to Remote Manipulator Control," Proc. 6th Annual Conference on Manual Control, April 1970.

19. N. J. Nilsson, "A Mobil Automaton: An Application of Artificial Intelligence Techniques," Proc. International Joint Conference on Artificial Intelligence, May 1969.

20. M. L. Minsky and S. A. Papert, "Research on Intelligent Automata," Status Report II, Project MAC, M.I.T., Sept. 1967.

21. K. Pingle, J. A. Singer and W. M. Wichman, "Computer Control of a Mechanical Arm Through Visual Input," Proc. IFIP Congress, Edinburgh, Vol. II, 1968.

22. P. H. Winston, "Learning Structural Descriptions from Examples," Rept. MAC TR-76, Project MAC, M.I.T., Sept. 1970.

23. W. G. Keckler and R. E. Larson, "Control of a Robot in a Partially Unknown Environment," Automatica, IFAC Journal, Vol. 6, May 1970.

ACKNOWLEDGMENT

This work was supported by National Science Foundation Grant GK-1970 and AFOSR Grant 69-1776.

HEURISTICS AND LEARNING CONTROL
(INTRODUCTION TO INTELLIGENT CONTROL)

K. Nakamura and M. Oda

Nagoya University

Nagoya, Japan

1. INTRODUCTION

In a phase of engineering development of cybernetics, researches on Artificial Intelligence go on increasing rapidly. Artificial Intelligence includes such pedagogical functions as learning, education, self-organizing, inductive inference, association, heuristics, concept formulation, creation, evolution and so forth, and plays the most serious roles in many problems such as pattern recognition, linguistic analysis, problem-solving, peak-searching, game-playing and so on.

Research on the introduction of Learning to control engineering which started about fifteen years ago has grown up favorably and now turned to one of the most remarkable and successful schemes of the trials to introduce the human intelligence to engineering fields. At present, there are found many researches and developments on learning and learning control, including fundamental studies of learning function [31]-[34], construction of learning control system with [35] or without identification [36], mathematical analysis of learning processes [37] and so on. Heuristics is one of the sophisticated artificial intelligences which are very difficult to be analyzed scientifically, but are most fundamental and important from the engineering view point. The authors have done some pioneering works [27]-[30] on heuristics and heuristic control (control with heuristics).

This paper first demonstrates definition of Heuristics, relation between learning, heuristics and intuition, and role of heuristics in control engineering, and next surveys several researches on heuristics and heuristic searching, and last discusses some future problems on heuristics.

2. DEFINITION AND MEANING OF HEURISTICS

2.1 Definitions of Heuristics

The pedagogical definition of Heuristics or heuristic learning is expressed in various forms [1], [2]. While the engineering definition of heuristics, even it is not yet so fixed at present, seems to stand on the fairly special concept. By heuristics is meant (1) to determine a principle of restricted generalization where by a given machine learns to apply a "generalized" form of a previously successful algorithm to a "similar" problem which is subsequently presented [3], (2) a principle or device that contribute, on the average, to reduction of search effort in problem solving activity [4],[5], (3) a method which helps in discovering a problem solution by making plausible but failable guesses as to what is the best thing to do next [6].

Now, the authors propose a definition that heuristics is (4) a method which solves a problem using an algorithm (heuristic algorithm) which utilizes a series of clues (heuristic elements) having a special order of priority [27]. The feature of definition (4) is to be divided into two main concepts of heuristic elements and heuristic algorithm. In Figure 1, the problem is to search an optimum path to reach at the end point (object, conclusion) from a starting point (initial state, premise). There are so many unknown branches on the way of any path that it is unfeasible to check the whole set of possible paths sequentially. In such a problem, heuristic elements mean the branches a_{ij}'s$(i = 1,2,\ldots;$ $j = 1,2,\ldots; J_i)$ which have a higher possibility of success of reaching to the end point among the possible branches b_{jk}'s in the i-th interval. On the other hand, a heuristic method means a method of making connection of a series of branches which can pass through from the starting point to the end point under some proper

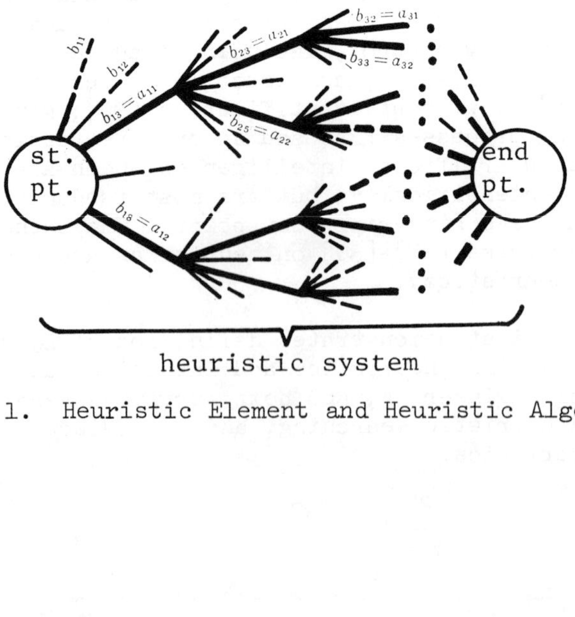

Figure 1. Heuristic Element and Heuristic Algorithm.

priority of such branches as a_{ij}'s with a high possibility of success. In practice, the following problems must be considered: What kinds of heuristic elements must be chosen? What kind of heuristic system must be adopted? From which side must the connection be started, from the starting point or the end point? and so on.

2.2. Relation Between Learning, Heuristics and Intuition

Learning, Heuristics and Intuition are three of the sophisticated native functions of human. The investigation of actual mechanisms of the above human intelligences concerns closely with the problem how does human use the gathered informations and/or the accumulated experiences. The engineering interpretation of the three intelligences may be as follows:* The feature of learning is the active usage of past experiences to improve system behavior. Then learning is a function relating to time actions like accummulation and utilization of experiences. On the other hand, heuristics is an useful procedure (heuristic algorithm) to extract several relevant factors (heuristic elements) from vast external informations and to make a best path (solution) connecting these heuristic elements in each step of the sequential decision process. In other words, heuristics is a function concerning spatial factors like the processing of enormous informations and the derivation of spatial solution. Repeating the heuristic processing, man becomes to be able to quicken the heuristic processing and to make a decision quickly, that is, heuristics could be quickened by learning. Moreover, experiences and knowledges obtained by learning will supply more available informations and lead to better probable solutions. Thus, learning can improve the heuristic processing both in speed and accuracy. Heuristics improved by learning may be called Learned Heuristics. The extremely quickened heuristics by learning may be able to say "Intuition". On the other hand, learning is also improved by applying heuristics. Namely, the accumulation and arrangement of informations and experiences can be rationalized by heuristic processing. So that the learning led by heuristic decision presents a considerable improvement in the speed of learning and precision of learned results. Learning improved by heuristics may be called Heuristic Learning. The few functions mentioned above, learning, heuristics, learned heuristics, heuristic learning, and intuition are, in all, important intelligences of human, and seem to be ordered as above in complexity and sophistication. Even if the actual mechanisms of human's intelligences, especially of intuition, are different from the above description, the above concepts of heuristics and intuition seem to be more convenient

*This is the opinion of one of authors induced from questions and discussions by Professor R. W. McLaren (University of Missouri) and Professor H. Sugiyama (Osaka University) at the seminar.

from the engineering view point of artificial intelligence. In a decision process, human will probably use these functions in proper way. We can easily guess that human will make the proper use of these functions in a peak-searching process. In game-playing, heuristics and intuition should be both in use to decide a move. A poor chess-player or beginner will be always in use of a slow heuristics to decide a moving policy, on the other hand, a good player will use intuition or learned heuristics more often than an average player. So a good-player can decide every move in game-playing very quickly or instantaneously.

3. ROLE OF HEURISTICS IN AUTOMATIC CONTROL

As been easily guessed from the definitions mentioned in the preceding section, heuristics seems to be a powerful real approach for the search of optimal solution for complex problems which have too many admissible solutions to be ordered one by one and of course to be checked analytically. We can find several applications of heuristic approach in such problems as the solution of mathematical equation [7], the mathematical proof problems [8]-[13], the search problem of best strategy in Games [14]-[18], the solution of pentomino puzzle [19],[20], the automatic design of printed network or IC circuit [21]-[23] and so on.

In the control problem for complex system with a lot of uncertainty, it is required to learn the association of control choices and a great number of measured samples, and to choose an optimal control from many allowable controls. There seems to be so many kinds of associations in this case that it is not easy to select one optimum control out of the various possible ones. In order to perform this difficult work efficiently, the heuristic approach should provide effective clues on the classification and decision-making which are important to realize recognition or learning. Then the introduction of heuristics or learned heuristics to control problem is coming to an important subject. Learning control directed by heuristic learning or learned heuristic may be called Heuristic Control. Sophisticated control schemed by the introduction of artificial intelligence may be named, in general, as Intelligent Control.

Major approaches to the design of learning processes include algorithmic approach, heuristic approach, and algorithmic heuristic approach. Most research efforts have been centered, so far, upon the first approach, but the second and third approach should not be ignored.* Now, the Heuristic Control is going to be highlighted as an interesting and worthy strategy in control.

*This paragraph is the discussion after the Professor J.T. Tou's comment in the Seminar.

4. BRIEF SURVEY OF SEVERAL RESEARCHES ON HEURISTICS AND HEURISTIC CONTROL

4.1. Heuristics Used for Selecting the Important Terms in a Self-Organizing Nonlinear Threshold Network

Mucciardi-Gose [24] proposed "weight size heuristic", which determines the specially important terms (and their weight values) from a lot of terms in a self-organizing network for pattern recognition. The problem is as follows: In the nonlinear threshold network of Figure 2, assume the input is $\underline{X}_j = (x_{j1}, x_{j2}, \ldots, x_{jN})$ and choose M linearly-independent terms $f_m(\underline{X}_j)$, $m = 1,2,\ldots,M$. This is a problem of selecting M proper terms from the 2^N possible linearly-independent terms $(x_{ji}, x_{ji}\, x_{jk}, x_{ji}\, x_{jk}\, x_{jl}, \ldots, x_{ji}\, x_{jk}\, x_{jl} \ldots x_{jN})$. The total number of possible combination is $\binom{2N}{M}$ and it grows very rapidly as the number N increases. Therefore, it is very difficult to select the M proper terms $f_m(\underline{X}_j)$ among them.

In the method of weight size heuristic proposed by Mucciardi-Gose, M terms $f_m(\underline{X}_j)$ are first chosen at random, then the system is trained by a series of J training samples. Comparing the values of w_m's after the training, the M_1 terms with higher values of weight w_m are kept in the network, while the $M_2 = (M - M_1)$ terms with lower values of weight are discarded to exchange for the M_2 randomly-chosen new terms. Then, the new network of M terms is trained again. As stated above, if such a process of both training and selecting phases is repeated, then the M desired terms $f_m(\underline{X}_j)$ can be discovered, and at the same time their weight values can also be decided.

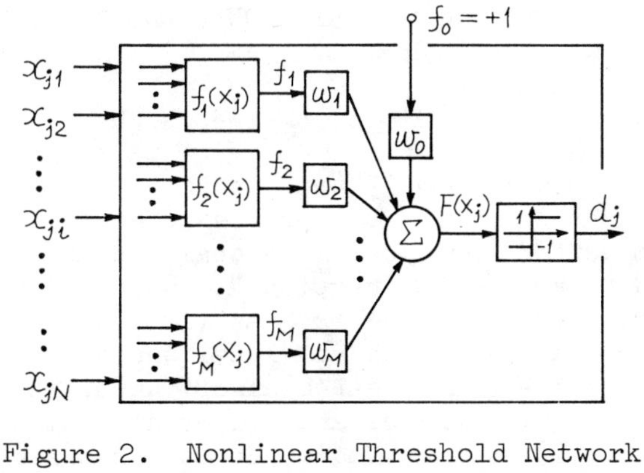

Figure 2. Nonlinear Threshold Network.

Such a serial process called an evolutional process by Mucciardi-Gose includes three important ideas (or subprocesses) of teaching or learning, weight size heuristic, and evolution. In other words, by teaching or learning the network, the weights can be modified; and it leads to the decision of the M terms $f_m(\underline{X}_j)$ into the M_1 terms to remain and M_2 terms to discard and exchange with M_2 newly chosen terms; after repeating the above procedure, the system attains a gradual evolution.

4.2. Heuristics Used for Approximately Solving an Optimal Control

Thomas-Tou [25],[26] have researched the heuristic method of solving an optimal control. The problem here is "to find an optimum control $\{y_k\}$ in order to shift a system from initial velocity V_0 to a desired final velocity in the specified period N under the condition of minimum fuel consumption". In their experiments, the subject is not quite informed about the plant character. The subject read the current values of six variables (control choice, velocity, difference between the desired and current velocity, fuel consumption and so on) on meters, and choose heuristically a current optimum control y_k or a series of control $\{y_k\}$ so as to minimize the fuel consumption at the terminal point.

As a result of the experiment, a new mode of heuristic search which a subject evolves was extracted as follows: "A subject pays his attention to the six indications of meters in order to detect if there are any invariants in the nine kinds of factors which are selected beforehand. If there is any one of invariants, a control y_k is chosen according to the invariant. On the other hand, if there are two (or more than two) invariants, y_k is chosen according to the priority designated in advance."

5. AUTHOR'S RESEARCH ON HEURISTICS USED FOR SEARCHING AN OPTIMUM POINT ON MULTI-DIMENSIONAL AND MULTI-MODAL HILL

5.1. Aim of the Research

The authors have recently tried to make clear the heuristics evolved by a human searcher in a process of searching the highest point (optimum vertex) on two-dimensional and multi-modal hills [27]. The work aimed to solve the question "to which of the several conventional searching techniques such as (1) random procedure, (2) mixed process of random search and local search, and (3) model simulation search,is the human search algorithm similar?" or "is the human's procedure quite unique and different from any of them?" Where the scheme (1) is a method which at first searches n points randomly, then selects the highest trial point among them. Although there are several varieties of the scheme (2), the basic frame of search is as follows. First a point x_i, i=1,2,..., is selected randomly and decided as a base point for a successive local search, which will get an extreme point x_i^m with a value f_i. Next another point x_{i+1} is selected, and the same procedure is repeated till times of fail ($f_{i+1} \leqq \max_i \{f_i\}$) are counted. The highest (optimum) point f^0 is got as $\max_i\{f_i\}$. There are also several varieties of (3). An example of them is as follows. First, a global model of a polynomial approximate equation is made for a criterion function from data of local search; then an optimum point of the model is conjectured. After jumping over to a conjectured optimum point, some correcting trials of search are added to stick a true optimum point (Polynomial Conjecturing Method). In another example, first local search points of four adjacent trials make a local approximate criterion function of a third-order polynomial equation; then a global criterion function is constructed by a piecewise connection of such local functions. A next trial is made at the maximum point of the global function. After the trial, the global function is reconstructed by an addition of the new trial result, and another trial point is selected: in this way, the search process is repeated (Method of piecewise cubic approximation). There are also a method of probabilistic model which approximates by probability the intervals between adjacent two points, and so on.

5.2. Method and Conditions of Experiment

In this research, it is planned to extract some heuristics of human through the analysis of experimental search by human subjects performed in the scene shown in Figure 3. The human subject (searcher) and the experimenter sit down at the opposite sides of

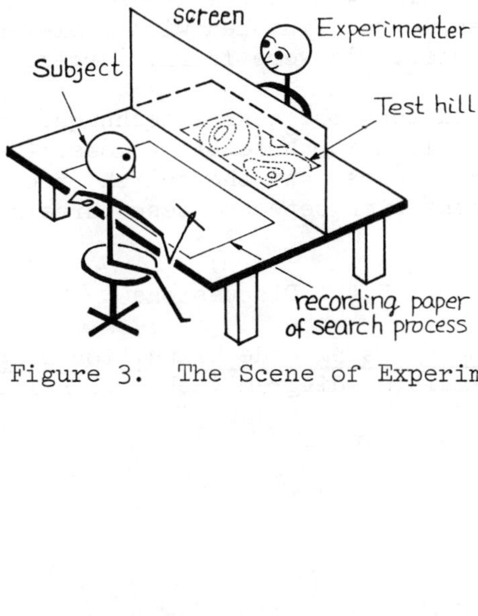

Figure 3. The Scene of Experiment.

a screen, which can avoid the subject's seeing a test hill (criterion function) in front of the experimenter. When the subject selects a trial point $x = (x_1, x_2)$, the experimenter tells him the height of the point on the test hill. The subject writes down the height on a recording paper, and selects a next trial point. Repeating the above procedure, the subject gets an optimum point which is drawn on the experimenter's test hill paper, but hidden for the searcher by the screen, by using the experience of search which recorded successively on his searching paper. Main goal is to search an optimum point $f(x^m)$. Incidental goal is to keep the value of

$$\overline{f}(x_i) = \frac{1}{N} \sum_{i=1}^{N} f(x_i)$$

as big as possible, where $f(x_i)$ is the height of the trial point $x_i (i=1,2,...N)$ and $\overline{f}(x_i)$ is the average value of $f(x_i)$. There is no restriction on the selection of trial points x_i, trial step width Δx_i, and trial number N. All a priori informations told to the subject about a test hill are that $f(x)$ is a step-wisely continuous and one-valued function on x. The number and location of peaks on a test hills are not, of course, informed to the subject.

There are provided twenty kinds of test hills in which some regularities (rules) in the number, arrangement, and shape (sharpness) of peaks. Then, the key point for the subject to get the highest peak quickly is to discover the hidden rules as quickly as possible.

5.3. Results of Experiment

One example of experiment is shown in Figure 4, in which actual equipotential curves are shown by real line and trajectry traced by the human searcher is drawn by dotted line. The searcher (YF) tried firstly four global searches (G-mode search), secondly five local searches (L-mode search) around the highest point of the group of first trials, thirdly two convergent searches (C-mode) to confirm the optimum point with a smaller step width. The search process of this example was finished within eleven trials (N=11), while, in general cases, N's were fairly larger.

It must be noted that the search technique presented in the examples can be dissolved in three basic modes (G-mode, L-mode, and C-mode) proposed by authors, and the search process evolved is expressed by the transition between these three modes.

5.4. Discussions

<u>(A) Discussion by Using Mode Transition Diagram</u>. Here is proposed a mode transition diagram (abb., transition diagram), which

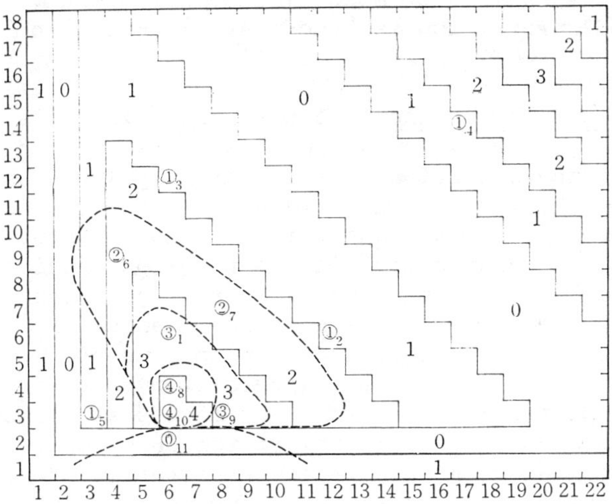

Figure 4. An Example of Searching Experiment (In Case of YF-16)

is useful to investigate a transition process of a search mode. Some examples of the diagram are illustrated in Figure 5, which indicates any transition between the two search modes by an arrow, on which is written the transition order (1, 2, ...). For instance, the transition of Figure 4 is illustrated as Figure 5(a) since it is formulated as Eq. (5.1):

$$\text{Transition of the Search Mode of YF-16} = G^{4}\underset{1}{\curvearrowright}L_{1}^{5}\underset{2}{\curvearrowright}C_{1}^{2}\ldots \quad (5.1)$$

As to the notation of $G_i^{m_i}$, $L_i^{m_i}$, $C_i^{m_i}$, upper suffixes show the number of trials and lower suffixes show the number of group of trial searches. The diagrams of Figure 5(b) through (f) correspond to another example. Figure 5 shows some of typical examples of the transition diagrams. In general, however, the mode transition by a human searcher has "a complete flexibility of changing from a

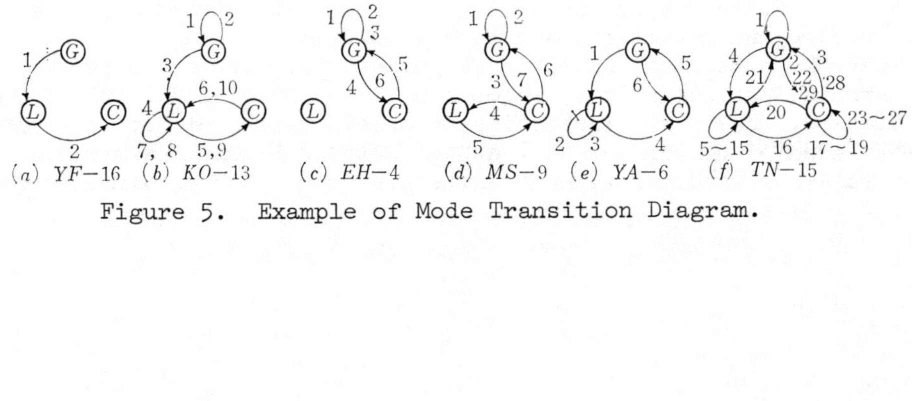

Figure 5. Example of Mode Transition Diagram.

specific mode to any other modes including itself". We call such a type of mode transition as "the most flexible type". (cf. Figure 5 (f)).

(B) Core of Heuristics. As the results of this experimental research, the following few heuristics are extracted. They are heuristics in the selection of first trial point (base point), and heuristics in the transition between searching modes. If we think that heuristics is a conscious jump of logical thought, then the transition of the search mode in the previous section is just corresponding to the conscious jump. Heuristics appeared in the selection of base point may be considered as heuristics in a kind of transition from starting condition to base point. The transition to a local search area is especially worthy of being called a heuristic jump, two kinds of which extracted as follows: One is a jump (transition) to the vicinity of the base point, which is selected from the G-mode search in the expectation of giving a higher value of the hill. The other is a jump to a point in an unsearched open local area with a comparable extent to that of the most plausible highest peak (a candidate of the optimum point) obtained in the previous process of search.

(C) Heuristics and Learning. In a practical process of search, the transition and/or the arrangement of the search mode are deformed variously according to the searching results (trial experiences), in other words, by learning of a subject. This deformation by learning is just correspond to the learned heuristics defined in 2.2. The study of such a deforming mechanism from a microscopic point of view is now under investigation at our laboratory.

6. CONCLUSION

Research of learning control has entered into the second stage where more serious effort should be devoted for the completion of systematic analysis and synthesis of learning process as well as for the high-grade development of the methodological or conceptual research of control.

This paper emphasized that the courageous work for the introduction of more sophisticated intelligence to control engineering should be started just now, and demonstrated author's opinion on definition and meaning of Heuristics, Heuristic Learning, Learned Heuristic and Intuition, and furthermore introduced authors' research on heuristics evolved by human searcher in the peak-searching process on a two-dimensional and multi-modal criterion function. There remains various problems worthy of future works: What kind of difference will appear between heuristics evolved by subjects with and without any sense organ such as visual organ, tactual organ, and so on? What is the adaptability of heuristics for the case of sudden changes of a hill? Although researches on heuristics and heuristic control may be very complex and difficult, they will bring a epochal aspect on the control of unknown system.

It is sincerely hoped that this brief paper will excite the successive progression of researches in this pioneering field.

REFERENCES

1. M. Minsky, "Steps Toward Artificial Intelligence", Proc. IRE, 49, 8/30 (Jan. 1961).
2. M. Minsky, "A Selected, Descriptor-Indexed Bibliography to the Literature on Artificial Intelligence", IRE Trans. on HFE, 2-1, 39/55 (Mar. 1961).
3. E. A. Feigenbaum and J. Feldman (eds.), Computers and Thought, McGraw (1963).
4. R. J. Solomonoff, "Some Recent Work in Artificial Intelligence", Proc. IEEE, 54-12 (Dec. 1966).
5. A. L. Chernyavskii, "Computer Simulation of the Process of Solving Complex Logical Problems (Heuristic Programming)", Automation and Remote Control, No.1, 145/167 (Jan. 1967).
6. E. L. Nappel'baum, "International Symposium on Machine Simulation of the Higher Functions of the Brain", Automation and Remote Control, No.8, 1247/1249 (Aug. 1967).
7. J. R. Slagle, "A Heuristic Program that Solve Symbolic Integration Problems in Freshman Calculus", in 3), 191/203.
8. H. Gelernter, "Realization of Geometry-Theorem Proving Machine", in 3), 153/163.
9. H. Gelernter, J. R. Hansen and D. W. Loveland, "Explorations of the Geometry-Theorem Proving Machines", in 3), 153/163.
10. A. Newell, J. C. Shaw and H. A. Simon, "Empirical Explorations with the Logic Theory Machine--A Case Study in Heuristics", in 3), 109/133.
11. F. V. Anufriev, V. V. Fedyurko, et al., "Algorithm for Proving Theorems in Group Theory", Kibernetika, No. 1 (1966).
12. A. Newell and H. A. Simon, "GPS, A Program that Simulates Human Thought", in 3), 279/296.
13. I. G. Wilson and M. E. Wilson, Information, Computer and System Design, Wiley (1965).

* in Japanese

14. A. Newell, J. C. Shaw and H. A. Simon, "Chess Playing Programs and the Problem of Complexity", 3), 39/70.

15. A. L. Samuel, "Some Studies in Machine Learning Using the Game of Checkers", in 3), 71/108.

16. A. L. Samuel, "Some Studies in Machine Learning Using the Game of Checkers (II - Recent Progress)", IBM J., 11-6, 601/917, (Nov. 1967).

17. E. B. Carne, Artificial Intelligence Techniques, Spartan (1965).

18. K. Nakano and F. Watanabe, "Simulation of Thinking Process in Game-Playing", Rept. at Research Meeting of Automata, Inst. Elect. Commu. Engrs. Japan, No. A68-21 (July 1968).

19.* T. Ikeda (ed.), Introduction to Electronic Computers, 260/263, Ohm Books Co. (June 1968).

20.* T. Ikeda and T. Tsuchimoto, "Pentomino and Computer", bit, 1-1, 84/85 (Mar. 1969).

21.* T. Kitamura, "Automation of Design of IC Logical Network Package", Electronic Science, 7-8, 73/79 (Aug. 1967).

22.* T. Hayashi, et al., "Automation of Designing IC Computer", FUJITSU, 19-1, 35/53 (1968).

23.* K. Migami, "Present State of Design Automation", Electronics, 12-12, 88/96 (Nov. 1967).

24. A. N. Mucciardi and E. E. Gose, "Evolutionary Pattern Recognition in Incomplete Nonlinear Multithreshold Networks", IEEE Trans. on Elec. Computer, 15-2, 257/261 (Apr. 1967).

25. J. T. Tou, R. E. Thomas and R. J. Cress, "Development of a Mathematical Model of the Human Operator's Decision-Making Functions", Battelle Memorial Inst. Rept. (Oct. 1967).

26. R. E. Thomas and J. T. Tou, "Evolution of Heuristics by Human Operators in Control Systems", IEEE Trans. SSC, 4-1, 60/71 (Mar. 1968).

27.* M. Oda and K. Nakamura, "A Heuristic Method of Searching an Optimum Point of a Two-Dimensional Multimodal Criterion Function", Instrument and Control Engrs., 7-12, 16/22 (Dec. 1968) and Res. Rept. of Auto. Contr. Lab. Nagoya Univ. 16 (1969).

28. M Oda, "Fundamental Discussion of Heuristic Function and its Introduction into Learning Control", Memoir of Faculty of Eng. Nagoya Univ., 20-1, 327/344, (May, 1968).

29. M. Oda and B. F. Womack, "A Unified Discussion on Heuristics in Artificial Intelligence", Proc. of Asilomar Conference, 387/399 (Dec. 10-12, 1969).

30. M. Oda, B. F. Womack and K. Tsubouchi, "A Pattern Recognizing Study of Palm Reading", Proc. of Allerton Conference, (Oct. 1970).

31. J. E. Gibson and K. S. Fu, "Philosophy and State of the Art of Learning Control Systems", TR-EE, 63-7, Purdue University, (Nov. 1963).

32.* K. Nakamura and M. Oda, "Learning Control and Learning Control Machine", Information Science Series, Korona-sha (1966)

33. J. T. Tou and J. D. Hill, "Steps Towards Learning Control", Preprint of JACC, P. 12, 12/26 (Aug. 1966).

34. K. Nakmura and M. Oda, Fundamental Principle and Behavior of Learntrols, Computer and Information Sciences-II, Academic Press (1967).

35. Ya. Z. Tsypkin, "Adaptation, Training and Self-Organization in Automatic Systems", Automatic and Remote Control, 27-1, 16/51, (Jan. 1967).

36. K. S. Fu, "Learning Control Systems--Reviews and Outlook", IEEE Trans. on Automatic Control, 210/221 (April, 1970).

37. J. M. Mendel and K. S. Fu, Adaptive, Learning and Pattern Recognition Systems, Academic Press (1970).

ADAPTIVE MODEL CONTROL APPLIED TO REAL-TIME BLOOD-PRESSURE REGULATION

Bernard Widrow

Stanford University

Stanford, California, U.S.A.

ABSTRACT

A real-time computer-control system for regulating the blood pressure of an animal in a prolonged state of shock has been successfully developed and is being theoretically analyzed. The computer controls the rate of infusion of a vaso-constrictor drug inputed to the animal, and monitors the blood-pressure output. An adaptive model of the animal's drug response is used to derive the required input for control of future blood-pressure values. A transversal-filter model is used, and control is derived by forward-time calculation including the known internal states of the model.

There is a great need for learning control systems which can adapt their control laws to accomodate the requirements of plants whose characteristics may be unknown and/or changeable in unknown ways. A principal factor that has hampered the development of adaptive controls is the intrinsic difficulty of dealing with learning processes embedded in feedback loops. Interaction beteeen the feedback of the learning processes and that of the signal flow paths greatly complicates the analysis which is requisite to the design of dependable operating systems.

An elementary form of adaptive control system employing an adaptive series compensator is shown in Figure 1. This system is simple in conception but is rather inefficient and difficult to deal with from the point of view of adaptation. The compensator could be easily adapted if one had available in real time an optimal output or plant driving signal corresponding to the particular real-time compensator input signal. The optimal compensator output

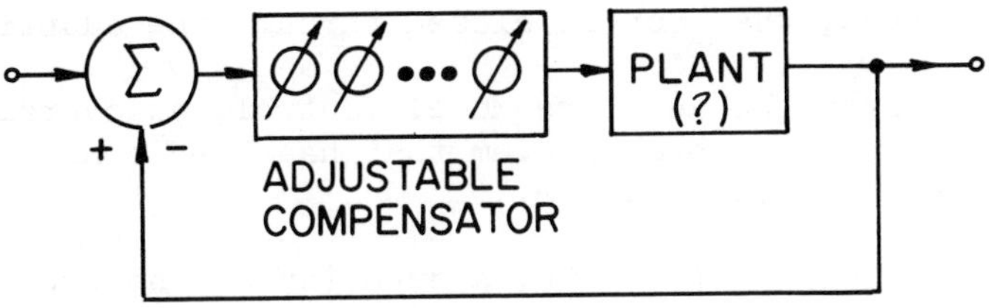

Figure 1. Adaptive Series-Compensated Control System.

signal could serve as a training signal for an adaptive compensator. This signal is very difficult to obtain when the plant is unknown however. If this signal were available, the compensator and the feedback loop would be unnecessary.

Another approach to the adaptation of the system of Figure 1 is the following. Suppose that the purpose of adaptation is the minimization of the servo error in the mean square sense. Gradient components could be measured by perturbing the compensator adjustments. The mean square error could be minimized by using a gradient method such as the method of steepest descent. There are two difficulties here that limit the technique. Regardless of the method used in perturbing the adjustments, whether one at a time or all at once, the system settling time must be waited before measurements can be taken each time the compensator adjustments change or the plant parameters change.[1]. Furthermore, assuming that the gradient can be successfully measured, the mean-square-error performance function is known to be irregular, non-parabolic, and containing relative optima [2]. Hill-climbing techniques for such functions still are in primitive stages of development.

The techniques proposed in this paper have been successfully tested in a limited number of medical-electronic experiments and they represent a different approach to plant control which circumvents many of the difficulties typified by the adaptive-system example of Figure 1. These techniques are still in development, so this paper should be regarded as a preliminary report. In some ways, these techniques are related to those of Powell [3] who used an adaptive model to determine a feedback-loop compensator.

The techniques proposed here will be referred to as Adaptive-Model Control (AMC). The principle operates as follows. Form a model of the plant, and continually update the model by an adaptive process. Using the model and its internal states, do a forward-time analysis to determine inputs to the model which will cause desired future model outputs, thereby controlling the model very closely. Apply the same control to the actual plant, and if the model behaves similarly to the plant, the output of the plant will be closely controlled. The control of the plant is in a sense open-loop, but in fact, the loop is closed through the adaptive process.

To illustrate the adaptive-Model Control, an overall diagram of a blood pressure control system that has been constructed and tested is presented in Figure 2.

At the beginning of a test, a quantity of the powerful drug Arfonod is injected into the animal (a dog). This drug has the effect of disabling the natural blood pressure regulating system of the animal inducing a prolonged state of shock. If left alone, the blood pressure would drop close to zero and irreversible damage would be done to the animal. A vasoconstrictor drug, Norepinephrine, is infused slowly over many hours to compensate and to support the blood pressure. The computer continually monitors blood pressure and regulates the rate of infusion of the vasoconstrictor. The ultimate purpose is to develop computer controls for human intensive care systems.

Typical dynamic responses of the mean animal blood pressure readings to step changes in rate of infusion of the vasoconstrictor drug are sketched in Figure 3. The type of response resulting depends upon the size, type, and especially upon the condition of

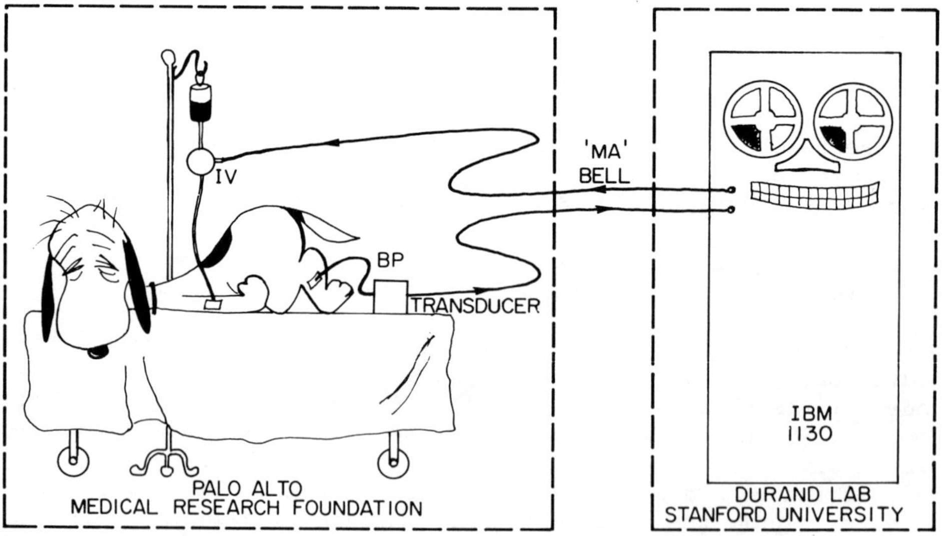

Figure 2. The Experimental Set-up.

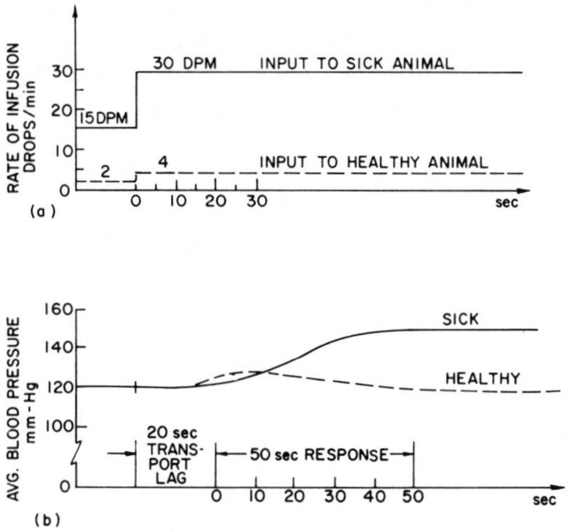

Figure 3. Typical Average Blood Pressure Time Responses to Step Changes in Vasoconstrictor Infusion Rates.

the animal. An animal in good health will respond to small increases in drug flow by eventually settling the blood pressure back to the original set point level. A sick animal will not be able to compensate for even moderate increases in vasoconstrictor inputs and hence the blood pressure will increase in a predictable manner and then settle at a higher level. Tremendous variations in animal responses to the vasoconstrictor have been observed. Typically, there is a transport lag of 10 to 20 seconds before the animals respond, and settling times are usually about 50 to 100 seconds.

The system illustrated in Figure 2 gives the appearance of being an ordinary feedback control system. But this is not the case. The dynamic response of an animal (including transport lag) is too variable to be managed by a conventional feedback control. A block diagram of the actual system is shown in Figure 4.

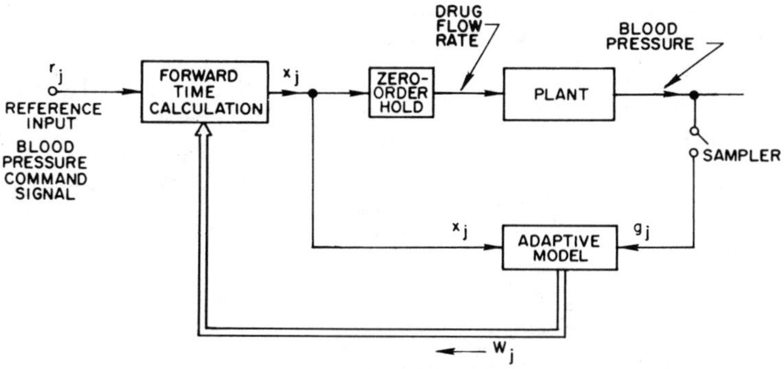

Figure 4. Block-Diagram of an Adaptive-Model Control System.

The functions labeled "forward-time calculation" and "adaptive model" are accomplished by a Hewlett-Packard 2116 computer, as are many data logging and data display function which are not shown but which are necessary in a laboratory set-up. The "plant" is the dynamic response of the blood pressure system to the drug. The zero-order hold is part of the electronic system interfacing the computer to the drug-flow solenoid valve. A cycle time occurs every five seconds. Once per cycle, the adaptive model is updated and a new drug rate (drops per minute) is calculated.

The adaptive model is a 20-tap transversal filter covering a total real-time window of 95 seconds. A bias weight is included to represent the ambient average blood pressure when the input drug rate is zero. The details of the adaptive model are shown in Figure 5.

The adaptive model would be linear if the weights were fixed or if their values were not functions of the input-signal characteristics. The adaptive process automatically adjusts the weights so that for the given input-signal statistics, the model provides a best minimum-mean-square-error fit to a sampled version of the combination of the zero order hold and the plant. The adaptive process utilized is the LMS algorithm, presented first in refs. [4] and [5] and presented more completely in the context of applications to pattern recognition [6] and applications to spatial and temporal filtering (adaptive antenna arrays) [7].

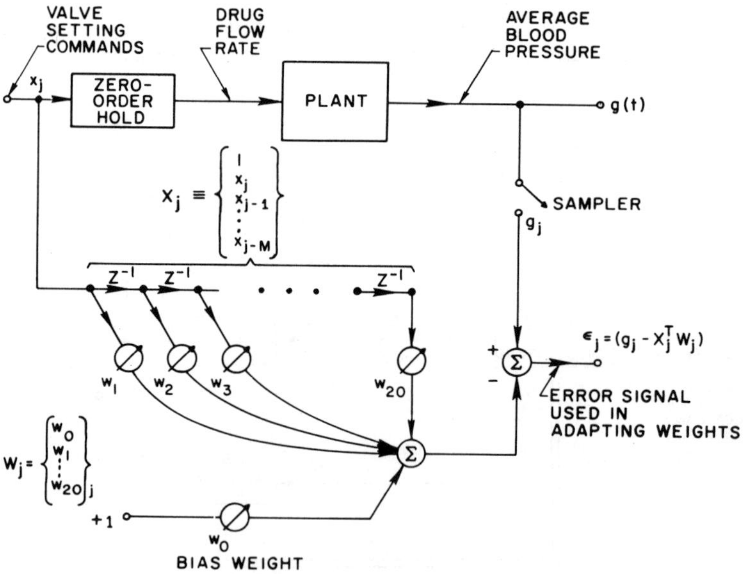

Figure 5. Details of "Adaptive Model" Box, a 20-tap Adaptive Transversal Filter with Bias Weight.

The adaptive algorithm is

$$W_{j+1} = W_j + 2\mu\varepsilon_j X_j \ ,$$
$$\varepsilon_j = d_j - X_j^T W_j \ . \tag{1}$$

The error ε_j is the difference between the desired response d_j and the model response $X_j^T W_j$. The desired response d_j is obtained by sampling the actual plant response $g(t)$. Therefore $d_j = g_j$, and the error is $\varepsilon_j = g_j - X_j^T W_j$. (Refer to Figure 5.)

With a stationary input and a stationary plant, the LMS algorithm is known to be convergent when the convergence factor μ is chosen in the range

$$\frac{1}{\lambda_{max}} > \frac{1}{\text{trace } R} > \mu > 0 \ . \tag{2}$$

The input autocorrelation matrix R is defined below. Its largest eigenvalue is λ_{max}. Note that trace $R = E[X_j]^2$. The factor μ controls stability and rate of convergence of the algorithm. The expected value of the weight vector converges to the optimal or "Wiener" solution W^*.

$$\lim_{j\to\infty} E[W_j] = W^* = R^{-1}P,$$

where

$$R \equiv E[X_j X_j^T] \quad \text{and} \quad P \equiv E[d_j X_j] \ . \tag{3}$$

A fundamental mathematical question is raised by this approach. The input cannot be stationary, and it will be shown that this input is partly determined from the weight values themselves (via the "forward-time calculation"). Yet, the LMS algorithm behaves stably, and in all cases in practice, always converges rapidly to a close model of the unknown plant. There seem to be no practical problems with the approach, only mathematical problems.

Refer now to the block diagram of the entire system shown in Figure 4. The plant control X_j is derived from the box labeled "forward-time calculation." This box generates X_j from the reference input r_j and from the weight vector W_j and the input vector X_j of the model. We shall now consider the operation of this box.

The objective is to derive a driving function x_j so that $X_j^T W_j = r_j$. Each iteration cycle, the model weight vector W_j is updated, and then x_j is calculated taking into account W_j and

$x_{j-1}, x_{j-2}, \ldots, x_{j-n+1}$. If the model behavior is essentially the same as that of the plant, application of x_j to the plant will cause its output response to be close to the reference command signal r_j. Let us choose x_j according to [1].

$$x_j w_1 + \sum_{i=2}^{n} x_{j-i+1} w_i + w_0 = r_j$$

$$x_j = \frac{1}{w_1} [r_j - w_0 - \sum_{i=2}^{n} x_{j-i+1} w_i] \tag{4}$$

Choosing x_j according to this formula will allow the model to be perfectly controlled, and applying the same input to the actual plant will result in a mean-square control error $E[r_j - g_j]^2$ equal to the mean-square error $E[\varepsilon_j^2]$ of the modelling process.

Everything goes well using this method as long as w_1 has substantial value. When there are transport delays however, w_1 tends to be small and noisy. The values of x_j computed by the above formula could be very large and erratic, since division by w_1 is required. This could create problems, particularly in the blood-pressure control system where massive doses of drug are undesirable and negative doses are impossible. Because of transport delays, two somewhat different approaches have been taken.

The first of these approaches constrains the first several weights of the adaptive model to be zero. The number of zero-constrained weights corresponds to the transport lag of the plant, which would be obtained from a priori knowledge.

Details of the functional box "Forward-time Calculation" of Figure 4 are shown in Figure 6a, illustrating how x_j is calculated in the situation when the first two model weights are zero. All the weights shown in Figure 6 are copied from the values derived by the appropriate modeling process. The particular values shown are for illustration only.

Each cycle, the value of x_j is calculated to cause the output of the summer y_j to be equal to r_j. Since the plant is driven by x_j, its sampled output g_j will closely approximate r_{j-2}, depending on the closeness of fit of the model to the plant. The delay in the response is an inevitable result of the plant transport delay. The values of x_j are calculated according to

$$x_j = \frac{1}{w_3} [r_j - w_0 - \sum_{i=4}^{n} x_{j-i+3} w_i] . \tag{5}$$

a) first two weights zero

b) first two weights small, finite

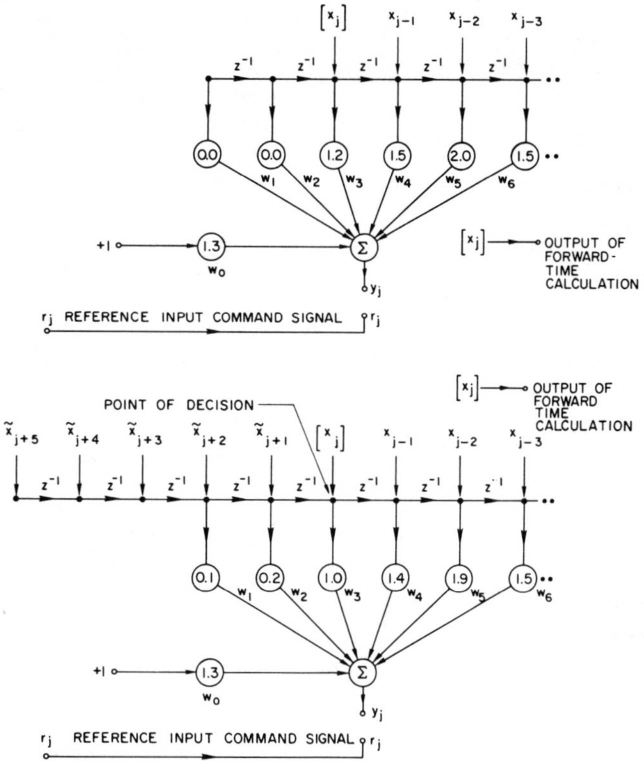

Figure 6. Details of "Forward-Time Calculation" Box.

It should be mentioned that to start one of these systems as quickly as possible, initial weight values in the modeling process are usually taken from the previous run. Initial values are not critical, but if they are close to correct, there will be very little start-up transient.

The second approach for dealing with plant transport delay does not require a decision constraining a certain number of model weights to zero. There are many cases where the leading weights are small in magnitude, but non-zero. Such a set of weights is illustrated in Figure 6b.

In this case, the values of x_j cannot be calculated to perfectly match the output y_j with r_j. Future tentative values of x such as $\tilde{x}_{j+1}$, $\tilde{x}_{j+2}$... are calculated so that the control signal x_j can be deduced. <u>The tilde indicates that the values are</u>

tentative. Absence of the tilde means that the value is decided and is used in controlling the plant. The "point of decision" in the calculation is indicated in Figure 6b. The position of this point of decision along the tapped delay line of the adaptive model is chosen a priori by the system designer to correspond to the plant transport delay. Choosing this position has some effect on system performance, but is not critical.

The calculation of x_j at the time of the jth cycle is accomplished according to eq. (6). The point of decision is taken as in Fig. 6b.

$$w_0 + w_1\tilde{x}_{j+2} + w_2\tilde{x}_{j+1} + w_3 x_j + \sum_{i=4}^{n} w_i x_{j-i+3} = r_j$$

$$w_0 + w_1\tilde{x}_{j+3} + w_2\tilde{x}_{j+2} + w_3\tilde{x}_{j+1} + w_4 x_j + \sum_{i=5}^{n} w_i x_{j-i+4} = r_{j+1}$$

$$w_0 + w_1\tilde{x}_{j+1} + w_2\tilde{x}_{w+3} + w_3\tilde{x}_{j+2} + w_4\tilde{x}_{j+1} + w_5 x_j + \sum_{i=6}^{n} w_i x_{j-i+5} = r_{r+2}$$

The number of equations is generally determined by the number of future values of the reference command signal $r_j, r_{j+1}, r_{j+2}, \ldots$ that may be available. These equations may be rearranged according to (7).

$$w_3 x_j + w_2\tilde{x}_{j+1} + w_1\tilde{x}_{j+2} = r_j - w_0 - \sum_{i=4}^{n} w_i x_{j-i+3}$$

$$w_4 x_j + w_3\tilde{x}_{j+1} + w_2\tilde{x}_{j+2} + w_1\tilde{x}_{j+3} = r_{j+1} - w_0 - \sum_{i=5}^{n} w_i x_{j-i+4}$$

$$w_5 x_j + w_4\tilde{x}_{j+1} + w_3\tilde{x}_{j+2} + w_2\tilde{x}_{j+3} + w_1\tilde{x}_{j+4} = r_{j+2} - w_0 - \sum_{i=6}^{n} w_i x_{j-i+5} \quad (7)$$

The numerical values of the right-hand sides of the equations (7) can be calculated since r_j, $r_{j+1}, r_{j+2}, \ldots$ are known, the weights are known, and the previously-decided driving function values $x_{j-1}, x_{j-2}, \ldots$ are known. Let the right-hand sides be calculated.

These equations cannot be solved yet, since there are too many "unknowns" for the number of equations. Since w_1 and w_2 are relatively small, a solution can be obtained by letting two adjacent distant-future values of $\tilde{x}$ take arbitrary values, such as zero. When only three values of the reference signal r_j, r_{j+1}, r_{j+2} are known, we have three simultaneous equations to solve. We let $\tilde{x}_{j+4} = \tilde{x}_{j+3} = 0$. It is then possible to solve for $x_j, \tilde{x}_{j+1}$, and $\tilde{x}_{j+2}$.

Although we only need the decided value x_j at the time of the jth cycle for direct control purposes, the future tentative values a interesting to have also.

At the j+1-th cycle, the entire process is repeated. The tentative value of $\tilde{x}_{j+1}$ calculated on the j-th cycle should agree closely with the decided value of x_{j+1} calculated on the j+1-th cycle. The agreement will not be perfect because of setting $\tilde{x}_{j+4}$ and $\tilde{x}_{j+3}$ to zero. Let us call this effect "truncation error." By using additional future values of the reference command signal, tentative values of x's can be determined further into the future and truncation error can be reduced.

Using x_j as the plant driving function will cause the plant sampled output g_j to agree closely with r_{j-2}. The error in the system response will be due partly to imperfection in the modeling fit and partly to truncation error.

Note that when the transport delay mechanism is such that the first k weights of the model are relatively small, solving the equations determining x_j and future $\tilde{x}$-values requires assuming that a sequence of k distant-future $\tilde{x}$-values are zero.

Also note that knowledge of future values of the plant driving function, although they are tentative, could be quite useful in modifying the goals of the control system (i.e., modifying r_j) in cases where demands are made on the driving function that would exceed limits, go negative where this is not possible, etc. For example, the sequence r_j could be modified by not insisting that the system settle in the minimum time achievable with an unrestricted r_j, etc. It is possible to state and to have the system respond to very sophisticated computer-directed goals. Since inexpensive modern computers can operate much faster than real time, various goals and control objectives can be practically explored each calculation cycle.

The AMC techniques have already been used a number of times in experiments to regulate and control average blood pressure in animals. In these experiments, the standard deviation of the noise in the blood-pressure sensing instrumentation has been about 5 to 10 mm Hg. The mean blood pressure is typically regulated to within about 2 to 4 mm Hg in steady state and could be off about 5 to 10 mm Hg temporarily under extreme transient conditions. The typical start-up settling times are of the order to two minutes, somewhat longer than the total time window spanned by the adaptive plant model. The Appendix presents data from an actual run.

APPENDIX - AN EXPERIMENTAL RUN

Figures 7-10 present results developed during an experimental run while controlling an animal's average blood pressure.

The beginning of the run is shown in Figure 7. The dog was healthy and normal until the Arfonod was injected, whereupon his blood pressure plummetted, as seen in the figure.

In this experiment, the adaptive-model weights began to form, starting from initial settings, at the very outset before the Arfonod was injected. The two upper tracings show the actual, average blood pressure of the animal and the output of the model respectively. Note how they stay closely together. They stay moderately close together even in periods of great stress such as just after the Arfonod was injected.

At the beginning of the run, the flow rate of the vasoconstrictor (the "drug rate") was manually set at 10 drops per minute. This was manually raised to 20 drops per minute after the Arfonod took hold. Raising the drug rate checked the blood pressure decline. Soon thereafter, as indicated by the cross on the drug-rate tracing, the control of drug rate was turned over to the automatic system and remained automatic thereafter. A pressure set point was entered through the computer keyboard, and this level was indicated by the cross near the upper two tracings. The control system then had the job of getting the animal blood pressure up to the set point and holding it there in spite of natural disturbances in the animal. Changes in the set point were inserted from time to time as part of the system test. The middle curve shows a running average mean square error (on a log scale) between the plant and the adaptive model.

The total memory time of the adaptive model was 100 seconds. The model contained 20 taps with 5 second delays between taps of the transversal filter. Once automatic control was established, the system took about 5 minutes to settle the blood pressure close to the set point. Thus the system settling time was about 3 times as long as the memory time of the model. This represents rather fast settling for an adaptive control system.

In this test, the computer controlled the blood pressure during several hours with the animal under different degrees of influence to Arfonod. The results were uniformly good, and the response data of Figures 8 and 9 typical. The data records were long, but the data of Figures 7, 8 and 9 are contiguous with slight time overlaps. Settling responses to changing set-point values are illustrated. In each case, approximately 5 minute settling times were evident.

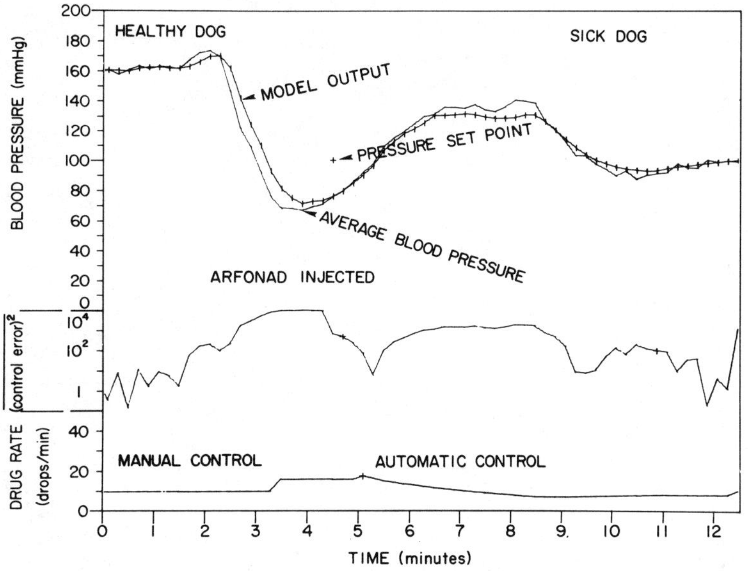

Figure 7. Actual run: Transition from Healthy to Sick, Manual to Automatic Control.

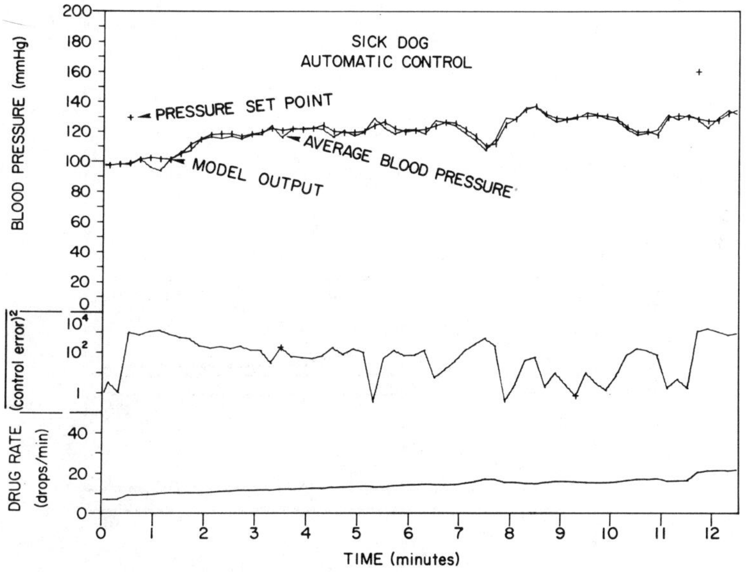

Figure 8. Actual run: Control of Sick Dog Blood Pressure.

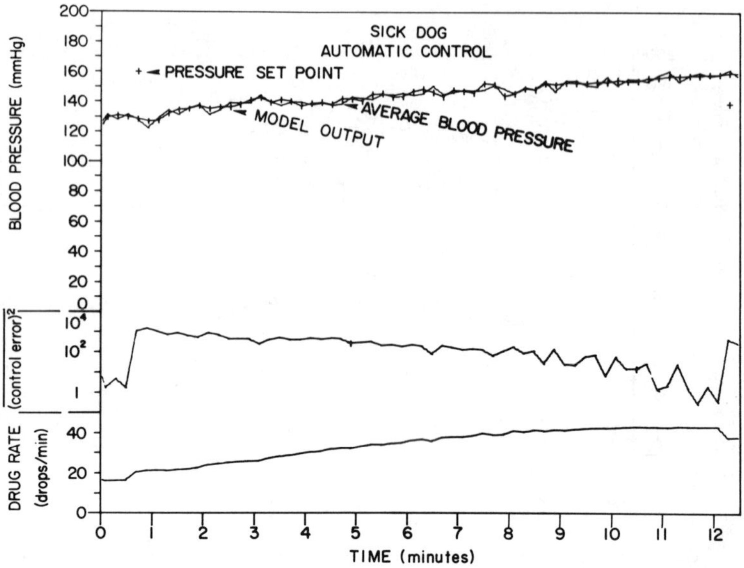

Figure 9. Actual run: Control with Raised Blood-Pressure Set Point.

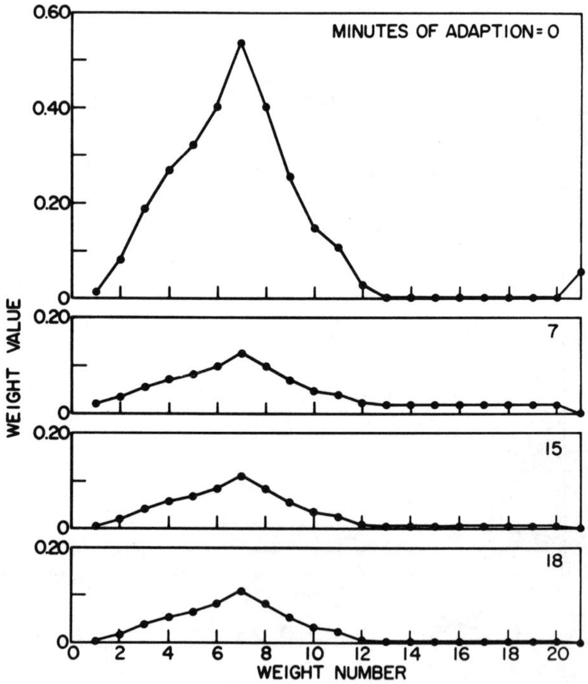

Figure 10. Model Weights (Impulse Response).

The tapped-delay-line-model weights at several times during the run were recorded and they are plotted in Figure 10. The weight values are arranged chronologically along the line, and so represent the model's view of the animal's impulse response. The twenty-first weight is the bias weight, (see Figure 5). The impulse response of the top frame was taken before the Arfonod was injected and here the animal was very sensitive to the vasoconstrictor drug. The next frame was taken after the Arfonod was injected and had taken hold, just before the automatic control was turned on. The shape of the response was changed somewhat, but the sensitivity level changed greatly. As time went on, the animal regulating system was disabled due to the Arfonod. Changes in the animal impulse response took place, but they were not drastic changes. One can see that the amount of transport delay was not very clear-cut, illustrating the type of plant behavior indicated in Figure 6b.

Although this system is simple in conception, making the hardware and software work reliably has taken considerable effort. Many runs over the past year or so were made in order to perfect the algorithms, the system software, and the interface hardware. Experimental results have been uniformly good using the algorithms and procedures outlined here. The development of mathematical analysis of the AMC adaptive control technique is progressing nicely.

REFERENCES

1. K. S. Narendra, and L. E. McBride, "Multiparameter Self-Optimizing Systems Using Correlation Techniques", IEEE Transactions on Automatic Control, Vol. AC-9, No. 1, January 1964, p. 31.

2. P. E. Mantey, "Convergent Automatic-Synthesis Procedures for Sampled-Data Networks with Feedback", Stanford Electronics Laboratories, SEL-64-112, TR 6773-1, Stanford, Calif., Oct. 1964.

3. F. D. Powell, "Predictive Adaptive Control", IEEE Transactions on Automatic Control, Vol. AC-14, No. 5, October 1969, pp. 550-552.

4. B. Widrow and M. E. Hoff, "Adaptive Switching Circuits," 1960 IRE Wescon Convention Record, Part 4, pp. 96-104.

5. B. Widrow, "Adaptive Sampled-Data Systems," Proceedings, 1st International Congress IFAC, Moscow, 1960, pp. 406-411.

6. J. Koford and G. Groner, "The Use of an Adaptive Threshold Element to Design a Linear Optimal Pattern Classifier," IEEE Transactions on Information Theory, Vol. II-12, No. 1, January 1966, pp. 42-50.

7. B. Widrow, et. al., "Adaptive Antenna Systems," Proceedings, IEEE, Vol. 55, No. 12, December 1967, pp. 2143-2159.

ACKNOWLEDGEMENTS

The author would like to thank Henry Crichlow, Fred Mansfield, Jeffrey Bower, Thomas Stibolt, Ulf Strom, and especially Cristy Schade who was the leader of this group of graduate students at Stanford University who have designed, assembled, and brought to operation the real-time experimental hardware and software. Special thanks go to Dr. Noel M. Thompson of the Palo Alto Medical Research Foundation who introduced the problem, made working with experimental animals possible, and who was able to evaluate performance from the point of view of both a Physician and an Engineer.

REAL-TIME DISPLAY SYSTEM OF RESPONSE CHARACTERISTICS OF MANUAL CONTROL SYSTEMS

Jin-ichi Nagumo

University of Tokyo

Tokyo, Japan

INTRODUCTION

Extensive research has been devoted to the response characteristics of human operators engaged in manual tracking tasks, but little work has been done on the changes in response characteristics with the lapse of time.

In the present paper, a new method is proposed that can determine the time-varying characteristics of manual tracking systems, and also a real-time display system of the characteristics using an on-line digital computer is presented.

In order to realize such a display system, the following three points should be settled. The first is to establish a method that can identify linear systems with random input signals in a short time. The second is, by making use of this method of identification, to construct a display system including an on-line computer that displays the slowly varying response characteristics of manual tracking systems, moment by moment, on a CRT screen. The third is to develop a software system for the computer that provides facilities for performing various kinds of tracking experiments.

The display system described in the following is regarded as having achieved our purpose.

LEARNING METHOD FOR SYSTEM IDENTIFICATION

The response characteristic of a manual tracking system with pursuit operation of the subject may be regarded as linear under certain circumstances. However, the response characteristic of the subject gradually changes with the lapse of time due to fatigue, proficiency, etc., so that it is desirable that identification of the response characteristic be completed in the shortest time possible. The response characteristic obtained by averaging over a long period of time is meaningless if changes take place during the course of the experiment.

In the case where the response characteristic of a system is linear and time-invariant, choice of the input signal is arbitrary since the output that corresponds to any input signal can easily be known from an input-output relation. In actual manual tracking systems, however, the response characteristic is apt to be influenced by the peculiarity of the input signal and, moreover, the normal condition of the subject is often disturbed by the input signal itself. Thus, for the identification of the response characteristics of manual tracking systems, a random input signal is better than those with particular waveforms, such as impulse, step function, or sinusoids. A well-known method of system identification, which makes use of a stationary random input signal and correlation functions, is nondisturbing but the time required for identification is, in general, rather protracted.

As mentioned above, the various methods for identifying linear systems known so far have their respective weaknesses. In this paper, a new method for linear system identification is proposed [1] that uses a random input signal and still has a short identification time for application to manual tracking systems whose response characteristics vary slowly with time.

In our method of system identification, linear systems are represented by weighting functions, and our consideration is restricted to discrete-time systems (sampled-data systems) for convenient application to computers.

The sampled weighting function of a linear time-invariant stable system is approximated by a finite set of values

$$W_1, W_2, W_3, \ldots, W_N \tag{1}$$

where W_i is the ith sampled impulse response of the system, N (named "span" of the weighting function) is chosen so that $N \cdot \Delta t$ covers the significant duration of impulse response, and Δt is the sampling period. Let an input sequence be

$$X_1, X_2, X_3, \ldots \tag{2}$$

and the corresponding output sequence of the system specified by (1) be

$$Y_1, Y_2, Y_3, \ldots \tag{3}$$

Then, for $j \geq N+1$, it follows that

$$Y_j = W_1X_{j-1} + W_2X_{j-2} + \dots + W_NX_{j-N}$$
$$= \sum_{i=1}^{N} W_iX_{j-i} \ . \qquad (4)$$

Our method of identifying the system specified by (1) is to find the sequence of the set of values

$$V_1^{(j)}, V_2^{(j)}, V_3^{(j)}, \dots, V_N^{(j)} \qquad j = N+1, N+2, \dots \qquad (5)$$

by an iterative procedure, such that each $V_i^{(j)}$ approaches W_i with the iteration step j (see Figure 1). Such a set of values is called the weighting function of the identifier, and the output Z_j of the identifier approaches Y_j with the iteration step j, where

$$Z_j = V_1^{(j)}X_{j-1} + V_2^{(j)}X_{j-2} + \dots + V_N^{(j)}X_{j-N}$$
$$= \sum_{i=1}^{N} V_i^{(j)}X_{j-i} \ . \qquad (6)$$

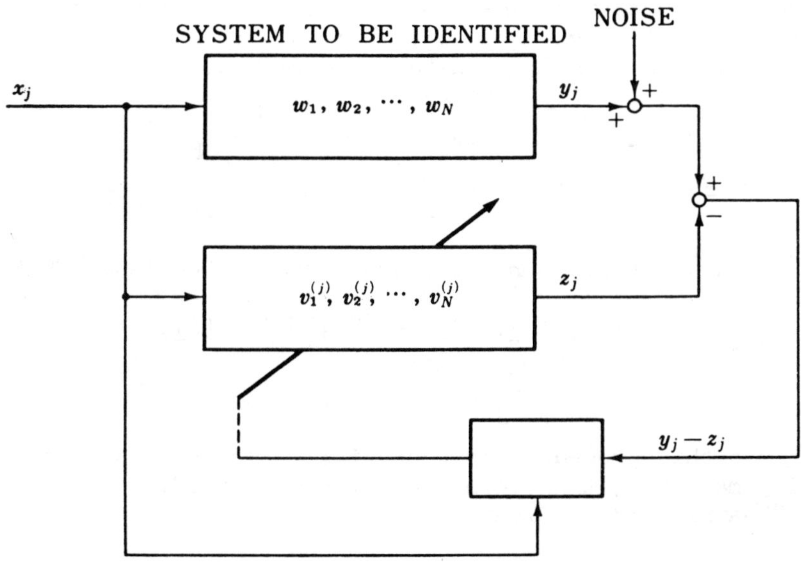

Figure 1. Learning Method for System Identification.

Define the following N-dimensional vectors:

$$\underline{W} = \begin{bmatrix} W_1 \\ W_2 \\ \vdots \\ W_N \end{bmatrix}, \quad \underline{V}_j = \begin{bmatrix} V_1^{(j)} \\ V_2^{(j)} \\ \vdots \\ V_N^{(j)} \end{bmatrix}, \quad \underline{X}_j = \begin{bmatrix} X_{j-1} \\ X_{j-2} \\ \vdots \\ X_{j-N} \end{bmatrix}.$$

The last vector is a modified input obtained by collecting N terms of the original input sequence (2). Hereafter, the vector input sequence

$$\underline{X}_{N+1}, \underline{X}_{N+2}, \underline{X}_{N+3}, \cdots \tag{7}$$

is used instead of (2). By the use of these vectors,

$$Y_j = (\underline{W}, \underline{X}_j), \tag{8}$$

$$Z_j = (\underline{V}_j, \underline{X}_j). \tag{9}$$

Now the adjustment procedure for the weighting function $\underline{V}_j$ of the identifier is as follows. The identification error for an input vector is allotted to each component of the weighting function vector of the identifier, proportional to the magnitude of the corresponding component of the input vector, so that the output of the adjusted identifier gives a correct output if the same input is applied at the next sampling instant. More precisely, at the jth step, $\Delta\underline{V}_j$ is added to $\underline{V}_j$, where

$$\Delta\underline{V}_j = \underline{V}_{j+1} - \underline{V}_j = (Y_j - Z_j)\frac{\underline{X}_j}{\|\underline{X}_j\|^2}, \quad \|\underline{X}_j\|^2 = \sum_{i=1}^{N} X_{j-i}^2 \tag{10}$$

which means an error-correcting procedure in the sense that

$$(\underline{V}_{j+1}, \underline{X}_j) = (\underline{V}_j + \Delta\underline{V}_j, \underline{X}_j) = (\underline{V}_j, \underline{X}_j) + (\Delta\underline{V}_j, \underline{X}_j)$$

$$= Z_j + (Y_j - Z_j) = Y_j. \tag{11}$$

Since this procedure is essentially the same as that of the learning machines, this method of system identification is referred to as "learning identification."

A geometrical interpretation of this adjustment procedure is as follows (see Figure 2). Define a hyperplane Π_{j-1} in the N-dimensional space by

$$\Pi_{j-1} = \{\underline{p} \mid (\underline{p}, \underline{X}_{j-1}) = Y_{j-1}\}. \tag{12}$$

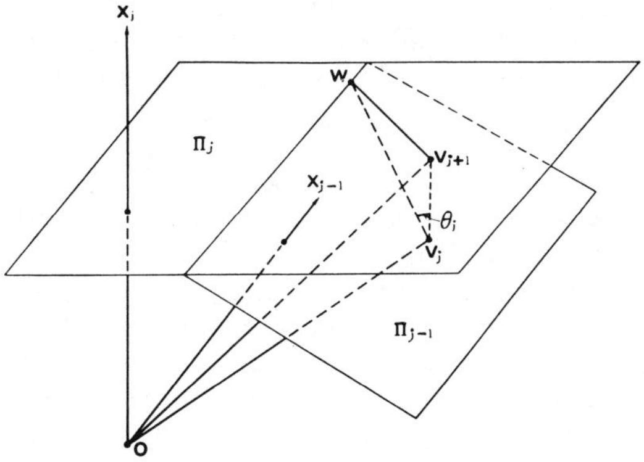

Figure 2. Geometrical Interpretation of the Adjustment Procedure When N=3.

The hyperplane Π_{j-1} is perpendicular to vector $\underline{X}_{j-1}$ and, from (8) and (11), it is known that points $\underline{W}$ and $\underline{V}_j$ are on Π_{j-1}. Similarly, hyperplane Π_j is perpendicular to $\underline{X}_j$, and points $\underline{W}$ and $\underline{V}_{j+1}$ are on Π_j. Since, from (10), vector $(\underline{V}_{j+1} - \underline{V}_j)$ is parallel to vector $\underline{X}_j$, the point $\underline{V}_{j+1}$ coincides with the foot of the normal from point $\underline{V}_j$ to hyperplane Π_j.

It is shown that the sequence $\{\underline{V}_j\}$ converges to $\underline{W}$ independent of the initial value $\underline{V}_{N+1}$ for a random input sequence $\{X_j\}$ or $\{\underline{X}_j\}$, which belongs to a very wide class of random sequences [1].

It should be noted that the adjustment procedure mentioned above can be slightly extended, and

$$\Delta\underline{V}_j = \underline{V}_{j+1} - \underline{V}_j = \alpha(Y_j - Z_j)\,\frac{\underline{X}_j}{\|\underline{X}_j\|^2} \tag{13}$$

can be employed instead of (10), where $\alpha(2>\alpha>0)$ is an error-correcting coefficient.

Now define the normalized error e_j of the identification at the jth step by

$$e_j = \frac{\|\underline{W} - \underline{V}_j\|}{\|\underline{W}\|} \qquad j \geqq N+1. \tag{14}$$

Computer simulation of the identification process shows that the expected value of e_j decreases at the rate of 1/e as the identification process proceeds 2N steps, regardless of the characteristics of the systems to be identified [1].

So far, noiseless measurements have been assumed. In what follows, the effect of noise in the system output on the process of learning identification will be considered [2].

In this case, the process of learning identification may be regarded as consisting of two stages. One is the transient state where the influence of the initial error of identification cannot yet be ignored, and the length of this state represents, roughly speaking, the rate of identification. The other is the limiting state where the adjustment procedure has been almost completed and is fluctuating in the neighborhood of the weighting function to be identified, and the amplitude of the fluctuation determines the achievable accuracy of the identification. Moreover, it should be noted that the range of α limited to $0<\alpha<1$.

Now, there is a distinct inconsistency between these two states: increase in the value of α shortens the transient state but decreases the accuracy at the limiting state. This incompatibility is utilized as a guiding principle in determining the error-correcting coefficient α, that is, in mediating between the two conflicting requirements of reducing the period of the transient state and of increasing the limiting accuracy.

Furthermore, it has been known that the time variation of the characteristic of the system to be identified causes prolongation of the transient state and reduction of the limiting accuracy.

APPARATUS

A block diagram of the experimental apparatus is shown in Figure 3. A bright vertical line is displayed on a CRT screen and

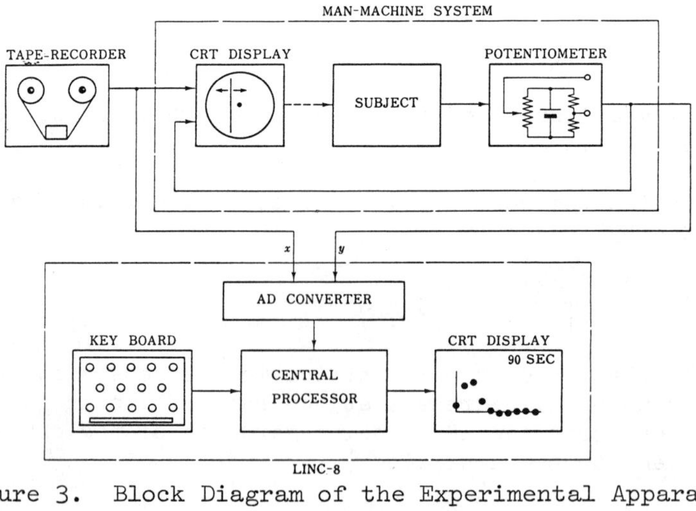

Figure 3. Block Diagram of the Experimental Apparatus.

is driven horizontally by a random signal that has been prepared beforehand and recorded on a magnetic tape. The random signal is an analog signal that is obtained by filtering the white noise through a filter with a cutoff frequency of about 0.1 Hz.

A subject seated in front of the CRT oscilloscope rotates the knob of a potentiometer, which generates a voltage proportional to the angle of rotation, and the voltage drives a bright spot on the same CRT screen horizontally. The subject controls the angle of rotation of the potentiometer knob so that the spot remains as close to the vertical line as possible, watching the movements of both on the CRT screen.

The results of the experiment are processed by an on-line digital computer LINC-8 (12 bits/word, cycle time 1.5 μS, 4K-word core memory) in real time. The input signal and the corresponding output signal of the pursuit tracking system are sampled and digitalized ($\pm$ 8 bits), and turn out computation data. The latter is treated by the learning method for system identification and the resultant weighting function is, moment by moment, displayed in graphic form, on a CRT oscilloscope, with the legends written in alphanumeric characters.

The sampling interval Δt and the span of the weighting function N can be assigned by the computer program within the range of 60-180 ms and 10-20, respectively. The error-correcting coefficient α is usually set at 1.

PROGRAM IDSYS

A software system called GUIDE has been provided for LINC-8 and users of the computer can conveniently utilize this system in writing, editing, registering, and carying out programs. However, in order to afford further facilities for the identification of linear systems and for making various kinds of manual tracking experiments, a program named IDSYS has been newly developed, to be registered as a part of the GUIDE system.

Program IDSYS is written by instruction words of the computer and takes the form of a question-answer system; that is, the operator can carry out the program by typing answers on the keyboard, to questions displayed on the CRT screen. Figure 4 is a flow chart of the program system IDSYS.

SOME EXPERIMENTAL RESULTS

Although pursuit tracking is a simple operation for the subject, it is rather difficult to maintain concentrated operation for

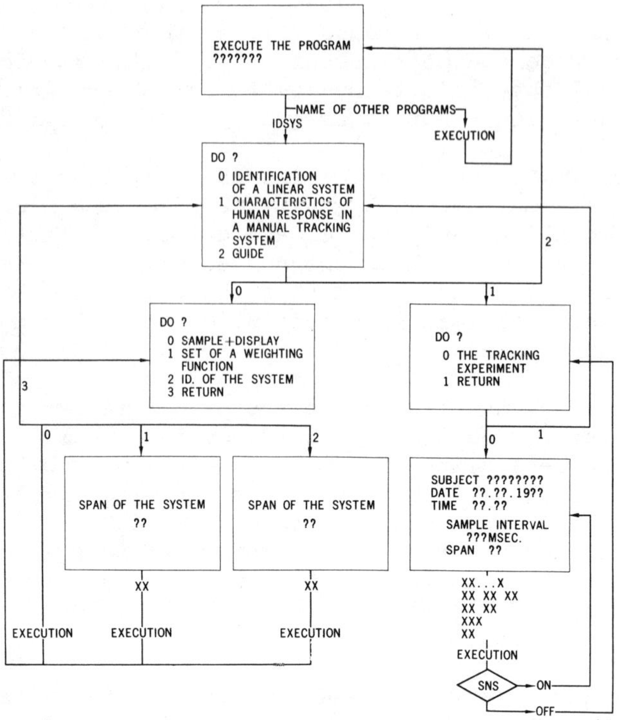

Figure 4. Flow-Chart of the Program System IDSYS.

a long period of time because of fatigue. Thus, each tracking experiment is limited to 3 minutes.

A series of off-line investigations on the same experimental data resulted in the following. To identify pursuit tracking systems with a certain degree of accuracy, it is necessary to choose Δt (sampling interval) and $N \cdot \Delta t$ (span of weighting function in time scale) so as to satisfy $\Delta t < 150$ ms, $N \cdot \Delta t > 1.6$ seconds. (Minimum sampling interval due to computer capacity is 40 ms.) However, it is not advantageous to choose too large a value of N, since the identification time increases in a direct ratio.

The shapes of weighting functions obtained by our experiments have, in general, relatively large positive values at the head and relatively small positive or negative values at the tail. Here the head implies the part of weighting function W_i ($i=1 \sim N$) that corresponds to small values of i (usually $i \leqq 4$), while the tail implies the other part of the weighting function.

When a smooth tracking operation is in progress, it is observed that the tail lies almost on the horizontal axis. If the subject is confused by having failed in tracking, however, the tail oscillates up and down until smooth tracking operation is resumed. For the sake of convenience, the former tracking operation is named the "normal mode" while the latter is the "abnormal mode." It is presumed that the normal mode appears when the human characteristic may be regarded as linear, while the abnormal mode appears when it cannot be so regarded.

Usually, the subject can keep to the normal mode throughout a tracking experiment of about 3 minutes. In certain cases, however, he falls into the abnormal mode once or twice during the course of the experiment. In such cases, the response characteristic obtained by averaging over the whole period of the experiment is meaningless.

The shapes of the heads of the weighting function in the normal mode are divided into two classes. One is trapezoidal (Figure 5(A)), the other bell-shaped (Figure 5(B)). The former appears to correspond to a fine tracking operation and the latter

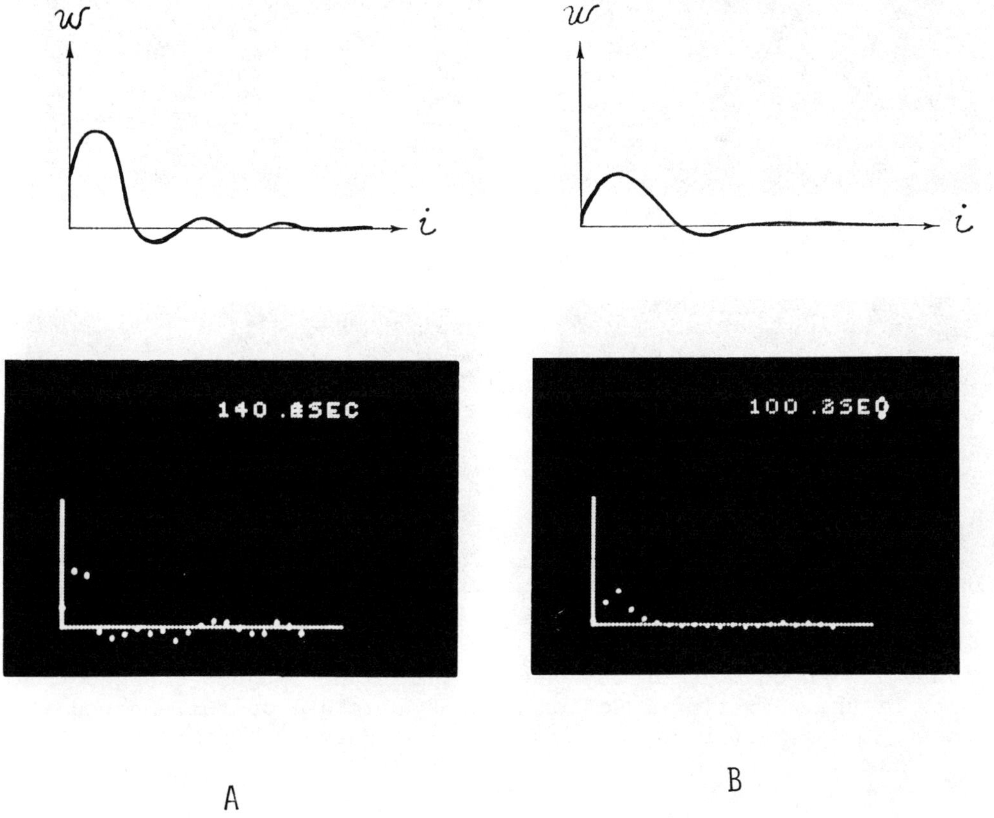

Figure 5. Two Typical Shapes of the Weighting Functions of the Pursuit Tracking Systems in the Normal Mode Operation of the Subject. A: trapezoidal, B: bell-shaped.

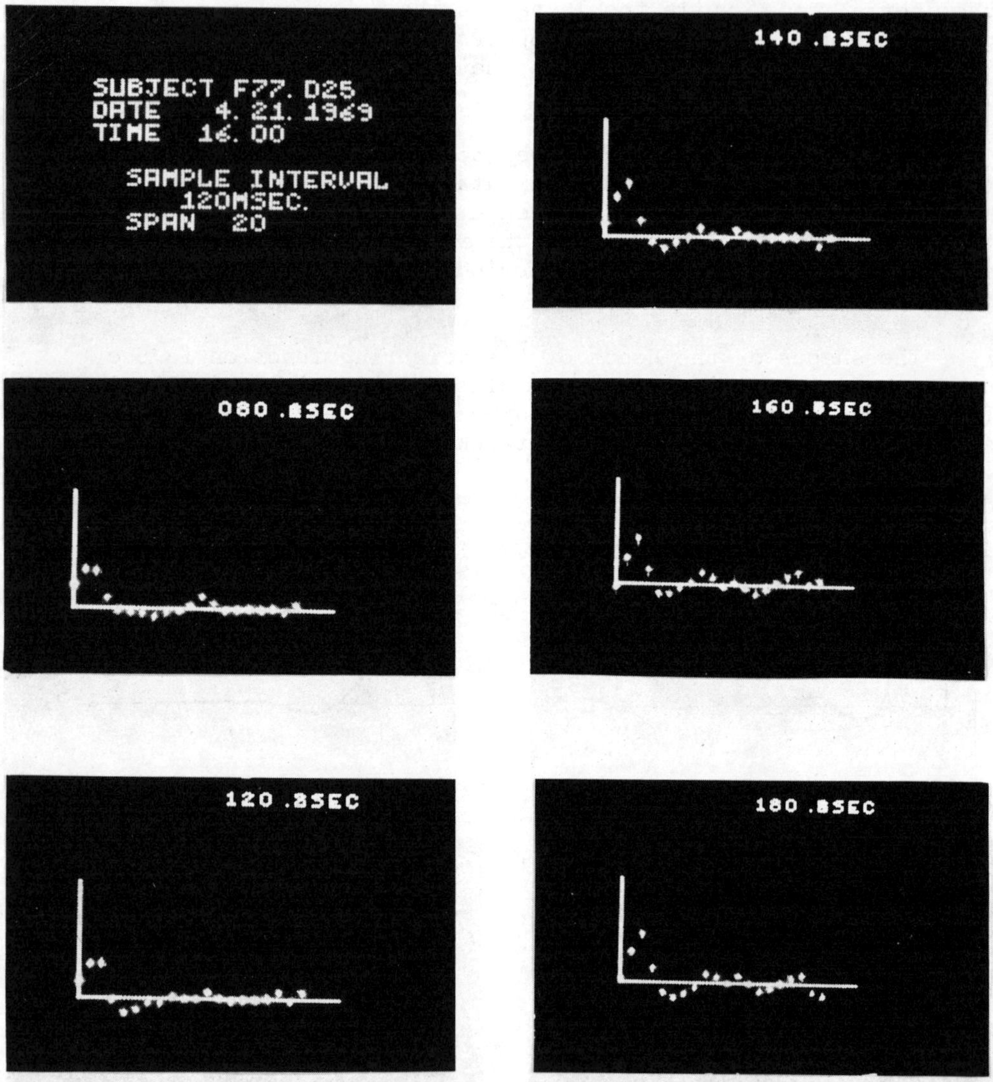

Figure 6. An example of the Case in Which the Subject Changed the Shape of the Weighting Function from Trapezoidal to Bell-Shaped at about 130 Sec. after Beginning a 3-Minute Pursuit Tracking Experiment.

to a rough one. Although the subject usually keeps to one of them throughout an experiment, in some cases the subject changes the shape of the weighting function he has chosen from the former to the latter, as shown in Figure 6.

Usually, the time required for convergence was about 20 sec. It is recommended that the initial value of the weighting function of the identifier should be chosen as close as possible to that of the identified system.

SUMMARY

First, a new method is proposed that can identify a linear system with a random input signal in a short time for application to the identification of manual tracking systems whose response characteristics vary slowly with time. Secondly, making use of this method of identification, a display system including an on-line computer is constructed. It displays the slowly varying response characteristics of manual tracking systems on a CRT screen in real time. Thirdly, a software system that facilities performance of various kinds of tracking experiments is developed. Lastly, some results of simple pursuit tracking experiments are described and the mode of the tracking operation is discussed.

REFERENCES

1. J. Nagumo and A. Noda, "A Learning Method for System Identification," IEEE Trans. on Autom. Contr., Vol. AC-12, pp. 282-287, June 1967.

2. A. Noda, "Effects of Noise and Parameter-Variation on the Learning Identification Method," J. Soc. Instr. Contr. Engrs. Japan, Vol. 8, pp. 303-312, May 1969.

LIST OF DISCUSSERS

"Some Studies on Pattern Recognition with Nonsupervised Learning Procedures" by Kokichi Tanaka
Discussers: G. J. McMurtry, Y. Suzuki, Y. T. Chien

"Linear and Nonlinear Stochastic Approximation Algorithms for Learning Systems" by Y. T. Chien
Discussers: S. Eiho, B. Widrow, K. S. Narendra, T. Kitagawa, G. N. Saridis

"Multi-Category Pattern Classification Using a Nonsupervised Learning Algorithm" by Iwao Morishita and Ryuji Takanuki
Discussers: L. A. Gerhardt, Y. Kinouchi

"A Mixed-Type Non-Parametric Learning Machine Without a Teacher" by Masamichi Shimura
Discussers: Y. T. Chien

"Recognition System for Handwritten Letters Simulating Visual Nervous System " by Katsuhiko Fujii and Tatsuya Morita
Discussers: L. A. Gerhardt, R. Suzuki, K. Tanaka

"Sequential Identification by Means of Gradient Learning Algorithms" by J. M. Mendel
Discussers: S. Sagara, K. S. Narendra, G. N. Saridis

"Stochastic Approximation Algorithms for System Identification Using Normal Operating Data" by Yasuyuki Funahashi and Kahei Nakamura
Discussers: G. N. Saridis, K. Kumamaru, T. Kitagawa, M. Aoki

"On Utilization of Structural Information to Improve Identification Accuracy" by M. Aoki and P. C. Yue
Discussers: T. Kitagawa, K. S. Narendra

"An Inconsistency Between the Rate and the Accuracy of the Learning Method for System Identification and Its Tracking Characteristics" by Atsuhiko Noda
Discussers: D. W. C. Shen, G. N. Saridis, Y. Funahashi

"Weighting Function Estimation in Distributed-Parameter Systems" by Henry E. Lee and D. W. C. Shen
Discussers: H. Takeda, T. Kitagawa, M. Aoki, M. Kimura

"System Identification by a Nonlinear Filter" by Setsuzo Tsuji and Kousuke Kumamaru
Discussers: J. Zaborszky, Y. Suzuki, H. Sugiyama

"A Linear Filter for Discrete Systems with Correlated Measurement Noise" by Tzyh Jong Tarn and John Zaborszky
Discussers:

"Stochastic Learning by Means of Controlled Stochastic Processes" by Seigo Kano
Discussers: M. Aoki, R. W. McLaren, M. Kimura, K. S. Narendra

"Learning Processes in a Random Machine" by Sadamu Ohteru, Tomokazu Kato, Yoshiyuki Nishihara and Yasuo Kinouchi
Discussers: B. Widrow

"Learning Process in a Model of Associative Memory" by Kaoru Nakano
Discussers: S. Ohteru

"Adaptive Optimization in Learning Control" by George J. McMurtry
Discussers: M. Ito, K. Kumamaru, J. M. Mendel, G. N. Saridis

"Learning Control of Multimodal Systems by Fuzzy Automata" by Kiyoji Asai and Seizo Kitajima
Discussers: K. S. Fu

"On a Class of Performance-Adaptive Self-Organizing Control Systems" by George N. Saridis
Discussers: K. Asai, J. M. Mendel, R. W. McLaren

"A Control System Improving Its Control Dynamics by Learning" by Kahei Nakamura
Discussers: K. S. Narendra, J. M. Mendel

"Self-Learning Method for Time-Optimal Control" by Hiroshi Tamura
Discussers: J. M. Mendel

"Learning Control Via Associative Retrieval and Inference" by Julius T. Tou
Discussers: B. Kondo

"Statistical Decision Method in Learning Control Systems" by Shigeru Eiho and Bunji Kondo
Discussers: G. J. McMurtry, M. Kimura, H. Takeda

"A Continuous-Valued Learning Controller for the Global Optimization of Stochastic Control Systems" by R. W. McLaren
Discussers: K. Tanaka, G. N. Saridis, S. Kitajima

"On Variable-Structure Stochastic Automata" by R. Viswanathan and Kumpati S. Narendra
Discussers: S. Kano

"A Critical Review of Learning Control Research" by K. S. Fu
Discussers: S. Tsuji, L. A. Gerhardt, J. M. Mendel, R. W. McLaren

"Heuristics and Learning Control (Introduction to Intelligent Control)" by K. Nakamura and M. Oda
Discussers: J. T. Tou, K. S. Fu, R. W. McLaren, H. Sugiyama

"Adaptive Model Control Applied to Real-Time Blood-Pressure Regulation" by Bernard Widrow
Discussers: J. Nagumo and L. A. Gerhardt

"Real-Time Display System of Response Characteristics of Manual Control Systems" by Jin-ichi Nagumo
Discussers: S. Ohteru, K. S. Narendra

INDEX